A scene from a second century A.D. sarcophagus in the Capitoline Museum in Rome. A shepherd is feeding his dog its evening meal of porage. The dog has the characteristics described as desirable in a sheepdog by Varro. (Photo: *Deutsches Archäologisches Institut, Rome.*)

VARRO
THE
FARMER

A Selection from the Res Rusticae

edited by

Bertha Tilly, Ph.D., M.A.

UNIVERSITY TUTORIAL PRESS

Published by University Tutorial Press Ltd
9/10 Great Sutton Street, London ECIV ODA

Published 1973

ISBN: 0 7231 0605 3

Printed in Great Britain by
University Tutorial Press Ltd
Foxton, near Cambridge

Preface

An easy unlaboured reading of the Latin text and a consequent heightening of interest in the subject matter and the way the author treats it should, I feel, be the first aim of an editor. In this edition I have tried to encourage such a response—by expansion and digression where this seemed desirable and helpful. I have tried to reconstruct the life of the farm Varro describes in *Res Rusticae*, and have emphasised the agricultural advice he offers. Scenes and practices of farm life therefore bulk larger than might have been expected of an edition of this kind.

There is no up-to-date English Commentary on any of Varro's work. Many of the authorities who must be consulted for a proper understanding of his literary merit and his scientific thought are quite inaccessible, being found only in foreign libraries. Though this edition can make no claims to being complete, let alone comprehensive, I have tried to take account in my editing of the best foreign work. Inevitably questions are opened which this edition cannot answer; I hope that areas for future study will be indicated by this edition, and that there will be a revival of interest in Varro in the English-speaking world, for his *Res Rusticae* at least is worthy to be mentioned in the same breath as the *Georgics*.

I have made this selection not with the professional scholar particularly in mind, but rather the cultured general reader, the Sixth Form or University student who is interested in and attracted to the background the history and literature of the period at the end of the Republic and the beginning of the Empire. Those whose knowledge of Latin is not yet very wide will be able to enlarge it with the help of the summaries, notes and full vocabulary. My selections comprise the greater part of the *Res Rusticae*, covering all the major topics. What I have omitted I have summarised in English. My choice of what to omit was a hard one: some will disagree violently, for example, with my omission of certain sections on stock-breeding (namely on pigs, asses and mules) while I include the management of the aviary (in which science Varro was an innovator), but it had to be made. I hope that dissatisfaction with it will lead to the appearance of other selections, and, ultimately, of a complete scholarly edition.

Textual comment has been almost wholly omitted. I have followed in its entirety the text of Goetz' second Teubner edition (1929) as it is used in the Loeb edition of 1960. My only change has been to print *es* for

PREFACE

is for the accusative plural of the third declension. I have kept references to other writers, ancient and modern, down for the sake of greater simplicity; those who wish to take the subject further will find the Select Bibliography a useful guide to the beginning of what is available. The recent welcome agricultural bibliography of K. D. White (1970) supplies many deficiencies.

My grammatical comments are made to facilitate reading. Varro's Latin is often unclear—a result of haste in composition ?—and the more frightening departures from normal usage have been noted: too much concentration on his grammar where the sense is nevertheless clear would have militated against my desire to explore and explain his agricultural theories and practice.

BERTHA TILLY.

Contents

Illustrations

Acknowledgements

The text is that of M. Terenti Varronis, *Rerum Rusticarum libri III*, edited Keil and Goetz (Stuttgart, 1870), and has been reproduced with the permission of B. G. Teubner. It has been followed in its entirety except for the substitution of *es* for *is* in the accusative plural of the third declension.

Grateful acknowledgement is also made to the following:

Professor E. T. Salmon for his help in the chapter on Varro's Life.

The Director of the American Academy in Rome, Professor Frank Brown, and his successor, Mr Bartlett Hayes, for kind permission to use the Library and the amenities of the Academy; also to the Librarian, Mrs Inez Langobardi, and her staff for unfailing help and kindness.

Dr Ernest Nash of the Fototeca Unione and his staff for help over the illustrations.

Mr E. C. Kennedy for advice and help.

The officers of University Tutorial Press who first suggested this work, setting me what seemed at first a colossal, and unattainable, task, but which ultimately proved delightful, stimulating and, though exacting, the occasion of long visits to Rome.

These acknowledgements cannot end without an expression of thanks to my friend, Mrs Marion Eastty Gregory, for her constant and important encouragement.

B. T.

Chapter I

I. VARRO'S LIFE

Varro was a man of vast learning and encyclopaedic knowledge which ranged over all the arts and sciences known to the ancient world, and at the same time a man of action in the political sphere. Almost 90 when he died, he was a giant among men. During his long life which covered the period of the civil wars, one of the most troubled times in Roman history, he saw a succession of men who, one after another, seized power merely to fall to ruin. Only after Octavian's victory at the battle of Actium in 31 B.C. was ultimate and complete peace established: and a few years of tranquillity were still left for ·Varro's old age. He is said to have died on the sixteenth day in the January of 27 B.C. when the emperor received an ovation in the Senate and was given the title of "Augustus".[1] The magnitude of Varro's achievement was acknowledged by his contemporaries and others. Cicero, his close friend, calls him πολυγραφώτατος, "a polygraphist", "a prodigious writer of books";[2] Plutarch honours him as βιβλιακώτατος,[3] "a man of vast book-learning", and Quintilian as *vir eruditissimus*,[4] "the most learned of the Romans." Perhaps the greatest tribute that could be paid to him was that of Petrarch, the Italian poet of the early Renaissance, who lived from 1304-1374: he placed him beside Cicero and Vergil and paid homage to him as *il terzo gran lume di Roma*,[5] "the third great light of Rome". Petrarch, more fortunate than we in this age, had once possessed a manuscript of Varro's *Antiquitates Rerum Humanarum et Divinarum* but lost it by a piece of misfortune: the result was that one of Varro's major works, of the utmost importance for knowledge of the ancient world, was lost for ever. Tragic it is that out of Varro's immense output, there survives today only one book in its entirety, the *Res Rusticae*, a

---

[1] Some scholars hold that Varro died in 28 B.C. in his eighty-ninth year.
[2] Cicero, *Ad Atticum*, XIII. xviii.
[3] Plutarch, *Romulus*, XII.
[4] Quintilian, *Institutio Oratoria* X. i. 95.
[5] Petrarch, *Trionfo della Fama* (*The Triumph of Fame*) III. 38.

lesser work. In addition we still have five out of the original twenty-five books of the *De Lingua Latina*, one of his major works. The rest are only known in fragments quoted by other writers. These are the pitiful remnants spared to us of some six hundred and ninety books attributed to him.

Ten years older than Cicero and Pompey and friend of both, seventeen years after the assassination of Tiberius Gracchus (which was itself an important milestone in Roman agrarian history),[1] Marcus Terentius Varro was born of an undistinguished family of equestrian rank, at or near the town of Reate (now Rieti) in 116 B.C. in the Sabine country. Some fifty miles from Rome it stands on an upland plain watered by copious streams which flow down from the central Apennines from the Monti Reatini which close it in on the east, and from the snow-covered heights of Terminillo. The largest stream among them, the river Velinus (now the Velino) meanders far and wide over the level land draining in its course several small lakes of great beauty, to join the Nar (now the Nera) some three miles above Terni, ancient Interamna Nahars, where the combined rivers reach the Tiber. The green waters of the Velinus, from translucent mountain sources and melting snow, cause the plain, the Rosea, to be exuberantly fertile. As a result of these unfailing copious streams so abundantly and rapidly grew the grass that it was said that as much as was grazed in one day would grow up again by the next.[2] When pleading a cause before the censors, Caesar Vopiscus, who once held the rank of aedile, spoke of the Rosea as the *sumen* ("udder") of Italy, saying that it was so fertile that a rod left lying on the ground overnight would be covered with grass the next morning.[3] In early times this level basin known as the *Ager Reatinus* was an extensive lake. After the Roman conquest of the Sabines in 272 B.C., however, the victorious general Manius Curius Dentatus drained the lake by tunnelling through the mountains and making an outlet for its waters to the north, into the Nar. As a result a large tract of land previously submerged, in the district of Reate, was reclaimed to become amongst the most productive areas of Italy. The name, Rosea or Rosia, probably connected with *ros* "dew", still persists in a place near Rieti called Le Roscie. Five miles to the east of Terni are the spectacular falls of the Velino, the Cascate Delle Marmore which take their origin from the artificial channel made by Dentatus. The drainage scheme subsequently became the subject of a lawsuit between

---

[1] See Chapter II. *Latifundia* and Transhumance.
[2] Servius, *Ad Aen.* VII, 712.
[3] Varro, *R. R.* I. vii. 10. Pliny, *Naturalis Historia*, XVII, iii. 32.

the citizens of Reate and those of Interamna because waters diverted from the Rosea flooded the territory of the Interamnates. About the year 54 B.C. Cicero himself was called to defend the Reatini: in order to be briefed for the case which he subsequently won, he visited the locality and was so impressed with the fertility which he saw there that he compared it with the vale of Tempe in Thessaly in Greece, also famous for its fresh green growth. He wrote to Atticus at the time,[1] "the people of Reate took me to their Tempe to plead for them against the Interamnates before the consuls and ten assessors because the Lake of Velinus flows into the Nar through a cutting in the mountains made by M'. Curius as a result of which the plain of Rosea was drained, yet retains sufficient moisture . . . I stayed with Axius. More than that he took me to see the Seven Waters." Axius is one of the speakers in Book III of the *Res Rusticae*: it seems that he showed Cicero the beautiful lake scenery of the Rosea and perhaps the falls. In Roman times the Rosea was wholly given over to pasture.[2] All kinds of flocks and herds, fine horses, mules and asses were bred and flourished on the lush grass.[3] Asses called the Rosean breed, *Roseani*, were as much in demand as the prized asses of Arcadia in Greece.[4] Any man who owned land in such a well-watered district in Italy was exceedingly fortunate. Axius,[5] Cicero's friend, drew wealth from his property and owned villas in this favoured Sabine country, as did Varro's own family. It was in this environment that Varro's childhood and boyhood were spent.

Reared after the stern Sabine tradition Varro learned lessons of hardihood and endurance. He has left a glimpse of his frugal life: "When I was a child I had a modest tunic and toga, shoes without laces, a horse without a saddle; I did not take a bath every day and rarely did I play at dice."[6] He found too that to one who is thirsty

---

[1] Ad *Att*. IV. xv. 5. *Reatini me ad sua* Τέμπη *duxerunt ut agerem causam contra Interamnates apud consules et decem legatos, qua lacus Velinus a* M'. *Curio emissus interciso monte in Nar(em) defluit; ex quo est illa siccata et umida tamen modice Rosea. vixi cum Axio; quin etiam me ad Septem Aquas duxit.* Cicero's word *modice* means that the land retained an adequate amount of moisture to ensure the good grazing for which it was famous. See also *Pro Scauro*, XII. 27, where Cicero also mentions his visit and his desire to be fully briefed by meeting his clients and seeing the place and the lakes for himself.

[2] Pl. *N.H.* XVII. iii. 32.
[3] Varro, *R.R.* II. i. 17; III. ii. 10.
[4] Varro, *R.R.* II. i. 14, II. vii. 6; II. vi. 1, II. viii. 3. (text omitted). In II. vi. 1, Varro states that he has several times sold colts even to Arcadians.
[5] Varro, *R.R.* III. ii. 9-10.
[6] Varro, *Logistoricon*, fr. 19 Bol.

3

thin wine was as good as honeyed wine, to one who is hungry black bread is as good as that made of fine milled flour, and sleep is sweet to one tired out with toil.[1] He worked with his father's herdsmen with the flocks and herds. The sheep grazed on the Rosea in summer but were driven to Apulia on the south-eastern coast for winter pasture[2]; perhaps he too went on the long trek with them along the old grassy droves over the Sabine mountains and learned by experience the economy of the ranching system. He eventually inherited his father's estates and enjoyed the prosperity produced by stock-rearing, but his horizon was not to be limited to his native Sabine country. In time he entered and became a leading figure in the wide Roman world, and in the course of his military and political career acquired still greater wealth with which he purchased villas in fashionable surroundings, at Tusculum in the Alban hills outside Rome, at Baiae[3] on the gulf of Naples, and at Casinum in southern Latium, that part called by the Romans *Latium Adiectum*, near the borders of Campania. On his Tusculan estate he preserved in the woods deer and boar which were trained to come at feeding times at the sound of a horn.[4] At Casinum he had an aviary built[5] for pleasure where guests could dine to the song of blackbirds and nightingales.[6]

In contrast to his simple upbringing, in Rome his parents gave him the best education of the time. His first master, Lucius Accius, the first Roman philologist, gave the young Varro an interest in letters

[1] Varro, *Logistoricon*, fr. 28. Bol. These two fragments are thought to be autobiographical.

[2] Varro, *R.R.* II. ii. 9.

[3] This was his *Cumanum* called after the neighbouring city of Cumae: Cic. *Ad Familiares*, IX. i, 2; ii, 3, Tyrrell and Purser, 1918, 456 and 461: it is the imaginary setting of the second edition of Cicero's *Academica*: (see below p. 14 and note).

[4] *R.R.* III. xiii. 1-2 (text omitted). Varro bought his estate at Tusculum from Marcus Pupius Piso: wild boar and deer would gather for food when a horn was sounded, at a regular time; mast was thrown to the boars and vetch to the deer. Appius reports in the conversation that at Quintus Hortensius' place near Laurentum, there was an extensive forest enclosed with a wall. Inside there was a place for dining. As the guests sat at dinner Quintus would call for Orpheus. A man would appear wearing a robe and carrying a harp, and would be ordered to sing. When he blew a horn, a crowd of stags, boars, and other animals poured around them. *Ibid.* 2-3.

[5] Varro, *R.R.* III. v. 8-17. A full description is given in this passage.

[6] Although Ll. Storr-Best, *Varro on Farming*, London 1912, vii, was shown some remains of Varro's villa and aviary at Cassino, along the river now called the Rapido, about the year 1912, G. F. Carettoni reports that little or nothing remains of it. (*Casinum (Presso Cassino) Regio I—Latium et Campania Italia Romana*: *Municipie e Colonie serie I, vol. II.* 1940-48).

4

and words which was an important influence on his developing mind. He had as his second teacher Aelius Stilo, a distinguished grammarian and master of rhetoric: he too exercised a formative influence on his pupil whose whole intellectual life was to be devoted to research and erudition. His abiding interests in linguistics and antiquarian studies were thus early aroused. When about sixteen he began his military career which was to continue at intervals for more than sixty years. Few facts are known about these early years but some historians consider it certain that he served as *quaestor* under Cinna[1] in the invasion of Liburnia and crossed over in 85 B.C. This district lay on the north-eastern coastland of the Adriatic and extended to the borders of Dalmatia.[2] Even on campaign, in accordance with his practice of observing local customs, he recorded how the sturdy Liburnian women bore children while at work, only withdrawing for the birth and returning as if they had found, not borne them.[3]

In 88 B.C. two distinguished philosophers fled to Rome on the outbreak of the Mithridatic war, Philo of Larissa and his pupil Antiochus of Ascalon. Philo was the founder of the school of philosophy known as the fourth Academy; and Antiochus of the fifth Academy. It is certain that Varro continued his academic training under these teachers about this time and although some authorities hold that he was in Athens from 84–82 and attended their lectures when they had once more been able to return, the evidence for his presence there is slender. It is perhaps more probable that he studied under them during the time they took refuge in Rome. Their teachings in general were a blending of the philosophic systems of Plato and Aristotle, combined with Stoicism, which latter required of its adherents a virtuous life and stern devotion to duty.

Varro did not, as might have been expected of one so given to scholarship, keep himself withdrawn from the turbulent politics of the years of his early manhood, but as a partisan of Pompey, who, in his estimation, stood for the maintenance of the old Roman order, he fought during the four years of the war against Sertorius in Spain. He was able now and all through his subsequent career to combine action with literary studies and erudition, finding time to earn the title given him

---

[1] E. Badian, *Journal of Roman Studies*, vol. LII, 1962, p. 60. It is suggested that Varro crossed with the advance party from Ancona to organise the landing. Cinna never went to Liburnia himself.

[2] C. Cichorius,, *Römische Studien* (Berlin, 1922), V. "Historische Studien zu Varro: I zu Varro's Lebesgeschichte, 189–207", would put Varro's service in Liburnia in 78-7; in a later campaign he may have served as *legatus*.

[3] *R.R.* II. x. 8.

by Quintilian, *vir eruditissimus*. As Pompey's *legatus* from 76-72 B.C. he observed, with an agriculturist's eye, Spanish methods of cultivation such as the growing of the vine without props,[1] threshing with a sled fitted with toothed rollers (*plostellum punicum*)[2], ways of storing grain,[3] the double shearing of sheep.[4] He saw too horses in their wild state[5] and brought back a species of rabbit (*cuniculus*)[6] which he reared in his own preserves. Pompey's return to Rome and his election to office as consul together with Crassus in 70 B.C. brought hopes of an established peace. Varro, then in his forty-sixth year, was tribune of the plebs, and could have expected that his military service was finished.

His tranquillity was not to last for long. In 67 B.C. Pompey was called on to rid the seas of the pirates who were strangling trade in the Mediterranean. Various attempts had been made to dislodge them, beginning as far back as 102 B.C., but all failed; they now had control of some forty harbours and could openly advertise their presence with gilded masts and purple sails. Pompey divided the Mediterranean into thirteen commands to each one of which he assigned a *legatus navalis*. As one of these naval lieutenants Varro was given the sphere of operations which extended from Sicily to the North African coast. By Pompey's masterly strategy the pirates were expelled from the western seas, then driven to their eastern bases in Cilicia and quickly reduced to surrender. Varro subsequently received the rare naval award of the *corona rostrata*[7] for his services.

As in Spain, Varro profited from observation of the agriculture of the countries he saw. He visited Epirus,[8] undoubtedly was shown over the estates of Atticus,[9] his own and Cicero's friend and one of the speakers in the conversation in Book II of the *Res Rusticae*. Varro was impressed with the slave families,[10] the sheep-rearing,[11] and the fine oxen[12] and dogs.[13] The date of this meeting is thought to belong to the time of his naval command, possibly 67 B.C.

---

[1] *R.R.* I. viii. I.

[2] *Ibid.* I. lii. I. This type, called *Punicum*, is thought to have been introduced into Spain by the Carthaginians.

[3] *Ibid.* I. lvii. 3. (Text omitted.)

[4] *Ibid.* II. xi. 8. (Text omitted.) The sheep were sheared every six months.

[5] *Ibid.* II. i. 5. (Text omitted.)

[6] *Ibid.* III. xii. 7. (Text omitted.) The *cuniculus* is described as burrowing into the ground so it is to be recognised as a species of rabbit: it is so-called because it makes tunnels (*cuniculi*) in the fields in which to hide.

[7] Some authorities think the proper title for this award was the *corona navalis*.

[8] *R.R.* II. Praef. 6.   [9] *Ibid.* II. ii. 1.   [10] *Ibid.* I. xvii. 5.   [11] *Ibid.* II. ii. 20.

[12] *Ibid.* II. i. 2. (Text omitted.) II. v. 10.   [13] *Ibid.* II. ix. 5.

Varro remained a partisan of Pompey and although the facts of his life are not clear at this stage it is thought by some that he took part in the third war against Mithridates, King of Pontus, under Pompey's leadership. If this be so, Varro may have drawn added wealth from the spoils of this war and further enriched his knowledge and experience by travel in Asia through countries until then little known. In 59 B.C., however, the facts are clearer. Caesar, Pompey, and Crassus had by that time formed the coalition usually called the First Triumvirate. The three men with armed forces at their command destroyed the old republican constitution and seized power which at first they shared, but eventually Pompey and Caesar fought between themselves for supreme command. One of their first acts was the passing of the agrarian law called after Caesar the *Lex Iulia* to enable the *Ager Campanus*, a fertile district in west central Italy between the Apennines and the coast, to be parcelled out in small holdings to ex-service men. A commission of twenty (*vigintiviri*) was appointed to administer the details; among them was Varro and alongside him on the committee was Gnaeus Tremelius Scrofa, the leading agriculturist and agronomist of the time who appears as the chief speaker in Book I. By this means landless peasants, displaced, and unemployed persons were removed from Rome and restored to the land and to a healthy working life. At this time, 59-58 B.C., Varro's prestige was so high and his standing with Pompey so sure that Cicero turned to him for help in his misfortunes: he sought to secure Varro's support in a reconciliation with Pompey. This move was unsuccessful and did not prevent Cicero from going into exile. In a series of letters to Atticus[1] who was also a close friend of Varro he expresses his hopes that Varro will intercede for him, but eventually there are tones of disappointment. When, however, Cicero was able to return to Rome in the year 57 he was reconciled with Varro and the two were on friendly terms.[2]

---

[1] *Ad Atticum*, II. 20. 1. Cicero gives expression to thoughts clothed with anxiety: *Varro satisfacit nobis*, "I am very pleased with Varro." II. 21. 6, *Varro mihi satisfacit*, "I am very pleased with Varro". II. 22. 4, Cicero's hopes seem to rise: *multa per Varronem nostrum agi possunt*: "much can be done through our friend Varro". But in II. 25. 1, his hopes seem to have been dashed: he reproaches Atticus for not having let Varro himself know how much Cicero appreciated his good offices, not because Varro had as yet done anything on his behalf but in the hope that he might in this way be encouraged to do so: he adds *mirabiliter enim moratus est*: "he is wonderful at procrastination".

[2] *Ad Familiares*, IX. i-viii.

In the next ten years it is thought that Varro held various magistracies, as *aedilis, triumvir*, and *tribunus plebis*. Now he might have felt that at last his days of campaigning were ended, but again this was not to be his lot. In 49 B.C. when Caesar attacked the Pompeian party in Spain, Varro, as a loyal Pompeian, was given command as *legatus propraetore* of two of the five legions stationed there. He was not able to withstand Caesar's victorious march south and surrendered to him without a struggle at Corduba (modern Cordova), handing over to him the legions and supplies under his command. Although his actions were inevitably under duress, he appears either now or later to have changed his political sympathies. Varro received a ready pardon from Caesar and was allowed to leave Spain. As a man of letters wholly devoted to learning, he was no menace on the field and had no forces at his command. Caesar perhaps thought that as the most distinguished scholar of his day he could help much in his intended reforms, and add prestige to his cause by having him on his side; Caesar's usual practice was, however, to let Pompeian officers who surrendered go unharmed.

At this point Varro, now only a private citizen, could have been expected to retire from the struggle; but, still a supporter of Pompey, he rejoined Pompey who was in the eastern theatre of the war with an army and a fleet at Dyrrachium on the south-eastern Adriatic coast off Macedonia: he also held another base for some of his forces and his storeships on the small island of Corcyra (now Corfu) nearby. Varro found Cicero in Pompey's camp as apprehensive as he was himself about the outcome of the war. While Pompey waited here events led on to the battle of Pharsalia in 48 B.C., one of the decisive encounters of the Civil War in which Pompey, utterly defeated, fled from the field only to be murdered in Egypt. At some time during the campaign, either before the battle or more probably after the rout of the Pompeians when many of the defeated took refuge in Corcyra, the forces were attacked by the illnesses common to overcrowding and the place was full of sick and dying men. Varro saved the army and the fleet. Acting on his knowledge of medicine and the winds, he opened new windows to let in fresh breezes and blocked others to keep out the infected air.[1]

The year of Pharsalia, 48 B.C., marked the end of his military career. Some twenty years were still before him which were to be given to writing—the life which was his true predilection although his tranquillity was twice disturbed. After Pharsalia, Antony, in command in

---

[1] *R.R.* I. iv. 5.

the west, proposed to proscribe Varro among others as a former supporter of Pompey. Such an action would have meant that Varro became a stateless person, that anyone who assassinated him would be rewarded, and that all his possessions could be confiscated and sold under the hammer. Caesar, now in Alexandria, heard of Antony's intentions, and immediately sent a dispatch forbidding the proscription and ordering immediate restitution. Varro was saved from a violent death, but was obliged for a time to live in retirement at his villa at Tusculum. In gratitude to Caesar Varro dedicated the second part of the *Antiquitates Rerum Humanarum et Divinarum* to him. Some authorities hold that it was on this occasion that Antony plundered Varro's estate at Casinum, but it is more likely to have happened after Caesar's assassination in 44 B.C. when Varro's defences were down. During his enforced stay at Tusculum he and Cicero were on affectionate terms: a series of letters from Cicero[1] reflect these feelings and even go to show that he envies Varro the tranquillity he now enjoys. In the year after Pharsalia, however, Varro's position was secure: Caesar honoured him, with rare magnanimity, as the greatest man of learning of his day. In 47 B.C. he was commissioned to supervise the assembling of Greek and Latin books for the formation of a great public library in Rome. This would have been equivalent to a modern university library and could have made Rome the greatest centre of learning in the ancient world. Caesar may have been inspired to attempt to rival the great library at Alexandria which he saw during his campaigns in Egypt or during his affair with Cleopatra. Caesar's project was not to materialise: two years afterwards he fell dead, assassinated, at the feet of Pompey's statue. Later, however, in 39 B.C. when Asinius Pollio set up the first public library in the *Atrium Libertatis*[2] in Rome, Varro's bust was erected alongside those of other great men, his the only one ever to be put up of a living person.

In 45 B.C. Cicero dedicated to Varro the second edition of the *Academica* and included him as one of the speakers in the discussion on philosophy. He writes of his intention to do this in a letter to Atticus[3] and says that Varro has already promised to dedicate to him the *De*

---

[1] Cicero: *Ad Familiares*, IX. i-vii.

[2] The *Atrium Libertatis* was originally a building containing the offices of the censors, both laws and records. It was restored in 194 B.C. and later, very lavishly, by Asinius Pollio, from the spoils of his Parthian campaign. It became the first public library. It contained Greek and Latin books with portrait busts of authors. The exact location is not known but it was certainly not far from the forum of Caesar which itself was adjacent to the old forum.

[3] *Ad Atticum*, XIII. xii. 3; xix. 5.

Lingua Latina which he was writing at the time. When the latter work was eventually published in 44 B.C. Books V to the end were dedicated to Cicero. The first quarter and fragments of the remainder survive of this second version of Cicero's work; in it Varro speaks to us over the ages, in a philosophical and not political vein. The letter of dedication addressed to Varro precedes the text in which Cicero says that he has cast him for the part of the champion of the philosopher Antiochus whose pupil and follower he was. Cicero writes too of the solace to him of his present studies—without them life would have little meaning.[1]

Two months after the assassination of Caesar in 44 B.C. Varro's peaceful life was disturbed for the second time. Antony regarded any previous partisan of Caesar and now of Octavian as an enemy. When leading a military colony to Capua in Campania he took possession of Varro's villa at Casinum when he was turned away from Capua, plundered the library and gave himself up to feasting and debauchery. The scene is described (perhaps rhetorically exaggerated) by Cicero in the Second Philippic which was a speech directed against Antony but never delivered.[2] In the next year, 43 B.C., Varro was proscribed for the second time by Antony but was given refuge in the house of a neighbour, Calenus, and so escaped for a second time a violent end, more fortunate than his friend Cicero who was murdered towards the end of that same year.

Varro's last years were serene, passed as he must always have wished in the quiet of his country estates. We have a glimpse of him in the *Res Rusticae*[3] at eighty years of age in the villa at Casinum, cared for by his wife Fundania, younger than himself, reading and writing in his study beside the stream,[4] and enjoying the beauties of the birds and birdsong of his aviary[5] close by. He died at the age of ninety in the year 27 B.C., quietly in his bed, differently from his great contemporaries who fell victim to the changing fortunes and savageries of the Civil Wars.

---

[1] Cicero: *Ad Familiares*, IX. viii.
[2] Cicero: *Philippic* II. xl-xli.
[3] *R.R.* I. i. 1-2.   [4] *Ibid.* III. v. 10.   [5] *Ibid.* III. v. 9-17

II. VARRO'S WORKS

It has been calculated by both ancient and modern scholars that Varro's output amounted to some six hundred and thirty books in all; yet out of all this vast achievement, there remains, today, only one in its entirety, a small treatise on agriculture, the *Res Rusticae*, which is among his least important works. In addition six books have survived out of the total of twenty-five of his treatise on the Latin language, the *De Lingua Latina*, and some thousand fragments from this and other writings quoted by other authors. Since it is not possible to arrive at a chronological sequence of Varro's works, nor to enumerate here all those of which we have knowledge, the most helpful approach is perhaps to group them according to their content, and to describe the major works.

Varro's earliest writings were in verse: these were one hundred and fifty books of satire, the *Saturae Menippeae* so-called after Menippus of Gadara, the Cynic philosopher of the third century B.C. Some six hundred fragments remain which show that they were written in various metres and interspersed with prose passages. The aims of the Cynic teaching were to have few wants, to renounce possessions, and to keep aloof from the troubles of the world. The true follower should lead a simple life, and be independent, serene even, in the midst of war and disorder. One of the best known of its exponents was the philosopher Diogenes who was content to live in a tub and who when asked by Alexander the Great if he needed anything answered only that his questioner would do well to stand out of the sun. The Cynic attitude to life is reflected in the Menippaean fragments: philosophical truths are presented in a kindly genial manner so as to draw out a smile, not to reproach or blame. The *Logistoricon*, in seventy-six books, contained passages on philosophical subjects illustrated and supported with examples from history: it is thought to have been composed towards the end of his life, and if so the publication is to be attributed approximately to the year 44 B.C. In content and tone it is to be regarded as being of the nature of satire.

The *Imagines* or *Hebdomades* were among his last works, and appeared in 39 B.C., two years only before the publication of the *Res Rusticae*. There were fifteen books which contained word pictures of Greek and Roman characters together with a portrait of each. The whole work was grouped in sevens, which gave rise to the second title, which is derived from the Greek "ἑπτά" "seven". The *Imagines* was probably the first illustrated book to be published in the western world.

Varro's preoccupation with the literature of Rome extended to the theatre and especially to the plays of Plautus. He wrote several treatises on subjects connected with the theatre, such as its origin and history, on masks and characters, and recognised twenty-one comedies as genuinely Plautine, thereby establishing a canon which is still accepted today.

Another deep interest was centred on the study of the national language. His principal work in this field was the *De Lingua Latina* in twenty-five books, which was published in 44 B.C. Only the books from V to X are extant. His main linguistic studies were contained in this, a work of major importance. There were sections on etymology containing histories of words and suggested derivations such as are met with in the *Res Rusticae*. They are often naïve, fanciful, even impossible but the work is a rich mine of countless quotations from old Latin, from the older poets, and from ancient documents of all kinds such as laws, ritual formulae, and many others. The importance of the work is in its content, not so much in Varro's theories about language, and in the material which has been preserved for us. Perhaps more famous than those already mentioned, and again an outcome of Varro's predilection for all things Roman was the *Antiquitates Rerum Humanarum et Divinarum* in forty-one books, "Antiquities of Things Human and Divine". St Augustine,[1] who had it before him as he wrote, has given us a detailed description of the work: the first part treated of things human and the second of things divine: both parts were schematised and tabulated in the manner which is apparent also in the *Res Rusticae*. This work appeared in 47 B.C., the second part dedicated to Julius Caesar who was then *Pontifex Maximus*, in gratitude for Varro's escape from proscription. The *Humanae* was published rather earlier. Varro's patriotism was perhaps nowhere more strongly expressed than in this book: he was at pains especially to see that the ancient cults were not lost in oblivion and as a result laid a foundation for the Augustan revival of the old Roman religion. The *Antiquitates* was widely read in the first and second centuries A.D. and later by Christian writers, among them Tertullian. A smaller work in four books, the *De Gente Populi Romani* had affinities with the *Antiquitates*: in writing this Varro's aim was to find the origin of the Roman people, tracing it to remote Greek antiquity. Another smaller book, *De Vita Populi Romani* was dedicated to his close friend Atticus, and contained an historical picture of public and private Roman life.

[1] *De Civitate Dei*, VI. iii.

VARRO'S WORKS

The nine books of the Disciplines, *Disciplinarum Libri IX* made up an encyclopaedia of the liberal arts, that is, the subjects studied in schools and universities, and comprised the knowledge which an educated person should acquire: the subjects treated of were grammar, dialectic (the art of reasoning), rhetoric (the art of speaking), geometry, arithmetic, astrology, music, medicine, architecture. The first seven were those which were included in the medieval systems of the *trivium* and the *quadrivium*[1] which are in fact a direct outcome of his teaching. The *Disciplinae* were the foundation of education through the Middle Ages. This work, in itself a tremendous achievement and yet only one of the several already mentioned, was produced in his old age. The *Res Rusticae* finds a place too among the writings of his old age, perhaps composed alongside some of far greater import, and written hastily as being of no great consequence.[2]

For the immense mass of work completed, for his patriotic fervour, his high moral sentiments, for versatility in forms of writing and in subjects, for the vast range of material, Varro towers above all his contemporaries and his successors: he was distinguished for learning as no other man had ever been or was to be. Writers of his own age and many that came after drew on his works. His style, heavy and unattractive, caused in time the reading of his books to lapse, but it was reanimated by Christian writers who drew on his works for their own purposes. Varro's books and doctrine pervaded all learning after his time: his great merit is to be found in the vast mass of facts and data which he preserved from sources of every kind. Without what we know and possess of him (even in spite of the loss of nearly all his works), our knowledge of Roman antiquity would be greatly impoverished.

It was as if the wisdom and knowledge of the Greeks and Latins of former ages were suddenly gathered into one man's mind and then, so gathered, given to the world for education and enlightenment. His works marked a gigantic forward step in the world of learning; his method of working in collecting information for comprehensive treatment in encyclopaedic form fulfilled a need felt by his own and all ages.

In his praises of Varro's achievement Cicero sums up his major

---

[1] In mediaeval educational systems the curriculum which included grammar, rhetoric, and logic made up the *trivium*, and that together with the *quadrivium* (arithmetic, music, geometry, and astronomy), made up the seven liberal arts.

[2] *R. R.* I. i. 1. It is allowable to conjecture that Varro was writing the *Disciplinae* at this time.

13

works in the *Academica*:[1] "for we were wandering and straying about like visitors in our own city and your books led us, so to speak, right home and enabled us at last to realise who and where we are. You have revealed the age of our native city, the chronology of its history, the laws of its religion and its priesthood, its civil and its military institutions, the topography of its districts and its sites, the terminology, classification, and moral and rational basis of all our religions and secular institutions, and you have likewise shed a flood of light upon our poets and generally on Latin literature and the Latin language, and you have yourself composed graceful poetry of various styles in almost every metre and have sketched an outline of philosophy in many departments."

[1] Cicero, *Academica* I. iii. This book as we have it belongs to the second edition of this work, and is sometimes called the *Academica Posteriora*: translated by H. Rackham, Loeb Classical Library; Cicero, *De Natura Deorum et Academica*, 1933.

Chapter II

LATIFUNDIA[1] AND TRANSHUMANCE

From earliest times, perhaps from as far back as the neolithic age, the land of Italy was worked by peasant proprietors who tilled with their families their own ancestral farms, content with humble circumstances and drawing enough food from the soil for the maintenance of man and beast. Throughout the peninsula were self-supporting smallholdings where agriculture and stock-breeding were practised together, where the father and his family worked "with black hands" and were not ashamed. The parcels of land were small—in fact it was said that by a law attributed to Romulus, every citizen had received two *iugera* as a *haeredium,* and this became his own private possession which he could bequeath to his heirs. Undoubtedly there must have been in addition some common land available for grazing. This scene is admittedly legendary, but is valuable in that it reflects the conditions of a primitive and sound economy on which the livelihood of the early Italians was based. From the first, however, there was always a tendency (and often opportunities) for the owner to add to his parcel of land with the result that an original allotment of land did not remain stable. During the conquest of Italy in the early centuries, after Rome's victories over the neighbouring tribes and gradually over the whole peninsula, her custom was to leave to the conquered a portion of their land, but to annex the rest to herself to become *ager publicus,* the property of the Roman people. Portions of this might be sold, rented, or leased, as the government decided, to Roman citizens. There was another course too by which land could be acquired and one that cost little. Sometimes the government took steps to use the public land and in this case anyone who wished might cultivate it and pay a small rent. Roman citizens often took advantage of this chance to acquire more property: at the same time it could happen that the original holders were not dispossessed, although their land had become technically speaking *ager publicus.* Those who occupied land belonging

---

[1] For a discussion on the meaning of *latifundia* see K. D. White, *Latifundia.* (University of London Institute of Classical Studies Bulletin, 1967.)

to the government were called *possessores*, a word almost equal in meaning to "squatters". They could be ejected by the government if it was thought necessary but often they had held the land so long that they were almost regarded as owners. It was easy for the great landowners, who were also those who controlled the state, to annex for themselves much of the public land and to enlarge their own estates.

Another and far more serious factor was at work, also an outcome of Rome's wars, which ceased with Marius' reforms of the army after 100 B.C. A great change took place when he recruited volunteers to be soldiers, mercenaries who would make a career of military service. Until his time the landed peasantry had been subject to conscription and they alone were regarded as fit to fight for Rome. They were torn from the soil which they and their forebears had tilled for generations. Conscripts could not know even where military service would take them, nor for how long their absence under duress would last. Manifold in number must have been those small family farms which lapsed to weeds and desolation through absence of the master. The owners might be too long away ever to be able to repair the damage of neglect on their return, nor in many cases will they have had the means to restock their holdings when as veterans they returned home. They became landless, unemployed, and drifted into the towns there to eke out what livelihood they could. Wars and conquests were at the expense of a sound agricultural economy: they were the peasant's ruin but the profiteer's source of wealth. The tendency for land to be concentrated in the hands of great landowners was becoming so harmful and alarming that as early as 367 B.C., long before Rome was mistress of all the peninsula, a law of restraint was proposed by Caius Licinius Stolo and called after him the *Lex Licinia*. A descendant of his is one of the speakers in Book I of the *Res Rusticae*.[1] By this law possession was to be limited to five hundred *iugera* (about three hundred and forty acres): in addition not more than two sons in a family were allowed to hold two hundred and fifty *iugera* each. Land held in excess of these amounts was to be transferred to the state and allotted in holdings to the poor. It is remarkable to see sympathies being expressed for the needy as early as the fourth century, more than two hundred years before the time of Gracchi, and to see evidence of land hunger among the masses at so early a stage in history. Some authorities have in fact cast doubt on the date and content of this law,

---

[1] *R.R.* I. ii. 9.

assigning it to a later age, but the fourth century coincides with the acquisition by Rome of extensive tracts of *ager publicus* and the need for a large army of conscripts with which to continue her career of conquest.

In spite of this early law and later legislative[1] attempts to prevent the increase of large estates the trend towards larger estates was not halted: it was even hastened by agrarian conditions after the wars with Hannibal and the later wars and conquests in Greece, the East and in Gaul. The need for larger and larger armies increased, accompanied by greater harm to the agricultural economy. Many times over, the fate of the peasants' ancestral plot was to be bought with capital which some war lord had perhaps obtained by plunder—or merely annexed. The luckless peasant served in hard fought campaigns only to his own ruin. In the second century B.C., after the Hannibalic wars, agrarian conditions combined with several other factors to favour the growth of the large estates which fell now into the hands of the rich aristocracy, that is, of those who could make a capital outlay. The *Lex Licinia* had been powerless to prevent this process which continuously assumed greater proportions. As a result of war many of the nobility who had had command profited from payments of large gratuities and from plunder, which was sometimes acquired on a vast scale: at the same time Rome was in possession of more *ager publicus* than could be efficiently controlled. In addition there was a source of cheap labour: the thousands, even tens of thousands of war captives from Carthage, Greece, Syria, Gaul, and elsewhere, together with those from the later wars, were there to be bought in the market place to become workers on the land. As an encouragement to agriculture on a large scale markets for produce were available in the towns which had increased in population with the drift from the land as the old economy collapsed. Capital to invest, cheap land and labour, and a ready market inevitably resulted in the growth of large estates.

A further development which aided the growth of these large estates, sometimes called *latifundia* (literally "wide farm-lands"), was ranching, which now became an integral part of the farming system. Although sheep and cattle had always been reared on the small farms it was now found that grazing on a large scale with cheap slave labour could be very profitable and could make a large contribution to the produce of the estate. Tillage could be exercised on the land in the immediate

---

[1] The question of the date of the *Lex Licinia* is discussed at length by A. J. Toynbee, *Hannibal's Legacy*, Vol. II (Oxford, 1965), pp. 557-61, and the probability of other legislation before the age of the Gracchi.

neighbourhood of the farm while ranching could be practised quite separately, but yet as part of it. When the whole of peninsular Italy was under the control of the Roman State ways of communication for summer and winter pasture were open and accessible. The flocks and herds, raised in great numbers, became transhumants—that is, they changed their grazing grounds in summer and winter and in many cases never saw the home farm. The owner, usually an absentee landlord, living perhaps in Rome, and rarely visiting his land, entrusted the working of the estate to a *vilicus* (bailiff) and the care of the herds to a chief shepherd (*magister pastorum*). Work on the *latifundia*, whether at home or on the far-flung pastures, was done by the slaves, former prisoners of war who were considered lucky to have escaped a violent death by being sold into servitude.

The attempt of the tribune Tiberius Gracchus in 133 B.C. to make a more equitable distribution of land and to call a halt to the growth of *latifundia* ended in his assassination. He had proposed that the ownership of land should once again be reduced to what was still the legal limit of five hundred *iugera*, now rarely adhered to, and that the surplus land surrendered after reduction should be given back to the state and distributed in smallholdings to the poor. In spite of strong opposition legislation was passed whereby a commission of three, of whom one was Tiberius, was established and work was begun on the redistribution of land: after a brief period of activity it came to an end in 129 B.C. His brother Caius Gracchus, elected tribune in 123 B.C., proposed the distribution in Rome of corn at a cheap price, the founding of new colonies, especially one at Carthage, and the building of roads. Though brilliantly conceived, these reforms were not effective; his death in 121 B.C. ended all hope of their being implemented. The efforts of the Gracchi to improve the agrarian situation left conditions unchanged.

Rich men who had capital at their disposal were now free to speculate on a still more lavish scale in easily-obtained land, acquiring vast acres which they could farm as profitable business concerns. It is this type of economy which Varro describes in the *Res Rusticae* and which was by his time fully developed in spite of repeated attempts at restraint. He shows himself an agricultural tycoon, the ultimate product of Rome's years of war and conquest, of the blindness of statesmen and politicians to the need for a stable economy based on sound farming communities settled on the land. The *latifundia* with their ranching systems were harmful to the land in that vast tracts became over-pastured to the neglect of proper tillage. The soil was denuded and depleted, robbed of its proper renewal of growth from both wild and cultivated plants. Some years later than Varro's own time the elder Pliny wrote: *latifundia Italiam perdiderunt*: "large estates have ruined Italy".[1]

The lofty Apennines[2] with extensive plains at their foot make Italy excellently suited for sheep-rearing on a large scale, but the flocks must move twice yearly to find sufficient pasture. In the mountains there are many upland plateaux (called by the Italians *conche* or *alti piani*), some many miles across, which in summer are watered from clouds gathered about the heights and from the melting snow on the high peaks: the grass is green and plentiful, well-watered even during the hottest months; and there sheep can graze to their fill. On the plains, by contrast, there is green pasture during the mild, wet Mediterranean winter, but only then, for in summer it is dried up to stubble during the scorching months of June, July, and August: the burning sun is merciless and there is scant shelter for man and beast. The flocks with their shepherd must make the long trek twice yearly back and forth to find grazing grounds covered with fresh green grass: in September they are driven down to the plains, in May up to the mountains. They move along wide grassy droves (*calles*) often more than a hundred yards across wherever mountain and plain are naturally connected by trails—as Varro himself remarks,[3] just as a pair of buckets are held by their yoke. The migration of flocks probably dates from early, even from prehistoric times, since it is the outcome of climatic conditions: so it may be that the grassy trails which lead by devious ways following the conformation of the land from mountain to plain and back are the most ancient ways of communication in Italy, far anterior perhaps to the tracing of roads. In the early centuries the practice was probably only on a small scale, but after the Roman conquest of the peninsula of Italy the droves came under state control and were in consequence more accessible and unimpeded. With the rapid growth of *latifundia* after the Hannibalic wars, farmers found it profitable to pasture sheep away from, but yet as an integral part of, the farm. To ensure profits it was necessary for the owner to have land for pasturing in his possession, or to rent from the government public

---

[1] Pliny, *N.H.* XVIII. vii. 35.

[2] Useful accounts of transhumance are to be found in A. Grenier, *La Transhumance des Troupeaux en Italie et sa Rôle dans L'Histoire Romaine* (Mélanges de l'École Française de Rome XXV, 1905); E. T. Salmon, *Samnium and the Samnites* (Cambridge, 1967), pp. 65-70; A. J. Toynbee, *The Legacy of Hannibal* (Oxford, 1965), pp. 286-95 and authorities there cited.

[3] *R. R.* II. ii. 10.

grazing grounds at a distance and to have the use of the public trails (*calles*) which connected his estate with them. Samnium and Apulia where Varro had land were together ideally situated for transhumance. In Apulia (now called Puglia) the district on the east coast of Italy which lies south of the Gargano peninsula there stretches a vast fertile plain suitable for winter grazing. The chief towns in Varro's day were Luceria and Brundisium. Samnium, roughly equivalent to the modern district of the Abruzzi and Molise was the central plateau in the heart of the Apennines, a mass of limestone mountains, bounded on the east by the plain of Apulia.[1] Some of the highest peaks reach an altitude of eight thousand feet and afford summer pastures on their slopes. Many sheep tracks led through Samnium to the plains on the coast: one such has been traced leading by Bovianum, Saepinum, and Beneventum into Apulia along which in all probability Varro's flocks were driven for the rich winter pasture. Inscriptions[2] of the time of Marcus Aurelius on the arch which still stands on the site of Saepinum (now called Altilia) prove that it spanned a Roman sheep track and that in that place the nomadic flocks were registered.[3] The *calles* were the property of the state and supervised by agents (*publicani*) who collected the tolls for the use of public land. There were, too, grazing grounds along the route of the *calles* where the flocks halted to feed and rest during the long trek which took several days: payment was exacted also for these. The supervision of the public sheep tracks was in the hands of the censors (*censores*). Of such importance were the *calles* that they came to be regarded a *provincia*, a sphere of government allotted to a magistrate: in 59 B.C. the oversight of this *provincia* was assigned by the Senate to Julius Caesar in his first consulship, in an attempt perhaps to curb his growing ambitions.

Migrating flocks still move regularly in Italy along the ancient droves. They are still pastured for instance on the Roman Campagna in winter, even on the very outskirts of Rome: until lately, flocks of sheep were to be seen being driven through the City on changing pastures. Sheep are, for instance, still pastured from about October to the end of May in the grounds of the Villa Doria Pamphili on the Janiculum, and are taken for summer grazing to the Marche, in the central hills near the Adriatic coast. Some are brought to graze on the Roman Campagna from the district of L'Aquila in the Abruzzi, where is the shepherds' settled home, others from Visso in the Marche. The men live on the

---

[1] *R. R.* II. ii. 9-10: III. xvii. 9 (text omitted).
[2] *C.I.L.* IX 2438.   In Saepinum the sheep track formed the *decumanus* (main street) of the town.   [3] *R.R.* II. i. 17.

Campagna in straw huts which seem to recall a primitive age and must be very like what Varro knew for his own slave-herdsmen. The life of the shepherd who looked after the transhumanting flocks was one of endurance. Steepness and roughness of mountain and long distances trudged up rough paths, and nights spent out of doors exposed to the weather all contributed to the hardships they had to face: Varro recommends[1] that the slave purchased for this occupation should be robust, able to move quickly, to lift heavy loads on to pack animals, alert and with strength to grapple with robbers and wild beasts. On the chief shepherds who led the migrating flocks lay a great responsibility requiring careful calculation and foresight in addition to bodily strength. Lambing times had to be carefully watched for those born in one pasture would have to be strong enough to walk all the way by the time the return trek began, which might involve steep climbs and rough stony ground. Also there had to be careful calculation to ensure that the flock reached a grazing ground along the trail each night where a halt could be made in time for the evening pasture, for folding and the night's rest. The journeys were beset by dangers— storms might delay them, the lambs might be born too soon before reaching the summer or winter grounds, or a flock might reach grazing land only to find other herds already in possession. Since it was necessary to fold the sheep each night an enclosure was made with hurdles which were wooden frames filled with wicker work, or with nets supported on stakes set firmly in the ground. After the evening pasture the shepherds would assemble the flocks, with the dogs to assist, within the fold. They were then safe from wolves which in those days infested the mountainous and remote places, and were kept from straying. Their pasturage also could be restricted; in winter for instance they should not feed until the hoar frost had melted. Even when the flocks moved along the *calles* from summer grazing to winter and from winter to summer, wherever they halted for the night, a temporary fold would be made, for there were added dangers along the old tracks which traversed remote and wild places: robbers might ravage the flock and even other shepherds with flocks changing pasture by the same route might make depredations. There is no wonder that Varro describes these folds as being made *in solitudine*.[2]

The shepherds would have to take with them a great load of paraphernalia of many kinds. They would need everything necessary for sleeping rough and for their life in huts if they were leaving their settled homes on the plains or in the mountains; blankets and coverings

[1] *R.R.* II. x. 1-4. [2] *R.R.* II. ii. 9.

for tents, cooking pots, water jars, food and drink for the several days of the journey, even perhaps some fodder, were among the many things needed.   Pack animals, mules, and donkeys accompanied the migration carrying great bundles and loaded panniers.   Dogs were essential for the safety of the transhumanting flocks and herds and were their constant companions on the trail.   The ideal breed is described in detail by Varro.[1]   The Roman sheep-dog was primarily their defender against the marauding wolf which must have been a constant and real danger in the remote places in the mountains.   It had to be strong and fearless enough to attack a wolf and make it drop its prey.   The modern sheep-dog called by the Italians *canelupo* which is about the size of a large English collie, with a thick shaggy white coat and black eyes, corresponds very well to that described by Varro.

---

[1] *R.R.* II. ix. 1-5.

Chapter III

THE *RES RUSTICAE*

The *Res Rusticae* was published in 37 B.C., ten years before Varro's death. The literary form in which the book is written is that of the dialogue, or conversation, in which several speakers take part. By means of discussion, question, and answer, information is given on a set subject, or after reasoned argument some conclusion is reached. Often one person speaks at length and the others have insignificant parts. The talk is natural, easy, and reflects the manners of the day. A scene is set, a place and occasion described, and the participants who are real people take up the tale. A little living drama is represented, but without a stage. The dialogue originated among the Greeks from the conversations of the philosopher Socrates with his companions. His custom was to develop and expound his philosophic enquiries by means of question and answer; groups of hearers gathered round him in the Agora in Athens or other places where philosophers and students could reason together. The form became established in Plato's dialogues, in which Socrates is often the chief speaker. In them major philosophic questions are discussed, in the course of which philosophic truths are revealed and expounded. Among those who wrote in Greek on agricultural subjects, Xenophon wrote a treatise in dialogue form on the farming of a small estate, the *Economicus*. Latin writers used the form. Cicero employed it in many of his political and philosophic books, especially the *De Oratore* and the *Tusculanae Disputationes* and the *Academica* in which Varro himself takes part as one of the interlocutors. In adopting the conversational form for presenting his material Varro was only following a long tradition going back nearly four hundred years to the Greek thinkers.

The *Res Rusticae* contains three books, each of which has a different subject. The first treats of agriculture in general, the second of the rearing of cattle and sheep, the third of the breeding of smaller stock (*pastio villatica*) on the home-farm. Each has a different setting and dramatic date, each of the three scenes is moving and lively; each picture presented is like some *genre* painting (but one which comes alive) giving a glimpse of Roman life as vivid as anything that might

23

be acted on the stage. Most of the speakers have names to fit the subjects which are discussed and are living people, chosen from among Varro's own friends. Their names correspond with the subject they discuss, often in a punning way after Varro's favourite manner. The first book is set in Rome, on the occasion of the festival of Tellus, the goddess of the earth. Those who take part in the dialogue have been invited to dinner by Lucius Fundilius, the sacristan of the Temple of Tellus: they are Varro himself, Fundanius, his father-in-law, and four others. Just when they are arriving, the sacristan is called away to attend on the aedile, the magistrate who has charge of the temple. We see them then waiting for him in the hall in the Temple sitting on the benches and looking at a map of Italy painted on the opposite wall. They discuss agriculture, decide on its true limitations, and then divide the subject under headings: the location and building of the farm, the best land-usage, machines and tools, the different plantations and crops, and the care needed in all operations are the subjects on which they talk. The conversation is long and it seems as if they have forgotten about their host and the invitation to dinner, but it ends with the sudden appearance of the sacristan's freedman who is in a state of distress. To their consternation he brings the news that his master has been stabbed by mistake by some unknown person with a knife in the crowd and has died of the wound. Varro and his friends go their several ways home sorrowing more for the fate of the old man than for the temper of the times. In such a scene as this Varro shows us how acts of violence were a commonplace during the turbulent days of the Civil Wars. The dramatic date, that is the time at which Varro describes the dialogue as taking place, is not clear but it is thought by some scholars, judging from the allusion to Corcyra, to be some time after the battle of Pharsalia, which took place in 48 B.C.

The place, the occasion, and the date of the dialogue of Book II are difficult to decide, because some material appears to be wanting and we are plunged into a conversation which has already begun. It can be concluded, however, that Varro and his friends meet on the day of a festival, and that since the subject of the book is cattle- and sheep-rearing, the occasion is the *Parilia*, the shepherd festival on April 21st and the birthday of Rome. We are not in Rome for this festal day, but, it appears, in Epirus, the country of large flocks and herds. The time appears to be when Varro, in command of the fleet operating on the coast during the war against the pirates in 67 B.C., met certain of the landed proprietors of the country, one of whom was Atticus, Cicero's friend. Varro speaks first about the origin, history, and repute of stock-breeding, followed by Scrofa who speaks of the science of the subject. Atticus speaks next about sheep-rearing, and others follow on goats, pigs, cows, asses, horses: at this point there is an interruption when a freedman of Menates brings a message that the cakes have been offered and the sacrifice is ready and asks that the gentlemen come if they wish to take part. The discussions, however, continue with a discourse from Murrius on asses and one on dogs by Atticus. The last "animal" to be discussed is the herdsman: Cossinius describes the ideal type, which must be sturdy and able to endure hardship, and goes on to talk about profit from milk, cheese, and wool, giving instructions for sheep-shearing and goat-clipping. The conversation ends and the friends disperse when a freedman of Vitulus comes with a message asking them not to shorten the holiday but come in good time. Scrofa and Varro go on to Vitulus' house and the others either go home or to Menates' house.

The setting of the Third Book is the most vivid of the three, and full of dramatic event and movement. We see enacted as it were before our eyes a scene familiar in Rome from year to year. It is the day of the election of the curule aediles, about the year 54 B.C. Varro and his friend Axius of Reate have come to Rome for the election. After casting their votes, they go and shelter from the heat in the nearby *Villa Publica* while they wait for the results to be announced. They find there Appius Claudius who as augur is ready to assist the presiding consul if he is needed, and sitting with him four friends whose names are also those of birds. The conversation which ensues during the waiting period ranges over the rearing of small stock on the home-farm, including aviaries and enclosures for all kinds of birds, peafowl, turtle-doves, fieldfares, geese, teal, ducks and hens, preserves for hare and boar, deer, and fish-ponds for both fresh- and salt-water fish. In the course of the discussion on birds Varro describes his aviary built for pleasure at his villa at Casinum. As the talk goes on there are interruptions: shouting is heard in the Campus and a messenger comes to say that a man has been caught tampering with a ballot-box. Pavo, one of the speakers, goes off. In a little while a message comes from the consul to say that the augurs are needed. Appius has to leave them, but is not long away. Eventually Pavo comes back to say that the lots are being cast and the herald is announcing the results. All hurry out, leaving Varro and Axius to finish the dialogue together. Axius talks about the rearing of fish until interrupting him the aedile elect comes wearing the broad stripe of a magistrate's toga. Varro and Axius escort him to the Capitol and then they all depart, he to his home and Varro and Axius to theirs. The dialogue in Book III is enlivened throughout by the events in the background: the election crowds are close at hand, there is frequent coming and going. Among the friends in the Villa, there is their anxiety (admittedly unexpressed) about their own candidate: it adds to the general restlessness and excitement of the day.

Varro's style is far from the elegant, unlaboured prose of Cicero or the simplicity and clarity of Caesar, his younger contemporaries. First and foremost he was an encyclopaedist, as one who with the industry of a great scholar, the greatest of his day, was always at pains to acquire knowledge and forever insatiable. His method was to read voraciously and widely, and as he read to make excerpts which were written on scrolls, probably not by himself but by a scribe. These may have been taken down unaltered, or in the form of digests or even notes. Of this we cannot be quite certain, but undoubtedly he never read a manuscript without extracting material which he wished to keep. His canvas was vast. He was a hasty reader and a hurried writer. His works covered all the then known arts and sciences, including philosophy, astronomy, history, rhetoric, law, music, linguistics, medicine, geography, architecture, antiquities, and literature; his range was so wide that only by working at high speed could he have achieved what he did. His aim was to collect material from all possible sources and to have it written down. There was no time for re-reading, for literary polish, or revision. His excerpts were probably kept in the raw just as they had been dictated.

Varro's prose has often been the subject of criticism in both ancient and modern times, and it is not to be denied that his language shows a lack of polish. This is due to several causes. His love of antiquity probably accounts for an archaic roughness in his expression, for instance in the placing of clauses and in grammatical constructions more characteristic of an earlier time such as that in the plays of Plautus and Terence, than of his contemporaries. His looseness of style may also be due to deliberate colloquialism. Many turns of expression seem to be such as are natural to conversation, where not overmuch attention is given to normal grammar, and sentences may even be disconnected. A group of friends are at ease and talk together: grammatical forms disappear in plain speech. Among the main characteristics of his inelegance of language is especially the frequent use of anastrophe, that is, the misplacing of a conjunction, phrase, or even a whole clause, sometimes in an anticipatory position, sometimes postponed. This is frequent in early Latin, and in the colloquial writing of Cicero's letters. Varro's treatment of relative clauses where for instance the relative is often placed before the antecedent, or is in an unnatural position, is to be noticed. There is repeated omission of the parts of *esse* even of the subjunctive: pronouns are often left to be understood. The generalising second or third person, not always logically followed out within the sentence, is common. A sudden change of subject is fairly frequent, and there are several instances of looseness of agreement which are sometimes difficult to resolve.

His use of the impersonal gerundive with an object is frequent; this is both an archaic and also colloquial use. It is still more remarkable when he uses the normal construction alongside the old. There are also several instances of anacolutha: sentences in which there is a break in the construction when the statement passes from one construction to another and the grammar is disconnected. This could be due to colloquialism, or just as easily to lack of revision. There are several inconsistencies in the use of the subjunctive, sometimes within the same sentence, and several departures from normal practice. There are, too, trickling sentences which seem to end with an afterthought. Three channels of thought are therefore found to be informing influences in his style: archaism arising from his antiquarianism, colloquialism the result of an effort to represent material as it might have been described in a conversation between friends, and haste in writing with consequent lack of revision and polish.[1] Varro is rarely obscure for all his faults and even in his departures from logical construction hardly ever is his meaning in doubt.

Varro takes his place in the long line of ancient agronomists some seven hundred years after the first Greek writers but early among the Romans. The tradition among the Greeks, exemplified in his own list of some fifty writers drawn up for the instruction of his wife Fundania,[2] shows that writing on agriculture, either in the form of general treatises or on specialised subjects,[3] was a well established practice, from as early as Hesiod's in the *Works and Days*, assigned to the eighth century. Among the Romans, as far as is known, Marcus Porcius Cato is the first to have produced a treatise in Latin. His *De Agricultura* is also the earliest connected Latin prose which we possess: the style is rough and heavy, the material more a collection

---

[1] The remarks on Varro's style owe much to E. Laughton, "Observations on the Style of Varro", *Classical Quarterly* (New Series, 1960), pp. 1-26.

[2] *R.R.* I. i. 1.

[3] *Ibid.* I. i. 8-11: (the text is omitted: see note).

of precepts and recipes than an ordered manual. The publication is to be dated about the year 160 B.C., some hundred and twenty-three years before the appearance of Varro's *Res Rusticae*. Cato's work antedates by some fourteen to twenty years one of the most important treatises of Roman times on which all the later writers drew and which must have exercised an incalculable influence on the practice of agriculture. After the fall of Carthage in 146 B.C. the great work of Mago the Carthaginian was translated from the Punic language into Latin by a committee under the direction of Decimus Silvanus:[1] the actual date of publication is not known for certain but it was in all probability not long delayed because the Senate regarded it as important for Roman agriculture. It was well known to Varro, who mentions Mago several times in the course of the *Res Rusticae*: Mago's works have all been lost (though they are sometimes quoted at length by other writers) as have those of the two Sasernae, father and son, which may also have appeared soon after Cato's works. They were known to Varro who was inclined to scoff at them. The writings of Cnaeus Tremellius Scrofa,[2] a contemporary and friend of Varro, who had large estates and was considered the leading authority on agriculture of his time, have not survived. A little earlier than the *Res Rusticae*, they are praised by Varro, and later writers pay tribute to him and to his good style.

The works of Mago, the Sasernae, and Scrofa have all perished:[3] we cannot assess Varro's debt to them but we can set his writing beside that of Cato to judge the progress that had been made. In the light of this comparison Varro's work has form, the content is logically set down, and is schematised throughout: it is put in a living background, the prose is fluent and though not polished the handbook is attractively written with even some wit and humour—and this is what it is, not an exhaustive treatise—far from the whole tenor of Cato's work. The invocation to the Muses and the many digressions on etymology and the history and derivation of words do not belong to a scientific treatise on agriculture, but add spice to Varro's material. Neither can he resist a pun or a sly dig at his contemporaries or the older writers. He seems to come to life in the pages of his book as a genial, good-hearted scholar with a certain sense of the humanity of one who has gone about amongst

---

[1] For the subsequent translations and abbreviations see notes on *R. R.* I. i. 10-11 (text omitted): also for references to Mago's books II. v. 18; III, ii. 13; and II. i. 27 (text omitted).

[2] Scrofa is one of the interlocutors in the conversation in Book I.

[3] This account of lost works owes much to H. Bardon, *La Littérature Latine Inconnue* (Paris, 1952), pp. 83, 84, 95, 318.

men and not withdrawn himself from the world: but then Cato never wrote poetry nor any treatise on philosophy.

The question of Vergil's debt to Varro in the composition of the *Georgics* is hardly within the scope of this book: it is only possible to outline the nature of the problem which is revealed in part by a consideration of dates. It is generally accepted that Varro's *Res Rusticae* was published in 37 B.C. This or the next year is the supposed date for the time when Vergil began to compose the *Georgics*: the task took seven years and was completed about 30 B.C. In both works the same agricultural subjects are treated of, such as (to follow Vergil's order) the soil, and tillage, propagation by slips and grafting, the tending of the vine, the rearing of farm animals, horses, oxen, sheep and goats, and bee-keeping; but in details there is a nearer relationship than this. In many passages both long and short the affinity is so close as to be startling; nowhere perhaps does it appear more vividly than in the descriptions of the pasturing of sheep,[1] and in bee-keeping.[2] How far Vergil was influenced by Varro and drew on the newly published *Res Rusticae*, or how far the two writers drew on a common source or sources cannot be determined with the scant knowledge we have. Detailed research might eventually throw a little light on this darkness but it is doubtful whether any useful conclusion could be reached. In the course of the notes attention has been drawn to some of the more noticeable comparisons with the *Georgics*, but the student is recommended to read them if possible in whole, or at least in part, alongside Varro. The reward will be ample, for beside the fascinating enquiries aroused in the mind, there will be vouchsafed an appreciation of the beauty of the Vergilian passages: by contrast with Varro's practical account with an eye to profit though not without some feeling for beast and man, there will shine out Vergil's true love of country life, and the sweetness of its solace, his compassion for the working beast, the flocks at pasture, the mare in foal, and his sympathy for all wild life and tender feelings for lesser creatures. In language Vergil's gentle touch, his sensitivity, the beauty which he sees with a poet's visionary eye in soil, field, vineyard, and olive grove and the inward joy of the countryman will be comprehended. Varro looks and calculates, Vergil regards and pities.

The *Res Rusticae* gave Varro an important place as an agronomist: in spite of its many shortcomings the work was prized from his own time onwards followed by all such writers after him and lasted in

---

[1] *Georgics III*, 295-393: *R.R.* II. ii. 7-12.
[2] *Georgics IV*, 8-115, 149-315: *R.R.* III. xvi. 4-30.

fame into the Middle Ages. The next treatise of importance after his time did not appear for nearly a hundred years. In A.D. 60 there was published the *Res Rustica* of Columella: far from being a handbook, it is a comprehensive and systematic treatise on farming in twelve books written for the information of landed gentry. He draws from Varro among others and speaks of Varro as a contemporary of his own grandfather. Columella's book is both practical and scientific, but Book X on gardens is written in hexameters after the manner of the *Georgics*. Unlike that of Varro his prose style is clear and at times to be compared with that of Cicero. Not until the fourth century is there to be found another agricultural treatise of importance. Palladius, who drew freely on Varro, wrote an *Opus Agriculturae* in fourteen books, twelve of which are arranged month by month, which was used extensively by medieval scientists. In the Middle Ages, after a long gap, is the major figure of Crescentius. He was born at Bologna in 1230 and was, by profession, a lawyer. After practising for thirty years he withdrew to his villa at Bologna in 1299. There he wrote a treatise on agriculture, usually known as the *Liber Ruralium Commodorum*, which was to make his name famous. The work is the first in an age when the sciences were being revivified: it was dedicated to Charles II of Anjou, King of Apulia. In compiling his treatise Crescentius drew largely on the ancient writers, among them Varro, whose importance during the Middle Ages never waned. Crescentius' work is among the greatest contributions which have come down to us from those times, and in it is the unbroken link with the Roman times of Varro. Later writers drew from him freely and his influence was continuous from his own to the Middle Ages.

Select Bibliography

It is not possible within the compass of this book to include the names of all the authorities which were consulted in the course of this edition. The ones listed below are those which are of greater importance and in some instances those more readily available to English readers: many others have been regretfully omitted.

EDITIONS OF THE *RES RUSTICAE*

The following are among the principal editions:

Gesner, J. M.   *Scriptores Rei Rusticae* (Leipzig, 1735). Vol. I.

Schneider, J. G.   *Scriptores Rei Rusticae* (Leipzig, 1794). Vol. I.

Keil, H.   *M. Porci Catonis de Agri Cultura Liber; M. Terenti Varronis Rerum Rusticarum Libri Tres* (Leipzig, 1884-95).

Goetz, G.   *M. Terenti Varronis Rerum Rusticarum Libri Tres* (Leipzig, 1929).

(The last named, by Goetz, is the second Teubner edition and is the text of this book of selections from the *Res Rusticae* in the Loeb edition of 1960.)

TRANSLATIONS OF THE *RES RUSTICAE*

Storr-Best, Ll.   *Varro on Farming*, Bohn Library (London, 1912).

Hooper, W. D., revised Ash, H. B.   *Marcus Porcius Cato on Agriculture, Marcus Terentius Varro on Agriculture*, Loeb Classical Library: 1st edition, 1934 (reprinted 1960). The text is that of the second Teubner edition (see above).

VARRO'S LIFE

Boissier, G.   *Étude sur la Vie et les Ouvrages de M. T. Varron* (Paris, 1861).

Cichorius, C.   *Römische Studien* (Berlin, 1922). V. Historische Studien zu Varro: 1. Zu Varros Lebensgeschichte, pp. 189-207.

Della Corte, F.   *Varrone: Il Terzo Gran Lume Romano* (Genova, 1954).

GENERAL STUDIES

Laughton, E.   "Observations on the Style of Varro": *Classical Quarterly* (New Series), X. 1, pp. 1-28.

Skydsgaard, J. E.   "Varro the Scholar. Studies in the First Book of Varro's De Re Rustica": *Analecta Romano Instituti Danici IV, Supplementum* (Copenhagen, 1968).

Wilkinson, L. P.   *The Georgics of Vergil. A Critical Survey* (Cambridge, 1969).

BEES

Billiard, R.   *Notes sur l'Abeille et L'Agriculture dans l'Antiquite d'apres les Ouvrages des Anciens Grecs et Latins:* (Lille, 1900).

SELECT BIBLIOGRAPHY

Royds, T. F.   *The Beasts, Birds and Bees of Vergil* (Oxford, 1930).
Tickner Edwardes   *The Lore of the Honey Bee* (London, 1908).
Zappi-Recordati, A.   *Apicoltura Moderna* (Roma, 1955).

GRAFTING
Marinucci, M.   *Innesto delle Piante da Frutto* [Roma (no date).]

HISTORY
Frank, T.   *Economic Survey of Ancient Rome*, Vol. I (Baltimore, 1933).
Michels, A. K.   *The Calendar of the Roman Republic* (Princeton, New Jersey, 1967).
Taylor, L. R.   *Roman Voting Assemblies* (University of Michigan Press; Ann Arbor, 1966).
Toynbee, A.   *Hannibal's Legacy*: Vols. I and II (Oxford, 1965).

LAW
Jolowicz, H. F.   *Historical Introduction to the Study of Roman Law* (Cambridge, 1952).

OLIVE-CULTURE
Marinucci, M.   *L'Impianto dell'Oliveto* (Roma, 1960).

SLAVES
Buckland, W. W.   *The Roman Law of Slavery* (Cambridge, 1908).
Heitland, W. E.   *Agricola: A Study of Agriculture and Rustic Life in the Greco-Roman World from the Point of View of Labour* (Cambridge, 1921).

FARM TOOLS AND MACHINES
Drachmann, A. G.   *Ancient Oil Mills and Presses* (Det Kgl Danske Veilenskabernes Selskuk Archaeologisk Kunsthistorischer Meddelelser I. 1). (Copenhagen, 1932.)
Forbes, R. J.   *Studies in Ancient Technology*, Vol. III (Leiden, 1955).
Moritz, L. A.   *Grain Mills and Flour in Classical Antiquity* (Oxford, 1958).
White, K. D.   *Agricultural Implements of the Roman World* (Cambridge, 1967).

TOPOGRAPHY
Carettoni, G. F.   *Casinum (Presso Cassinum) Regio I—Latium et Campania.* Serie I, Vol. II. (Istituto di Studi Romani, 1940-48.)
Semple, E. C.   *Geography of the Mediterranean Region: Its relation to Ancient History* (New York, 1931).

TRANSHUMANCE
Grenier, A.   *La Transhumance des Troupeaux en Italie et son Rôle dans l'Histoire Romaine.* (Mélanges de l'Ecole Française de Rome XXV, 1905, p. 293 *sq.*)
Salmon, E. T.   *Samnium and the Samnites* (Cambridge, 1967).

SELECT BIBLIOGRAPHY

THE VILLA

Gabriel, M. M.   *Livia's Garden House at Prima Porta* (New York, 1955).
White, K. D.   *Latifundia,* University of London Institute of Classical Studies. Bulletin No. 14 (London, 1967).
Grimal, P.   *Les Jardins Romains,* Deuxième Édition Revue, " Collection Hier " (Paris, 1969).

VITICULTURE

Billiard, R.   *La Vigne dans l'Antiquité* (Lyon, 1913).
Marescalchi, A.
    et Dalmasso, G.   *Storia della Vite e del Vino in Italia* (Milan, 1931-37).
Ricci, A.   *l'Allevamento della Vite* (1968² Roma).

GENERAL REFERENCE

Enciclopedia Italiana (1931).
Dizionario dell' Enciclopedia Italiana (1957).
Publications and Maps of the Touring Club Italiano.
Oxford Classical Dictionary (1949).

The following specialised bibliography recently published will be found indispensable. There is a useful introduction by the author (pp. vii-xxiii), together with short critical notes on the various books.

K. D. White, *A Bibliography of Roman Agriculture.* Institute of Agricultural History: *Bibliographies in Agricultural History; Number* 1 (University of Reading 1970).

PLATE I

A general view of Rieti (ancient Reate), where Varro was born. The Sabine Mountains are in the background. (Photo: *Gabinetto Fotografico Nazionale*.)

PLATE II

A relief from a sarcophagus of the third century A.D. in the National Museum of Rome, showing scenes on an estate. The sheep with thick fleeces and wool on their foreheads, the large-horned goats nibbling the trees, the heavy-dewlapped oxen, the sheepdog with large paws and wide mouth, all recall Varro's descriptions of farm animals. The man at the top holds a *falx arboraria* for lopping trees; those below are shepherds. One sitting outside a straw hut milks a goat; the other, with his dog, watches the flock. (Photo: *Deutsches Archäologisches Institut, Rome.*)

M. TERENTI VARRONIS
RERUM RUSTICARUM

LIBER PRIMUS

Introduction

I i, 1-4. I dedicate this work to you my wife Fundania, writing in haste because I have reached my eightieth year. Since you have just bought an estate, I wish to instruct you in methods of management which will make it profitable. Like the Sibyl I hope to leave behind me advice which will continue to be of help to you and others even after my death. My intention is to write three handbooks for your guidance. I shall first invoke the help of twelve Councillor gods, not those whose images stand in the Forum in Rome, but rustic gods who are the farmers' especial guardians.

    i. OTIUM si essem consecutus, Fundania, commodius tibi haec scriberem, quae nunc, ut potero, exponam cogitans esse properandum, quod, ut dicitur, si est homo bulla, eo magis senex. Annus enim octogesimus admonet me ut sarcinas conligam, antequam pro-
2 ficiscar e vita. Quare, quoniam emisti fundum, quem bene colendo fructuosum cum facere velis, meque ut id mihi habeam curare roges, experiar; et non solum, ut ipse quoad vivam, quid fieri
3 oporteat ut te moneam, sed etiam post mortem. Neque patiar Sibyllam non solum cecinisse quae, dum viveret, prodessent hominibus, sed etiam quae cum perisset ipsa, et id etiam ignotissimis quoque hominibus; ad cuius libros tot annis post publice solemus redire, cum desideramus, quid faciendum sit nobis ex aliquo portento; me, ne dum vivo quidem, necessariis meis quod prosit
4 facere. Quocirca scribam tibi tres libros indices, ad quos revertare, siqua in re quaeres, quem ad modum quidque te in colendo oporteat facere. Et quoniam, ut aiunt, dei facientes adiuvant, prius invocabo

35

eos, nec, ut Homerus et Ennius, Musas, sed duodecim deos Consentes; neque tamen eos urbanos, quorum imagines ad forum auratae stant, sex mares et feminae totidem, sed illos XII deos, qui maxime agricolarum duces sunt.

Invocation to the Gods of Country Life

i, 5-7. I invoke first Jupiter and Tellus, because the one is father, the other the mother: secondly Sun and Moon because they control the seasons, and thirdly Ceres and Liber who provide food and drink, the necessities of life: fourthly Robigo and Flora who guard the crops from disease: then Minerva and Venus, the one goddess of the olive groves, the other of gardens: lastly Lympha and Bonus Eventus, the first of whom ensures a supply of fresh water, and the second gives success and good yields to the farmer. I am now ready to give an account of conversations (or dialogues) I have recently held with various friends about agriculture. I shall also give you a list of writers in both Greek and Latin to whom you may refer for any more information you may need.

5   Primum, qui omnes fructos agri culturae caelo et terra continent, Iovem et Tellurem; itaque, quod ii parentes magni dicuntur, Iuppiter pater appellatur, Tellus terra mater. Secundo Solem et Lunam, quorum tempora observantur, cum quaedam seruntur et conduntur. Tertio Cererem et Liberum, quod horum fructus maxime necessari ad victum; ab his enim cibus et potio venit
6 e fundo. Quarto Robigum ac Floram, quibus propitiis neque robigo frumenta atque arbores corrumpit, neque non tempestive florent. Itaque publice Robigo feriae Robigalia, Florae ludi Floralia instituti. Item adveneror Minervam et Venerem, quarum unius procuratio oliveti, alterius hortorum; quo nomine rustica Vinalia instituta. Nec non etiam precor Lympham ac Bonum Eventum, quoniam sine aqua omnis arida ac misera agri cultura,
7 sine successu ac bono eventu frustratio est, non cultura. Iis igitur deis ad venerationem advocatis ego referam sermones eos quos de agri cultura habuimus nuper, ex quibus quid te facere oporteat animadvertere poteris. In quis quae non inerunt et quaeres, indicabo a quibus scriptoribus repetas et Graecis et nostris. Qui Graece scripserunt dispersim alius de alia re, sunt plus quinquaginta.

Writers on Agriculture
(the text is omitted: see notes on 8-11).

i, 8-11. *I give you a list of some fifty writers in Greek on agriculture
but Mago of Carthage who wrote a comprehensive work in 28 books is greater
than them all.   I myself will treat the subject in only three books, one on
agriculture, one on animal husbandry, and the third on the home farm
respectively.   I shall be guided by my own observations, knowledge, and
experience.*

The Setting and Occasion of the Conversation about Agriculture

ii, 1-2. *I had been invited to go to the temple of Tellus in Rome to
dine with the sacristan on the day of the festival, the* sementivae feriae.
*I found three of my friends there already, looking at a map of Italy:
they were Gaius Fundanius, Gaius Agrius, and Publius Agrasius.   Agrius
told me that we should have to wait for our host because he had been called
away.   It was suggested that we should sit on a nearby bench until he
returned.*

ii. Sementivis feriis in aedem Telluris veneram rogatus ab
aeditumo, ut dicere didicimus a patribus nostris, ut corrigimur a
recentibus urbanis, ab aedituo. Offendi ibi C. Fundanium,
socerum meum, et C. Agrium equitem R. Socraticum et P. Agrasium
publicanum spectantes in pariete pictam Italiam.   Quid vos hic?
inquam, num feriae sementivae otiosos huc adduxerunt, ut patres
2 et avos solebant nostros?   Nos vero, inquit Agrius, ut arbitror,
eadem causa quae te, rogatio aeditumi.   Itaque si ita est, ut annuis,
morere oportet nobiscum, dum ille revertatur.   Nam accersitus ab
aedile, cuius procuratio huius templi est, nondum rediit et nos uti
expectaremus se reliquit qui rogaret.   Voltis igitur interea vetus
proverbium, [*inquam*] quod est " Romanus sedendo vincit",
usurpemus, dum ille venit?   Sane, inquit Agrius, et simul cogitans
portam itineri dici longissimam esse ad subsellia sequentibus nobis
procedit.

The Geographical Situation of Italy

ii, 3-5. *Agrasius began the conversation by asking whether any land
was more widely cultivated than Italy.   Agrius said there was none, and*

recalled that according to Eratosthenes' geographical calculations Italy was in the northern and therefore the healthier and more productive part of the earth. Beyond were the arctic regions with perpetual light for six months of the year and a frozen sea. Fundanius exclaimed that nothing could grow there and that even in Italy where day and night alternated he could not go without his daily siesta.

3   Cum consedissemus, Agrasius, Vos, qui multas perambulastis terras, ecquam cultiorem Italia vidistis? inquit. Ego vero, Agrius, nullam arbitror esse quae tam tota sit culta. Primum cum 4 orbis terrae divisus sit in duas partes ab Eratosthene maxume secundum naturam, ad meridiem versus et ad septemtriones, et sine dubio quoniam salubrior pars septemtrionalis est quam meridiana, et quae salubriora illa fructuosiora, dicendum utique Italiam magis etiam fuisse opportunam ad colendum quam Asiam, primum quod est in Europa, secundo quod haec temperatior pars quam interior. Nam intus paene sempiternae hiemes, neque mirum, quod sunt regiones inter circulum septemtrionalem et inter cardinem caeli, ubi sol etiam sex mensibus continuis non videtur. Itaque in oceano in ea parte ne navigari quidem posse dicunt propter mare 5 congelatum. Fundanius, Em ubi tu quicquam nasci putes posse aut coli natum. Verum enim est illud Pacuvi sol si perpetuo sit aut nox, flammeo vapore aut frigore terrae fructos omnes interire. Ego hic, ubi nox et dies modice redit et abit, tamen aestivo die, si non diffinderem meo insiticio somno meridie, vivere non possum.

Praise of Italian Agriculture

ii, 6-8. Fundanius went on to say that every useful product was grown in Italy, and was of the highest quality. The best grain came from Campania, the best wheat from Apulia, the best wine from the Falernian district, the best oil from Venafrum. The produce of places in Greece famous for wines and corn could not be compared with hers. No land had a bigger yield of wine per iugerum. Marcus Cato recorded vast yields in the Ager Gallicus and Faventia. Fundanius reminded me that my friend the engineer used to say the same about Faventia. The Italian farmers, he said, had two guiding principles, yield in proportion to investment and the healthiness of the land.

6   Illic in semenstri die aut nocte quem ad modum quicquam seri aut alescere aut meti possit? Contra quid in Italia

utensile non modo non nascitur, sed etiam non egregium fit?
Quod far conferam Campano? Quod triticum Apulo? Quod
vinum Falerno? Quod oleum Venafro? Non arboribus consita
7 Italia, ut tota pomarium videatur? An Phrygia magis vitibus
cooperta, quam Homerus appellat ἀμπελόεσσαν, quam haec?
Aut tritico Argos, quod idem poeta πολύπυρον? In qua terra
iugerum unum denos et quinos denos culleos fert vini, quot quaedam
in Italia regiones? An non M. Cato scribit in libro Originum sic:
"ager Gallicus Romanus vocatur, qui viritim cis Ariminum datus
est ultra agrum Picentium. In eo agro aliquotfariam in singula
iugera dena cullea vini fiunt"? Nonne item in agro Faventino,
a quo ibi trecenariae appellantur vites, quod iugerum trecenas
amphoras reddat? Simul aspicit me, Certe, inquit, Libo Marcius,
praefectus fabrum tuos, in fundo suo Faventiae hanc multitudinem
8 dicebat suas reddere vites. Duo in primis spectasse videntur
Italici homines colendo, possentne fructus pro impensa ac labore
redire et utrum saluber locus esset an non. Quorum si alterutrum
decolat et nihilo minus quis vult colere, mente est captus adque
adgnatos et gentiles est deducendus. Nemo enim sanus debet velle
impensam ac sumptum facere in cultura, si videt non posse refici,
nec si potest reficere fructus, si videt eos fore ut pestilentia dispereant.

Gaius Licinius Stolo and Gnaeus Tremelius Scrofa

ii, 9-10. *Fundanius then said that Stolo and Scrofa were approach-*
ing: that an ancestor of Gaius Licinius Stolo was famous as the originator
of the lex Licinia: *and he himself had justified his own name Stolo*
because of his good farming. He was so thorough that all suckers were
cleared out from the roots as soon as they appeared: another ancestor of
his with democratic sympathies was the first to address the people in the
Forum on a political question rather than the patricians in the comitium.
Scrofa was a member of the Campanian Land Commission. I added that
his estates were so well managed that they had become show places for their
produce.

9   Sed, opinor, qui haec commodius ostendere possint adsunt.
Nam C. Licinium Stolonem et Cn. Tremelium Scrofam video
venire; unum, cuius maiores de modo agri legem tulerunt (nam
Stolonis illa lex, quae vetat plus D iugera habere civem R.), et qui

propter diligentiam culturae Stolonum confirmavit cognomen, quod nullus in eius fundo reperiri poterat stolo, quod effodiebat circum arbores e radicibus quae nascerentur e solo, quos stolones appellabant. Eiusdem gentis C. Licinius, tr. pl. cum esset, post reges exactos annis CCCLXV primus populum ad leges accipiendas
10 in septem iugera forensia e comitio eduxit. Alterum collegam tuum, viginti virum qui fuit ad agros dividendos Campanos, video huc venire, Cn. Tremelium Scrofam, virum omnibus virtutibus politum, qui de agri cultura Romanus peritissimus existimatur. An non iure? inquam. Fundi enim eius propter culturam iucundiore spectaculo sunt multis, quam regie polita aedificia aliorum, cum huius spectatum veniant villas, non, ut apud Lucullum, ut videant pinacothecas, sed oporothecas. Huiusce, inquam, pomarii summa sacra via, ubi poma veneunt contra aurum, imago.

Stockbreeding and Agriculture

ii, 11-14. Agrius assured Stolo and Scrofa that they were not too late for dinner because our host had not yet come: he suggested that until he came we should discuss the purpose of agriculture. Scrofa observed that the question was whether agriculture included sheep and cattle and stated that many writers had taken the wider view and included too much, but others had confined themselves to agriculture proper. Stolo insisted that stockbreeding was no real part of agriculture. He agreed that there were two headmen in charge of the two different occupations; the bailiff and the chief herdsman.

11   Illi interea ad nos, et Stolo, Num cena comessa, inquit, venimus? Nam non L. videmus Fundilium, qui nos advocavit. Bono animo este, inquit Agrius. Nam non modo ovom illut sublatum est, quod ludis circensibus novissimi curriculi finem facit quadrigis, sed ne illud quidem ovom vidimus, quod in cenali pompa solet
12 esse primum. Itaque dum id nobiscum una videatis ac venit aeditumus, docete nos, agri cultura quam summam habeat, utilitatemne an voluptatem an utrumque. Ad te enim rudem esse agri culturae nunc, olim ad Stolonem fuisse dicunt. Scrofa, Prius, inquit, discernendum, utrum quae serantur in agro, ea sola sint in cultura, an etiam quae inducantur in rura, ut oves et armenta.
13 Video enim, qui de agri cultura scripserunt et Poenice et Graece

et Latine, latius vagatos, quam oportuerit. Ego vero, inquit Stolo, eos non in omni re imitandos arbitror et eo melius fecisse quosdam, qui minore pomerio finierunt exclusis partibus quae non pertinent ad hanc rem. Quare tota pastio, quae coniungitur a plerisque cum agri cultura, magis ad pastorem quam ad agricolam pertinere
14 videtur. Quocirca principes qui utrique rei praeponuntur vocabulis quoque sunt diversi, quod unus vocatur vilicus, alter magister pecoris. Vilicus agri colendi causa constitutus atque appellatus a villa, quod ab eo in eam convehuntur fructus et evehuntur, cum veneunt. A quo rustici etiam nunc quoque viam veham appellant propter vecturas et vellam, non villam, quo vehunt et unde vehunt. Item dicuntur qui vecturis vivunt velaturam facere.

Does Stockbreeding belong to Agriculture ?

ii, 15-18. *Fundanius agreed that there was a difference between stockbreeding and agriculture like that between the right and left pipe in a double wind-instrument. To support this argument I made mention of Dicaearchus who taught that in primitive times only stockbreeding was practised, and that agriculture came in at a later stage, and played the accompaniment as did the left pipe. Agrius replied that without stockbreeding as a part of agriculture the owner could not have herds, nor slaves have the privilege of grazing and that there were even laws to regulate grazing. Fundanius, after reproaching him for misquoting the law, remarked that the goat was harmful to all newly planted trees.*

15   Certe, inquit Fundanius, aliut pastio et aliut agri cultura, sed adfinis et ut dextra tibia alia quam sinistra, ita ut tamen sit quodam modo coniuncta, quod est altera eiusdem carminis modorum
16 incentiva, altera succentiva. Et quidem licet adicias, inquam, pastorum vitam esse incentivam, agricolarum succentivam auctore doctissimo homine Dicaearcho, qui Graeciae vita qualis fuerit ab initio nobis ita ostendit, ut superioribus temporibus fuisse doceat, cum homines pastoriciam vitam agerent neque scirent etiam arare terram aut serere arbores aut putare; ab iis inferiore gradu aetatis susceptam agri culturam. Quocirca ea succinit pastorali, quod
17 est inferior, ut tibia sinistra a dextrae foraminibus. Agrius, Tu, inquit, tibicen non solum adimis domino pecus, sed etiam servis peculium, quibus domini dant ut pascant, atque etiam leges colonicas

tollis, in quibus scribimus, colonus in agro surculario ne capra natum pascat; quas etiam astrologia in caelum recepit, non longe 18 ab tauro. Cui Fundanius, Vide, inquit, ne, Agri, istuc sit ab hoc, cum in legibus etiam scribatur " pecus quoddam". Quaedam enim pecudes culturae sunt inimicae ac veneno, ut istae, quas dixisti, caprae. Eae enim omnia novella sata carpendo corrumpunt, non minimum vites atque oleas.

The Goat and the Conclusion of the Question whether Stock-breeding belongs to Agriculture

ii, 19-21. Fundanius went on to say that because of the harm it does it became customary for the goat to be sacrificed to one god, but not at all to another, although both had the same dislike of it. He-goats were regularly sacrificed to Liber, the discoverer of the vine, as a punishment for harm done to his plants, but none to Minerva because they were especially harmful to the olive tree. Goats might not be driven on to the Acropolis in Athens except for a particular sacrifice. I then made the suggestion that only those cattle which were of use on the land should be included in agriculture. Agrasius protested that herds provided manure which was profitable, but Agrius argued that by such reasoning many other things would have to be included which were no real part of it.

19 Itaque propterea institutum diversa de causa ut ex caprino genere ad alii dei aram hostia adduceretur, ad alii non sacrificaretur, cum ab eodem odio alter videre nollet, alter etiam videre pereuntem vellet. Sic factum ut Libero patri, repertori vitis, hirci immolarentur, proinde ut capite darent poenas; contra ut Minervae caprini generis nihil immolarent propter oleam, quod eam quam laeserit fieri dicunt sterilem; eius enim salivam esse fructuis venenum; 20 hoc nomine etiam Athenis in arcem non inigi, praeterquam semel ad necessarium sacrificium, ne arbor olea, quae primum dicitur ibi nata, a capra tangi possit. Nec ullae, inquam, pecudes agri culturae sunt propriae, nisi quae agrum opere, quo cultior sit, adiuvare, ut 21 eae quae iunctae arare possunt. Agrasius, Si istuc ita est, inquit, quo modo pecus removeri potest ab agro, cum stercus, quod plurimum prodest, greges pecorum ministrent? Sic, inquit Agrius, venalium greges dicemus agri culturam esse, si propter istam rem habendum statuerimus. Sed error hinc, quod pecus in agro

esse potest et fructus in eo agro ferre, quod non sequendum. Nam sic etiam res aliae diversae ab agro erunt adsumendae, ut si habet plures in fundo textores atque institutos histonas, sic alios artifices.

The Books of the Elder and Younger Saserna
(the text is omitted)

ii, 22-28. *Scrofa said that it was generally agreed to exclude stockbreeding from a discussion on agriculture. I asked how I was to regard the writings of the two Sasernae who held that even clay-pits are related to agriculture. Stolo reproached me for being jealous of them: Scrofa smiled at this because he thought little of their books. When Agrasius asked for an example from them Stolo told of recipes which had no real bearing on agriculture, such as a charm to cure gout; indeed, it was apparent that there were many marvels in their writings equally irrelevant.*

Agriculture both an Art and a Science

iii. *Agrasius suggested that now agreement had been reached on what should be excluded from agriculture, the decision should next be made whether agriculture was an art or a science. Stolo said that Scrofa was the one best suited to answer the question. Scrofa asserted that it was both, and especially was it a science since it taught choice of crops and methods of working the soil to bring in the best returns.*

iii. Igitur, inquit Agrasius, quae diiungenda essent a cultura cuius modi sint, quoniam discretum, de iis rebus quae scientia sit in colendo nos docete, ars id an quid aliud, et a quibus carceribus decurrat ad metas. Stolo cum aspexisset Scrofam, Tu, inquit, et aetate et honore et scientia quod praestas, dicere debes. Ille non gravatus, Primum, inquit, non modo est ars, sed etiam necessaria ac magna; eaque est scientia, quae sint in quoque agro serenda ac facienda, quo terra maximos perpetuo reddat fructus.

The Farmer's Aims

iv, 1-5. *Scrofa continued: "You must gain some knowledge of the four elements, before you begin cultivating your land. The farmer should keep two ends in view, profit and pleasure: the first is the more important, yet a piece of land improved by cultivation and planting will fetch more money than one just as rich but not as attractive to look at. Healthy land is more profitable than unhealthy even though the latter may be just as fertile. Great risks are taken in farming unhealthy land. By the application of knowledge we can alleviate such risks in cases of unhealthy marshes, unfavourable orientation, and direction of wind if some outlay is made. The siting of buildings is of great importance. With such knowledge Hippocrates saved towns during epidemics and Varro saved the army and the fleet at Corcyra."*

iv. Eius principia sunt eadem, quae mundi esse Ennius scribit, aqua, terra, anima et sol. Haec enim cognoscenda, priusquam iacias semina, quod initium fructuum oritur. Hinc profecti agricolae ad duas metas dirigere debent, ad utilitatem et voluptatem. Utilitas quaerit fructum, voluptas delectationem; priores partes
2 agit quod utile est, quam quod delectat. Nec non ea, quae faciunt cultura honestiorem agrum, pleraque non solum fructuosiorem eadem faciunt, ut cum in ordinem sunt consita arbusta atque oliveta, sed etiam vendibiliorem atque adiciunt ad fundi pretium. Nemo enim eadem utilitati non formosius quod est emere mavult pluris,
3 quam si est fructuosus turpis. Utilissimus autem is ager qui salubrior est quam alii, quod ibi fructus certus; contra in pestilenti calamitas, quamvis in feraci agro, colonum ad fructus pervenire non patitur. Etenim ubi ratio cum orco habetur, ibi non modo fructus est incertus, sed etiam colentium vita. Quare ubi salubritas non est, cultura non aliud est atque alea domini vitae ac rei familiaris.
4 Nec haec non deminuitur scientia. Ita enim salubritas, quae ducitur e caelo ac terra, non est in nostra potestate, sed in naturae, ut tamen multum sit in nobis, quo graviora quae sunt ea diligentia leviora facere possimus. Etenim si propter terram aut aquam odore, quem aliquo loco eructat, pestilentior est fundus, aut propter caeli regionem ager calidior sit, aut ventus non bonus flet, haec vitia emendari solent domini scientia ac sumptu, quod permagni interest, ubi sint positae villae, quantae sint, quo spectent porticibus, ostiis
5 ac fenestris. An non ille Hippocrates medicus in magna pestilentia non unum agrum, sed multa oppida scientia servavit? Sed quid ego illum voco ad testimonium? Non hic Varro noster, cum Corcyrae esset exercitus ac classis et omnes domus repletae essent aegrotis ac funeribus, immisso fenestris novis aquilone et obstructis pestilentibus ianuaque permutata ceteraque eius generis diligentia suos comites ac familiam incolumes reduxit?

The Divisions of Agriculture

(the text is omitted)

v, 1-4.   *Scrofa said that the number of divisions of the subject should be decided on.   Agrius stated his opinion, that according to the writings of Theophrastus, the divisions into which agriculture could be separated were innumerable.   Scrofa asserted that there were four main divisions of the subject: (1) a knowledge of conformation and soil, (2) of the equipment needed, (3) of the various operations on the farm, (4) of the proper season for the farm operations.   Each of these four could be divided into at least two sub-divisions: the first had to do with both the actual soil and the siting of the farm buildings, the second the workers, men and animals and the movable equipment; the third both the schemes of work which had to be made and the place where each operation had to take place; and the fourth the times determined annually by the sun and monthly by the moon. The four main divisions would first be discussed and after that the eight sub-divisions.*

The Soil

vi, 1-6.   *Scrofa said that there were two kinds of conformation, the natural and that produced by tillage, and that both could be good or bad. There were three simple types of land, plain, hill, and mountain; a fourth type could include two or three of these together.   Different systems of agriculture should be observed in these different regions because of the differences in climate.   The same crops were planted and harvested earlier in the lowlands than in the uplands.   Vegetation was also influenced by altitude and climate: grain did best on the plains, vines on the hills, and forests on the mountains.   On the plains work was easier in the winter, summer was a better time for work in the mountains.   Land in the lowlands which sloped was better than that which was level because the latter might become marshy.*

vi. Igitur primum de solo fundi videndum haec quattuor, quae sit forma, quo in genere terrae, quantus, quam per se tutus.   Formae cum duo genera sint, una quam natura dat, altera quam sationes imponunt, prior, quod alius ager bene natus, alius male, posterior, quod alius fundus bene consitus est, alius male, dicam prius de 2 naturali.   Igitur cum tria genera sint a specie simplicia agrorum, campestre, collinum, montanum, et ex iis tribus quartum, ut in eo fundo haec duo aut tria sint, ut multis locis licet videre, e quibus tribus fastigiis simplicibus sine dubio infimis alia cultura aptior

45

quam summis, quod haec calidiora quam summa, sic collinis, quod ea tepidiora quam infima aut summa; haec apparent magis ita esse 3 in latioribus regionibus, simplicia cum sunt. Itaque ubi lati campi, ibi magis aestus, et eo in Apulia loca calidiora ac graviora, et ubi montana, ut in Vesuvio, quod leviora et ideo salubriora; qui colunt deorsum, magis aestate laborant, qui susum, magis hieme. Verno tempore in campestribus maturius eadem illa seruntur quae in superioribus et celerius hic quam ilic coguntur. Nec non 4 susum quam deorsum tardius seruntur ac metuntur. Quaedam in montanis prolixiora nascuntur ac firmiora propter frigus, ut abietes ac sappini, hic, quod tepidiora, populi ac salices; susum fertiliora, ut arbutus ac quercus, deorsum, ut nuces graecae ac mariscae fici. In collibus humilibus societas maior cum campestri fructu quam 5 cum montano, in altis contra. Propter haec tria fastigia formae discrimina quaedam fiunt sationum, quod segetes meliores existimantur esse campestres, vineae collinae, silvae montanae. Plerumque hiberna iis esse meliora, qui colunt campestria, quod tunc prata ibi herbosa, putatio arborum tolerabilior; contra aestiva montanis locis commodiora, quod ibi tum et pabulum multum, quod in campis aret, et cultura arborum aptior, quod tum hic 6 frigidior aer. Campester locus is melior, qui totus aequabiliter in unam partem verget, quam is qui est ad libellam aequos, quod is, cum aquae non habet delapsum, fieri solet uliginosus; eo magis, siquis est inaequabilis, eo deterior, quod fit propter lacunas aquosus. Haec atque huiusce modi tria fastigia agri ad colendum disperiliter habent momentum.

The Quincunx *Method of Planting*

vii, 1-3. *Stolo quoted Cato's saying that the ideal farm faced south at the foot of a mountain. Scrofa asserted that profits would be in proportion to the good appearance of the farm: as for instance the new method of planting trees in* quincunces *which brought in more and better wine; and more grain was reaped by good planting from the same amount of ground than their ancestors had obtained.*

vii. Stolo, Quod ad hanc formam naturalem pertinet, de eo non incommode Cato videtur dicere, cum scribit optimum agrum esse, qui sub radice montis situs sit et spectet ad meridianam caeli

2 partem. Subicit Scrofa, De formae cultura hoc dico, quae specie fiant venustiora, sequi ut maiore quoque fructu sint, ut qui habent arbusta, si sata sunt in quincuncem, propter ordines atque intervalla modica. Itaque maiores nostri ex arvo aeque magno male consito et minus multum et minus bonum faciebant vinum et frumentum, quod quae suo quicque loco sunt posita, ea minus loci occupant, et
3 minus officit aliud alii ab sole ac luna et vento. Hoc licet coniectura videre ex aliquot rebus, ut nuces integras quas uno modio comprendere possis, quod putamina suo loco quaeque habet natura composita, cum easdem, si fregeris, vix sesquimodio concipere possis.
4 Praeterea quae arbores in ordinem satae sunt, eas aequabiliter ex omnibus partibus sol ac luna coquunt. Quo fit ut uvae et oleae plures nascantur et ut celerius coquantur. Quas res duas sequuntur altera illa duo, ut plus reddant musti et olei et preti pluris.

The Type of Soil
(the text is omitted)

vii, 5-10. *Scrofa continued by saying that the type of soil on the farm determined the kind of crops which could be planted. To prove that soil controlled the type of plants successfully grown he cited several plants growing only in particular places, such as an evergreen plane and oak trees and vines and apple trees which bore twice in the year. Some plants too could grow on marshy ground—reeds near Reate and alder trees in Epirus. He had seen places in Gaul where no vines and olives would grow, where a white chalk was used as a fertiliser and where there was no salt. Stolo quoted Cato's nine categories of soil: he put the vine first on the best soil but Scrofa retorted that he assigned the best to meadows and cited the instance of the plains of Rosea near Reate: because of its fertility Rosea had been dubbed the " udder " of Italy.*

Methods of Training the Vine

viii, 1-7. *I continued the discussion saying that the type of vineyard determined the expense of upkeep. In some the vines grew on the ground without props, others were trained on vertical and horizontal props. If the material for the props could be grown on the farm the expense was not heavy. Certain trees were natural props. The most economical vineyard was that which did not need props: one was the type in which the vines ran along the ground, the other in which only the grape-bearing branches*

47

*were propped on sticks. Where the soil was naturally moist the vine
should be trained higher so as to have as much sun as possible.*

viii. Contra vineam sunt qui putent sumptu fructum devorare.
Refert, inquam, quod genus vineae sit, quod sunt multae species
eius. Aliae enim humiles ac sine ridicis, ut in Hispania, aliae
sublimes, quae appellantur iugatae, ut pleraeque in Italia. Cuius
generis nomina duo, pedamenta et iuga. Quibus stat rectis vinea,
dicuntur pedamenta; quae transversa iunguntur, iuga; ab eo
2 quoque vineae iugatae. Iugorum genera fere quattuor, pertica,
harundo, restes, vites: pertica, ut in Falerno, harundo, ut in Arpano,
restes, ut in Brundisino, vites, ut in Mediolanensi. Iugationis
species duae, una derecta, ut in agro Canusino, altera compluviata
in longitudinem et latitudinem iugata, ut in Italia pleraeque.
Haec ubi domo nascuntur, vinea non metuit sumptum; ubi multa
3 e propinqua villa, non valde. Primum genus quod dixi maxime
quaerit salicta, secundum harundineta, tertium iunceta aut eius
generis rem aliquam, quartum arbusta, ubi traduces possint fieri
vitium, ut Mediolanenses faciunt in arboribus, quas vocant opulos,
4 Canusini in harundulatione in ficis. Pedamentum item fere
quattuor generum: unum robustum, quod optimum solet afferri
in vineam e querco ac iunipiro et vocatur ridica; alterum palus e
pertica, meliore dura, quo diuturnior; quem cum infimum terra
solvit, puter evertitur et fit solum summum; tertium, quod horum
inopiae subsidio misit harundinetum. Inde enim aliquot colligatas
libris demittunt in tubulos fictiles cum fundo pertuso, quas cuspides
appellant, qua umor adventicius transire possit. Quartum est
pedamentum nativum eius generis, ubi ex arboribus in arbores
traductis vitibus vinea fit, quos traduces quidam rumpos appellant.
5 Vineae altitudinis modus longitudo hominis, intervalla pedamentorum, qua boves iuncti arare possint. Ea minus sumptuosa vinea,
quae sine iugo ministrat acratophoro vinum. Huius genera duo:
unum, in quo terra cubilia praebet uvis, ut in Asia multis locis,
quae saepe vulpibus et hominibus fit communis. Nec non si parit
humus mures, minor fit vindemia, nisi totas vineas oppleris musci-
6 pulis, quod in insula Pandateria faciunt. Alterum genus vineti, ubi
ea modo removetur a terra vitis, quae ostendit se adferre uvam.
Sub eam, ubi nascitur uva, subiciuntur circiter bipedales e surculis

furcillae, ne vindemia facta denique discat pendere in palma aut funiculo aut vinctu, quod antiqui vocabant cestum. Ibi dominus simul ac vidit occipitium vindemiatoris, furcillas reducit hibernatum in tecta, ut sine sumptu harum opera altero anno uti possit. Hac 7 consuetudine in Italia utuntur Reatini. Haec ideo varietas maxime, quod terra cuius modi sit refert. Ubi enim natura umida, ibi altius vitis tollenda, quod in partu et alimonio vinum non ut in calice quaerit aquam, sed solem. Itaque ideo, ut arbitror, primum e vinea in arbores escendit vitis.

The Nature of the Soil determines the Crops
(the text is omitted)

ix, 1-7. *I went on to say that the use of the soil depends on its nature. The term* terra *itself needed definition. It could be " land " as in " the land of Italy ", " ground " as opposed to " air ", or " soil ". Of soil there were as many types as its constituents—rock, stones, sand, clay, loam, etc.—and within those types varying qualities depending on the proportion of those constituents. All soils divided again into dry, intermediate, or wet land. The general divisions could be made into poor, medium, or rich soils. In the thin soil of Pupinia nothing grew well, whereas in Etruria's rich soil crops and trees alike flourished, and the medium soil of Tibur allowed healthy growth. Stolo remarked that Diophanes of Bithynia gave useful guidance on judging the quality and nature of a soil from its appearance and consistency and from the natural vegetation.*

The Measurement of the Land
(the text is omitted)

x, 1-2. *Scrofa stated that different countries had different methods and standards. The amount ploughed in a day in Campania was the* versus, *Rome and Latium had the* iugerum *which contained two square* actus, *an area of 120 square feet. Two* iugera *made a* haeredium, *the amount which was that first allotted to each citizen by Romulus and that which could be transmitted by will to an heir. Then two* haeredia *were called a* centuria *which was an area of 2,400 square feet. Four* centuriae *were called a* saltus *in the distribution of public lands.*

4   49

The Size of the Farm Buildings, and the Water Supply

xi, 1-2. The size of the buildings should correspond to that of the estate, otherwise it would run at a loss. Storage space should be related to the farm produce, whether this was wine or grain. A supply of water should be close at hand, preferably a running stream: otherwise cisterns would have to be constructed.

xi. In modo fundi [*Scrofa inquit*] nonanimadverso lapsi multi, quod alii villam minus magnam fecerunt, quam modus postulavit, alii maiorem, cum utrumque sit contra rem familiarem ac fructum. Maiora enim tecta et aedificamus pluris et tuemur sumptu maiore. Minora cum sunt, quam postulat fundus, fructus solent disperire. 2 Dubium enim non est quin cella vinaria maior sit facienda in eo agro, ubi vineta sint, ampliora ut horrea, si frumentarius ager est.

Villa aedificanda potissimum ut intra saepta villae habeat aquam, si non, quam proxime; primum quae ibi sit nata, secundum quae influat perennis. Si omnino aqua non est viva, cisternae faciendae sub tectis et lacus sub dio, ex altero loco ut homines, ex altero ut pecus uti possit.

The Location of the Farm

xii, 1-4. Scrofa went on to say that it was important to locate the farm at the foot of a hill where there was extensive pasture land, facing healthy winds. The best orientation was towards the east. The farm buildings should not face a river nor be on swampy land. The latter could be unwholesome especially because of the invisible organisms which spread disease. Agrius stated his opinion that a farm so situated even if inherited should be sold or abandoned, but Scrofa suggested remedies: the farm should not face the infected wind and should be built on high ground. Good ventilation would destroy the disease-bearing organisms. Floods might occur in low-lying land and precautions there were more difficult against bandits.

xii. Danda opera ut potissimum sub radicibus montis silvestris villam ponat, ubi pastiones sint laxae, item ut contra ventos, qui saluberrimi in agro flabunt. Quae posita est ad exortos aequinoctiales, aptissima, quod aestate habet umbram, hieme solem. Sin cogare secundum flumen aedificare, curandum ne adversum eam ponas; hieme enim fiet vehementer frigida et aestate non salubris.

2 Advertendum etiam, siqua erunt loca palustria, et propter easdem causas, et quod crescunt animalia quaedam minuta, quae non possunt oculi consequi, et per aera intus in corpus per os ac nares perveniunt atque efficiunt difficiles morbos. Fundanius, Quid potero, inquit, facere, si istius modi mi fundus hereditati obvenerit, quo minus pestilentia noceat? Istuc vel ego possum respondere, inquit Agrius; vendas, quot assibus possis, aut si nequeas, relinquas.

3 At Scrofa, Vitandum, inquit, ne in eas partes spectet villa, e quibus ventus gravior afflare soleat, neve in convalli cava et ut potius in sublimi loco aedifices, qui quod perflatur, siquid est quod adversarium inferatur, facilius discutitur. Praeterea quod a sole toto die illustratur, salubrior est, quod et bestiolae, siquae prope nascuntur

4 et inferuntur, aut efflantur aut aritudine cito pereunt. Nimbi repentini ac torrentes fluvii periculosi illis, qui in humilibus ac cavis locis aedificia habent, et repentinae praedonum manus quod improvisos facilius opprimere possunt, ab hac utraque re superiora loca tutiora.

Directions for Farm Buildings

xiii, 1-5. *Scrofa went on to say that cowsheds should be in the place which was warmest in winter. Oil and wine should be stored in jars set on level ground, but dry products on raised floors. A place was needed for the farm hands to shelter in. The overseer's room should be built at the entrance to the farm and the kitchen should be close by because work started there in winter before daylight. Roofed sheds should be provided as cover from rain for carts and other equipment. Two farm-yards were needed, in one a pond with running water for the cattle and farm animals and in the other a pond for soaking lupins and other fodder crops. Chaff from the farm-yard provided valuable manure. Two pits were recommended for manure, both covered with branches and leaves to retain moisture. A shed to house the grain harvest was also necessary, open to the threshing floor.*

xiii. In villa facienda stabula ita, ut bubilia sint ibi, hieme quae possint esse caldiora. Fructus, ut est vinum et oleum, loco plano in cellis, item vasa vinaria et olearia potius faciendum; aridus, ut est faba et faenum, in tabulatis. Familia ubi versetur providendum, si fessi opere aut frigore aut calore, ubi commo-
2 dissime possint se quiete reciperare. Vilici proximum ianuam

51

cellam esse oportet eumque scire, qui introeat aut exeat noctu quidve ferat, praesertim si ostiarius est nemo. In primis culina videnda ut sit admota, quod ibi hieme antelucanis temporibus aliquot res conficiuntur, cibus paratur ac capitur. Faciundum etiam plaustris ac cetero instrumento omni in cohorte ut satis magna sint tecta, quibus caelum pluvium inimicum. Haec enim si intra clausum in consaepto et sub dio, furem modo non metuunt,

3 adversus tempestatem nocentem non resistunt. Cohortes in fundo magno duae aptiores: una ut interdius compluvium habeat lacum, ubi aqua saliat, qui intra stylobatas, cum velit, sit semipiscina. Boves enim ex arvo aestate reducti hic bibunt, hic perfunduntur, nec minus e pabulo cum redierunt anseres, sues, porci. In cohorte exteriore lacum esse oportet, ubi maceretur lupinum, item alia quae

4 demissa in aquam ad usum aptiora fiunt. Cohors exterior crebro operta stramentis ac palea occulcata pedibus pecudum fit ministra fundo, ex ea quod evehatur. Secundum villam duo habere oportet stercilina aut unum bifariam divisum. Alteram enim partem fieri oportet novam, alteram veterem tolli in agrum, quod enim quam recens quod confracuit melius. Nec non stercilinum melius illud, cuius latera et summum virgis ac fronde vindicatum a sole. Non enim sucum, quem quaerit terra, solem ante exugere oportet. Itaque periti, qui possunt, ut eo aqua influat eo nomine faciunt (sic enim maxime retinetur sucus) in eoque quidam sellas familiaricas

5 ponunt. Aedificium facere oportet, sub quod tectum totam fundi subicere possis messem, quod vocant quidam nubilarium. Id secundum aream faciendum, ubi triturus sis frumentum, magnitudine pro modo fundi, ex una parti apertum, et id ab area, quo et in trituram proruere facile possis et, si nubilare coepit, inde ut rursus celeriter reicere. Fenestras habere oportet ex ea parti, unde commodissime perflari possit.

The Aim in Building a Farm should be Profit not Luxury

xiii, 6-7. *Fundanius took up the argument, saying that if farm buildings were related to growing and storage as in former times, the farm would be more profitable, but men of the present days built luxurious country houses, spending more on them than on their town houses. In former times it was considered sufficient to have a kitchen, stables, and cellars for storing wine and oil, but now all efforts were concentrated on the dwelling house: the villas of Metellus and Lucullus had set a standard of luxury which were not for the public good. While nowadays the orientation of the dining-rooms was the owner's concern their forebears considered that of their cellars. If there was a hill on the estate, the house should be built there.*

6   Fundanius, Fructuosior, inquit, est certe fundus propter aedificia, si potius ad anticorum diligentiam quam ad horum luxuriam derigas aedificationem. Illi enim faciebant ad fructum rationem, hi faciunt ad libidines indomitas. Itaque illorum villae rusticae erant maioris preti quam urbanae, quae nunc sunt pleraque contra. Illic laudabatur villa, si habebat culinam rusticam bonam, praesepes laxas, cellam vinariam et oleariam ad modum agri aptam et pavimento proclivi in lacum, quod saepe, ubi conditum novum vinum, orcae in Hispania fervore musti ruptae neque non dolea in Italia. Item cetera ut essent in villa huiusce modi, quae cultura
7 quaereret, providebant. Nunc contra villam urbanam quam maximam ac politissimam habeant dant operam ac cum Metelli ac Luculli villis pessimo publico aedificatis certant. Quo hi laborant ut spectent sua aestiva triclinaria ad frigus orientis, hiberna ad solem occidentem, potius quam, ut antiqui, in quam partem cella vinaria aut olearia fenestras haberet, cum fructus in ea vinarius quaerat ad dolia aera frigidiorem, item olearia calidiorem. Item videre oportet, si est collis, nisi quid impedit, ut ibi potissimum ponatur villa.

Enclosures of the Farm

(the text is omitted)

xiv, 1-4. *Scrofa resumed by saying there were four types of enclosure for the protection of the farm, the natural, the rustic, the military, and the masonry types. The first was a quick-set hedge, the second a wooden palisade or fence, the third was a bank or ditch, often filled with water; this type was also used along the sides of public roads and streams for drainage purposes. In the district of Reate there were banks without trenches. The fourth kind was a wall of stone, brick, or moulded earth and gravel.*

M. TERENTI VARRONIS

Trees to be Planted on the Boundary
(the text is omitted)

xv. Trees might be planted where there were no enclosures, pines as *Fundania* had done on her Sabine farms, cypresses as he himself had done on *Vesuvius*, or elms. Elms were profitable because they also served as vine props, fodder, and fencing rails and firewood.

The Neighbourhood of the Farm

xvi, 1-3. Scrofa continued to lead the discussion saying that the safety of the neighbourhood, the possibilities for marketing in the district, the local means of transport by road and water, and the conditions on farms in the vicinity were essential considerations. There was much excellent land which could not be cultivated because of brigandage; opportunities for buying and selling with transport available made a farm profitable: some needed to import products, others had a surplus to send to market. Flowers could be grown on a large scale near a city, but not at a distance from a market. It was profitable to be able to buy necessities for the farm in the locality and in return to sell some of your own products.

xvi. Relinquitur altera pars, quae est extra fundum, cuius appendices et vehementer pertinent ad culturam propter adfinitatem. Eius species totidem: si vicina regio est infesta; si quo neque fructus nostros exportare expediat neque inde quae opus sunt adportare; tertium, si viae aut fluvii, qua portetur, aut non sunt aut idonei non sunt; quartum, siquid ita est in confinibus fundis, ut nostris agris 2 prosit aut noceat. E quis quattuor quod est primum, refert infesta regio sit necne. Multos enim agros egregios colere non expedit propter latrocinia vicinorum, ut in Sardinia quosdam, qui sunt prope Oeliem, et in Hispania prope Lusitaniam. Quae vicinitatis invectos habent idoneos, quae ibi nascuntur ubi vendant, et illinc invectos opportunos quae in fundo opus sunt, propter ea fructuosa. Multi enim habent in praediis, quibus frumentum aut vinum aliudve quid desit importandum; contra non pauci, quibus aliquid 3 sit exportandum. Itaque sub urbe colere hortos late expedit, sic violaria ac rosaria, item multa quae urps recipit, cum eadem in longinquo praedio, ubi non sit quo deferri possit venale, non expediat colere. Item si ea oppida aut vici in vicinia aut etiam divitum copiosi agri ac villae, unde non care emere possis quae opus

sunt in fundum, quibus quae supersint venire possint, ut quibusdam
pedamenta aut perticae aut harundo, fructuosior fit fundus, quam si
longe sint importanda, non numquam etiam, quam si colendo in tuo
ea parare possis.

Hired Labour: Profits Increased by Available Transport: The Neighbour's Boundary

xvi, 4-6. *Scrofa continued, saying that some farmers preferred to
employ skilled freemen from the neighbourhood rather than to keep them on
the farm: yearly contracts were made by the overseers who hired men
from a distance to live on the farm if none were available locally: in this
way the labourers were kept always at work. Saserna gave a rule that
only the overseer and certain others might leave the farm, but a better one
should require the overseer himself to have the master's permission to be
absent. Convenient means of transport in the vicinity made the farm more
profitable. What your neighbour planted on his boundary was also
important and might impoverish the soil.*

4 Itaque in hoc genus coloni potius anniversarios habent vicinos,
quibus imperent, medicos, fullones, fabros, quam in villa suos
habeant, quorum non numquam unius artificis mors tollit fundi
fructum. Quam partem lati fundi divites domesticae copiae
mandare solent. Si enim a fundo longius absunt oppida aut vici,
fabros parant, quos habeant in villa, sic ceteros necessarios artifices,
ne de fundo familia ab opere discedat ac profestis diebus ambulet
feriata potius, quam opere faciendo agrum fructuosiorem reddat.
5 Itaque ideo Sasernae liber praecipit, nequis de fundo exeat praeter
vilicum et promum et unum, quem vilicus legat; siquis contra
exierit, ne impune abeat; si abierit, ut in vilicum animadvertatur.
Quod potius ita praecipiendum fuit, nequis iniussu vilici exierit,
neque vilicus iniussu domini longius, quam ut eodem die rediret,
6 neque id crebrius, quam opus esset fundo. Eundem fundum
fructuosiorem faciunt vecturae, si viae sunt, qua plaustra agi facile
possint, aut flumina propinqua, qua navigari possit, quibus utrisque
rebus evehi atque invehi ad multa praedia scimus. Refert etiam
ad fundi fructus, quem ad modum vicinus in confinio consitum
agrum habeat. Si enim ad limitem querquetum habet, non possis
recte secundum eam silvam serere oleam, quod usque eo est

contrarium natura, ut arbores non solum minus ferant, sed etiam fugiant, ut introrsum in fundum se reclinent, ut vitis adsita ad holus facere solet.   Ut quercus, sic iugulandes magnae et crebrae finitimae fundi oram faciunt sterilem.

Farm Equipment: Slaves

xvii, 1-3.   *Scrofa continued, saying that the means by which land was tilled might be thought of in two parts, men and their aids, or in three, the voiced, the voiceless, and the mute, in other words, slaves, animals, and farm vehicles.   All agriculture was done by men, whether slaves or free; the freemen either tilled their own small holdings or worked for hire. Hired labour was more profitable on poor land and for heavier farm operations such as the vintage and harvest.   Cassius stated that hired men should be young and able to do the work.   This could be judged by observing and questioning them.*

Bipalium, *a spade with a foot rest (after K. D. White,* Agricultural Implements of the Roman World, *Cambridge* 1967).

xvii. De fundi quattuor partibus, quae cum solo haerent, et alteris quattuor, quae extra fundum sunt et ad culturam pertinent, dixi. Nunc dicam, agri quibus rebus colantur. Quas res alii dividunt in duas partes, in homines et adminicula hominum, sine quibus rebus colere non possunt; alii in tres partes, instrumenti genus vocale et semivocale et mutum, vocale, in quo sunt servi, semivocale, in quo sunt boves, mutum, in quo sunt plaustra. 2 Omnes agri coluntur hominibus servis aut liberis aut utrisque: liberis, aut cum ipsi colunt, ut plerique pauperculi cum sua progenie, aut mercennariis, cum conducticiis liberorum operis res maiores, ut vindemias ac faenisicia, administrant, iique quos obaerarios nostri vocitarunt et etiam nunc sunt in Asia atque Aegypto et in Illyrico 3 complures.   De quibus universis hoc dico, gravia loca utilius esse mercennariis colere quam servis, et in salubribus quoque locis opera rustica maiora, ut sunt in condendis fructibus vindemiae aut messis.

De iis, cuius modi esse oporteat, Cassius scribit haec: operarios parandos esse, qui laborem ferre possint, ne minores annorum XXII et ad agri culturam dociles. Eam coniecturam fieri posse ex aliarum rerum imperatis, et in eo eorum e noviciis requisitione, ad priorem dominum quid factitarint.

Farm Equipment: Slaves: Their Treatment and Supervision

xvii, 4-7. *In Scrofa's opinion overseers of slaves should be educated, reliable, and older than the others: they had to be experienced in farm work so that they could instruct their subordinates by example. The whip should only be used when the tongue failed. Too many slaves of the same nationality might make trouble. Foremen should have privileges such as that of having wives and children: these would endear them to the estate as was the case among the slave families of Epirus. They should be consulted about the work so that they might achieve some self-respect: better food and clothing, and certain other privileges would increase their interest in the farm and ensure their good will.*

Mancipia esse oportere neque formidulosa neque animosa.
4 Qui praesint esse oportere, qui litteris atque aliqua sint humanitate imbuti, frugi, aetate maiore quam operarios, quos dixi. Facilius enim iis quam qui minore natu sunt dicto audientes. Praeterea potissimum eos praeesse oportere, qui periti sint rerum rusticarum. Non solum enim debere imperare, sed etiam facere, ut facientem imitetur et ut animadvertat eum cum causa sibi praeesse, quod
5 scientia praestet. Neque illis concedendum ita imperare, ut verberibus coerceant potius quam verbis, si modo idem efficere possis. Neque eiusdem nationis plures parandos esse; ex eo enim potissimum solere offensiones domesticas fieri. Praefectos alacriores faciendum praemiis dandaque opera ut habeant peculium et coniunctas conservas, e quibus habeant filios. Eo enim fiunt firmiores ac coniunctiores fundo. Itaque propter has cognationes
6 Epiroticae familiae sunt illustriores ac cariores. Inliciendam voluntatem praefectorum honore aliquo habendo, et de operariis qui praestabunt alios, communicandum quoque cum his, quae facienda sint opera, quod, ita cum fit, minus se putant despici atque
7 aliquo numero haberi a domino. Studiosiores ad opus fieri liberalius tractando aut cibariis aut vestitu largiore aut remissione operis

concessioneve, ut peculiare aliquid in fundo pascere liceat, huiusce modi rerum aliis, ut quibus quid gravius sit imperatum aut animadversum, qui consolando eorum restituat voluntatem ac benevolentiam in dominum.

Farm Equipment: The Slave Personnel
(the text is omitted)

xviii, 1-8. *Scrofa went on to say that Cato said that the land and crop determined the number of slaves needed on an estate. An oliveyard of two hundred and forty* iugera *(one hundred and sixty acres) should have thirteen slaves, including an overseer, a housekeeper, five labourers, three herdsmen, one donkey-man, one swineherd, and one shepherd: a vineyard of a hundred* iugera *(sixty-seven acres) should have fifteen slaves. In the opinion of Saserna, however, one man was enough for eight* iugera *and should be able to dig over one* iugerum *in four days. Cato's rule could not be strictly applied because an overseer and housekeeper were needed even if the estate was small and were not duplicated if it were large. The standard unit was the century which contained two hundred* iugera. *It was therefore impossible if one sixth (forty* iugera) *were subtracted, at the same time to subtract one sixth from the slave personnel. Saserna's prescription made better sense. The neighbouring farms should be observed and from these the decision should be made as to how many labourers should be employed on your own land. Two principles should be followed in agriculture, imitation of others and systematic experiments, such as ploughing a second time, weeding a second and third time, and grafting figs in summer instead of spring.*

Farm Equipment: The Animals

xix, 1-3. *Scrofa next spoke of the farm animals, saying that with regard to the number of working beasts required for the farm Saserna prescribed two yoke of oxen for two hundred* iugera, *but Cato three yoke for an oliveyard of two hundred and forty* iugera. *Neither was to be regarded as entirely right because any given piece of land might be easier or more difficult than another to work. We had to be guided by the former owner's practice, that of the neighbouring owners, and by our own experiments. Cato added three donkeys for haulage and one for the mill: in addition only those animals should be kept which were beneficial to the land: sheep were preferred to swine for their manure: no farm was safe without dogs.*

RERUM RUSTICARUM

The Plough: (1) *Hand-grip,* manicula; (2) *the stilt,* stiva; (3) *share-beam,* dentale; (4) *the share,* vomer; (5) *plough-beam,* bara *or* buris; (6) *yoke-beam or pole,* temo (*after K. D. White,* Agricultural Implements of the Roman World, *Cambridge* 1967).

xix. De reliqua parte instrumenti, quod semivocale appellavi, Saserna ad iugera CC arvi boum iuga duo satis esse scribit, Cato in olivetis CCXL iugeris boves trinos. Ita fit ut, si Saserna dicit verum, ad C iugera iugum opus sit, si Cato, ad octogena. Sed ego neutrum modum horum omnem ad agrum convenire puto et

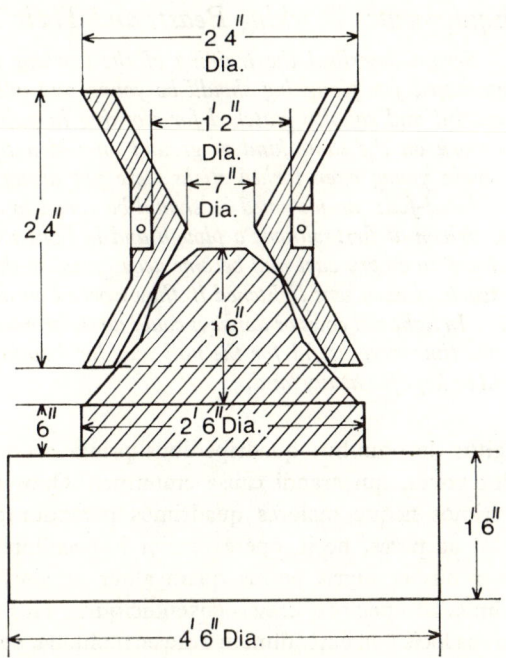

Section through a Donkey Mill from Pompeii (after L. A. Moritz, Grain Mills and Flour in Classical Antiquity, *Oxford* 1958).

59

utrumque ad aliquem. Alia enim terra facilior aut difficilior est; 2 aliam terram boves proscindere nisi magnis viribus non possunt et saepe fracta bura relinquunt vomerem in arvo. Quo sequendum nobis in singulis fundis, dum sumus novicii, triplici regula, superioris 3 domini instituto et vicinorum et experientia quadam. Quod addit asinos qui stercus vectent tres, asinum molarium, in vinea iugerum C iugum boum, asinorum iugum, asinum molendarium; in hoc genere semivocalium adiciendum de pecore ea sola quae agri colendi causa erunt et quae solent esse peculiaria pauca habenda, quo facilius mancipia se tueri et assidua esse possint. In eo numero non modo qui prata habent, ut potius oves quam sues habeant curant, sed etiam qui non solum pratorum causa habent, propter stercus. De canibus vero utique, quod villa sine iis parum tuta.

Farm Equipment: *Working Beasts and Their Training*

xx, 1-5. Scrofa described the training of the working beasts saying that oxen purchased for ploughing should be young and unbroken: they should be powerful and equally matched for working in pairs and should be bought to work on the same kind of ground on which they had been reared. To train young oxen forked sticks were put around their necks and they were hand-fed: an untamed ox should be yoked to a broken one, and should be driven at first without a plough and in light soil. Draught oxen should draw an empty cart and become accustomed to the noise of the town. The work of oxen was lightened if they worked on alternate sides of the plough. In light soil cows or donkeys could do the work of ploughing, and at the same time were useful for the mill and for hauling. Stronger animals should be kept for heavy land.

xx. Igitur de omnibus quadripedibus prima est probatio, qui idonei sint boves, qui arandi causa emuntur. Quos rudes neque minores trimos neque maiores quadrimos parandum; ut viribus magnis sint ac pares, ne in opere firmior imbecilliorem conficiat; amplis cornibus et nigris potius quam aliter ut sint, lata fronte, 2 naribus simis, lato pectore, crassis coxendicibus. Hos veteranos ex campestribus locis non emendum in dura ac montana, nec non contra si incidit, ut sit vitandum. Novellos cum quis emerit iuvencos, si eorum colla in furcas destitutas incluserit ac dederit cibum, diebus

paucis erunt mansueti et ad domandum proni.   Tum ita subigendum, ut minutatim assuefaciant et ut tironem cum veterano adiungant (imitando enim facilius domatur), et primum in aequo loco et sine aratro, tum eo levi, principio per harenam aut molliorem
3 terram.   Quos ad vecturas, item instituendum ut inania primum ducant plaustra et, si possis, per vicum aut oppidum; creber crepitus ac varietas rerum consuetudine celeberrima ad utilitatem adducit.   Neque pertinaciter, quem feceris dextrum, in eo manendum, quod, si alternis fit sinister, fit laboranti in alterutra parte
4 requies.   Ubi terra levis, ut in Campania, ibi non bubus gravibus, sed vaccis aut asinis quod arant, eo facilius ad aratrum leve adduci possunt, ad molas et ad ea, siquae sunt, quae in fundo convehuntur. In qua re alii asellis, alii vaccis ac mulis utuntur, exinde ut pabuli facultas est; nam facilius asellus quam vacca alitur, sed fructuosior
5 haec.   In eo agricolae hoc spectandum, quo fastigio sit fundus.   In confragoso enim haec ac difficili valentiora parandum et potius ea quae per se fructum reddere possint, cum idem operis faciant.

Farm Equipment: Dogs

xxi.   *A few good watch dogs should be kept trained to sleep in the daytime and to be on watch at night.   If an owner of meadows is without flocks of his own, he should allow a neighbour to pasture his flocks on the farm.*

xxi. Canes potius cum dignitate et acres paucos habendum quam multos, quos consuefacias potius noctu vigilare et interdiu clausos dormire.   De indomitis quadripedibus ac pecore faciendum; si prata sunt in fundo neque pecus habet, danda opera ut pabulo vendito alienum pecus in suo fundo pascat ac stabulet.

Farm Equipment: The ' Mute '
(the text is omitted)

xxii, 1-6.   *Scrofa stated that everything possible should be made on the farm from home-grown materials.   If any tools had to be bought, they were to be carefully priced and of good quality.   The number of tools needed was controlled by the size of the farm.   Stolo interrupted by saying that Cato prescribed for an oliveyard of two hundred and forty* iugera, *five presses and related equipment, three large carts, six ploughs and smaller items: for a vineyard of a hundred* iugera *three presses and other small equipment. He allowed an extra large number of wine jars for storage because old wine sold at a higher price than new. Scrofa then advised that the owner should keep a complete inventory of the farm equipment and that the overseer should store all tools near the home farm, keeping a special watch over those that were only needed at certain times such as the vintage.*

The Third Topic: The Crops

xxiii, 1-6. *Agrasius said that he was now ready to hear about the third topic. Scrofa continued, saying that the profit of the farm came from what was planted. Not all should be planted on rich soil: plants which did not need much nourishment should be planted in poorer soil, especially the legumes: those needing more food in rich soil such as cabbages, wheat, flax. Some were planted for green manure. Orchards and flower gardens were planted for pleasure. Willows and reeds needed*

MEDICAGO SATIVA, L. (*medica*), *now commonly called alfalfa, an important leguminous fodder crop in both the ancient and modern world.*

damp ground, grain and beans dry places. Some needed shade such as
the asparagus, but violets and gardens the sun. In other places you should
grow osiers for making farm equipment, and likewise other similar plants.
In young orchards garden crops and others might be planted in between
trees when they were young, but not later, to avoid damaging the roots.

xxiii. Suscipit Agrasius, Et quoniam habemus illa duo
prima ex divisione quadripertita, de fundo et de instrumento, quo
coli solet, de tertia parte expecto. Scrofa, Quoniam fructum, inquit,
arbitror esse fundi eum qui ex eo satus nascitur utilis ad aliquam rem,
duo consideranda, quae et quo quidque loco maxime expediat
serere. Alia enim loca apposita sunt ad faenum, alia ad frumentum,
alia ad vinum, alia ad oleum, sic ad pabulum quae pertinent, in
2 quo est ocinum, farrago, vicia, medica, cytisum, lupinum. Neque
in pingui terra omnia seruntur recte neque in macra nihil. Rectius
enim in tenuiore terra ea quae non multo indigent suco, ut cytisum
et legumina praeter cicer; hoc enim quoque legumen, ut cetera
quae velluntur e terra, non subsecantur, quae, quod ita leguntur,
legumina dicta. In pingui rectius quae cibi sunt maioris, ut holus,
3 triticum, siligo, linum. Quaedam etiam serenda non tam propter
praesentem fructum quam in annum prospicientem, quod ibi
subsecta atque relicta terram faciunt meliorem. Itaque lupinum,
cum minus siliculam cepit, et non numquam fabalia, si ad siliquas
non ita pervenit, ut fabam legere expediat, si ager macrior est, pro
4 stercore inarare solent. Nec minus ea discriminanda in conserendo
quae sunt fructuosa, propter voluptatem, ut quae pomaria ac floralia
appellantur, item illa quae ad hominum victum ac sensum delecta-
tionemque non pertinent neque ab agri utilitate sunt diiuncta.
5 Idoneus locus eligendus, ubi facias salictum et harundinetum, sic
alia quae umidum locum quaerunt, contra ubi segetes frumentarias,
ubi fabam potissimum seras, item alia quae arida loca secuntur; sic
ut umbrosis locis alia seras, ut corrudam, quod ita petit asparagus;
aprica, ut ibi seras violam et hortos facias, quod ea sole nutricantur,
sic alia. Et alio loco virgulta serenda, ut habeas vimina, unde
viendo quid facias, ut sirpeas, vallus, crates; alio loco ut seras ac
6 colas silvam caeduam, alio ubi aucupere, sic ubi cannabim, linum,
iuncum, spartum, unde nectas bubus soleas, lineas, restes, funes.
Quaedam loca eadem alia ad serendum idonea. Nam et in recentibus

pomariis dissitis seminibus in ordinemque arbusculis positis primis annis, antequam radices longius procedere possint, alii conserunt hortos, alii quid aliud, neque cum convaluerunt arbores idem faciunt, ne violent radices.

The Crops: Olives and Trees
(the text is omitted)

xxiv, 1-4. *Stolo quoted Cato (VI, 1-2) as saying that rich soil was to be used for grain, but if damp, it was suitable for root vegetables, millet, and panic-grass. Olives needed warm, heavy soil. The best land for olives faces west and is sunny. A* hostus *is the yield of oil from one* factus, *i.e. the amount pressed at one time: one* factus *is equal to one hundred and twenty or one hundred and sixty* modii *according to the equipment available. Cato's advice about planting elms and poplars on the borders of the farm so as to provide fodder and timber should be followed only on the north side.*

Scrofa quoted further advice from Cato: (VI, 3-4) " poplar cuttings and reeds should be planted in wet ground. The Greek willow planted along the border of the thicket will provide withes for tying vines ".

The Planting of Vines
(the text is omitted)

xxv. *Scrofa also quoted Cato, saying that when planting a vineyard care should be taken to see that different varieties were put in soil that is suited to them. The small Aminnian grape and the double* eugeneum *and the parti-coloured needed a good soil with a sunny position: the large Aminnian, the Murgentian, the Apician and the Lucanian would thrive in heavy and damp soil. Others, together with hybrids, would grow anywhere.*

The Vineyard
(the text is omitted)

xxvi. *Scrofa continued, saying that the vine should be protected on the north side by the prop. If live cypresses were grown to serve as props they should be planted in alternate rows and not near the vines because the trees were hostile to each other.*

Agrius now broke in and remarked to Fundanius that he was afraid the sacristan would return before Scrofa could begin on the fourth topic. He was eager to hear about the vintage. Scrofa reassured him, jokingly telling him to get the baskets and the wine jar ready.

The Farmer's Calendar: The Four Seasons

xxvii, 1-3. *Scrofa continued, explaining that there was both a solar and a lunar measurement of time. The sun's annual course was divided into four periods of about three months each and again into eight periods of a month and a half which represented the four seasons. In spring untilled ground should be broken up, weeded, and ploughed two or three times: the harvest was gathered in summer and the grapes in autumn: woods were to be cleared and trees trimmed; pruning was done in winter when there was no rain or ice.*

xxvii. Et quoniam tempora duorum generum sunt, unum annale, quod sol circuitu suo finit, alterum menstruum, quod luna circumiens comprendit, prius dicam de sole. Eius cursus annalis primum fere circiter ternis mensibus ad fructus est divisus in IIII partes, et idem subtilius sesquimensibus in IIX; in IIII, quod 2 dividitur in ver et aestatem et autumnum et hiemem. Vere sationes quae fiunt, terram rudem proscindere oportet, quae sunt ex ea enata, priusquam ex iis quid seminis cadat, ut sint exradicata; et simul glaebis ab sole percalefactis aptiores facere ad accipiendum imbrem et ad opus faciliores relaxatas; neque eam minus binis 3 arandum, ter melius. Aestate fieri messes oportere, autumno siccis tempestatibus vindemias, ac silvas excoli commodissime tunc, praecidi arbores oportere secundum terram; radices autem primoribus imbribus ut effodiantur, nequid ex iis nasci possit. Hieme putari arbores dumtaxat his temporibus, cum gelu cortices ex imbribus careant et glacie.

The Eight Divisions of the Solar Year

xxviii, 1-2. *Scrofa continued his discussion on the calendar, saying that the first day of each of the four seasons was the twenty-third day after the sun had entered the corresponding sign of the zodiac. These could be divided again into eight periods calculated by the rising of the west wind, the equinox and solstice, the rising of the Vergiliae and the Dog Star.*

xxviii. Dies primus est veris in aquario, aestatis in tauro, autumni in leone, hiemis in scorpione. Cum unius cuiusque horum IIII signorum dies tertius et vicesimus IIII temporum sit primus et efficiat ut ver dies habeat XCI, aestas XCIV, autumnus XCI, hiems XXCIX, quae redacta ad dies civiles nostros, qui nunc sunt, primi verni temporis ex a. d. VII id. Febr., aestivi ex a. d. VII
2 id. Mai., autumnalis ex a. d. III id. Sextil., hiberni ex a. d. IV id. Nov., suptilius descriptis temporibus observanda quaedam sunt, eaque in partes VIII dividuntur: primum a favonio ad aequinoctium vernum dies XLV, hinc ad vergiliarum exortum dies XLIV, ab hoc ad solstitium dies XLIIX, inde ad caniculae signum dies XXVII, dein ad aequinoctium autumnale dies LXVII, exin ad vergiliarum occasum dies XXXII, ab hoc ad brumam dies LVII, inde ad favonium dies XLV.

The Farmer's Calendar: The First Period

xxix, 1-3.   Scrofa began his account of the farmer's calendar, stating that the first period fell between the rising of the west wind and the vernal equinox.   The farm operations included the planting of nursery-beds, pruning, manuring, trenching of vines, clearing of meadows, planting of willows, and weeding of grain crops.   The derivations of the words seges, arvum, *and* novalis *were explained.   There were three stages of ploughing new or fallow land, " breaking up", " breaking down", and ridging.   The third operation was done with mould boards attached to the ploughshare by means of which the seed was covered in ridges, and ditches were cut between.   In some districts, any clods left on the ridges were broken down with hoes.   The hollow made by the plough was a " furrow ".   The ridge between the furrows was a* porca *from* porricere.

xxix.   Primo intervallo inter favonium et aequinoctium vernum haec fieri oportet.   Seminaria omne genus ut serantur, putari arbusta, stercorari in pratis, circum vites ablacuari, radices quae in summa terra sunt praecidi, prata purgari, salicta seri, segetes sariri.   Seges dicitur quod aratum satum est, arvum quod aratum necdum satum est, novalis, ubi satum fuit, antequam secunda
2 aratione novatur rursus.   Terram cum primum arant, proscindere appellant, cum iterum, offringere dicunt, quod prima aratione glaebae grandes solent excitari; cum iteratur, offringere vocant.

Tertio cum arant iacto semine, boves lirare dicuntur, id est cum tabellis additis ad vomerem simul et satum frumentum operiunt in porcis et sulcant fossas, quo pluvia aqua delabatur. Non nulli postea, qui segetes non tam latas habent, ut in Apulia et id genus praediis, per sartores occare solent, siquae in porcis relictae grand-3 iores sunt glaebae. Qua aratrum vomere lacunam striam fecit, sulcus vocatur. Quod est inter duos sulcos elata terra dicitur porca, quod ea seges frumentum porricit. Sic quoque exta deis cum dabant, porricere dicebant.

The Farmer's Calendar: The Second Period

xxx. *Scrofa recommended that during the second period which fell between the vernal equinox and the rising of the constellation* vergiliae, *crops should be weeded, the ground broken up, willows cut, and meadows fenced. Work left over from the former period should be finished before the plants began to bud. Olives were to be planted and pruned.*

xxx. Secundo intervallo inter vernum aequinoctium et vergiliarum exortum haec fieri. Segetes runcari, id est herbam e segetibus expurgari, boves terram proscindere, salicem caedi, prata defendi. Quae superiore tempore fieri oportuerit et non sunt absoluta, antequam gemmas agant ac florescere incipiant, fieri, quod, si quae folia amittere solent ante frondere inceperunt, statim ad serendum idonea non sunt. Oleam seri interputarique oportet.

The Farmer's Calendar: The Third Period

xxxi, 1-5. *Scrofa stated that in the third period between the rising of the constellation* vergiliae *and the solstice, the young vines were to be dug or ploughed and then forked to break up clods. They should be thinned by a professional vine-dresser. The strongest shoots only were to be left on the vine so that they might have all the sap. In the nursery the young vine was to be cut back so as to form a sturdier stock. The vine grasped the support by means of tendrils and for this reason a tendril was called* capreolus (*from* capere). *Fodder crops were to be cut;* farrago *was a mixture of barley and legumes sown for green feed:* vicia, *vetch, was so called because it bound other plants with its tendrils. Meadows should be irrigated: if there was drought, grafted fruit trees should be watered every evening.*

xxxi. Tertio intervallo inter vergiliarum exortum et solstitium haec fieri debent. Vineas novellas fodere aut arare et postea occare, id est comminuere, ne sit glaeba. Quod ita occidunt, occare dictum. Vites pampinari, sed a sciente (nam id quam putare maius), neque

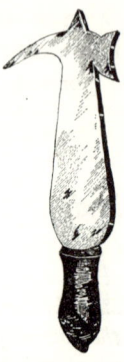

A falx vinitoria *or* vineatica, *a vinech dressing knife* (*after* K. D. *White,* Agricultural Implements of the Roman World, *Cambridge* 1967).

2 in arbusto, sed in vinea fieri. Pampinare est e sarmento coles qui nati sunt, de iis, qui plurimum valent, primum ac secundum, non numquam etiam tertium, relinquere, reliquos decerpere, ne relictis colibus sarmentum nequeat ministrare sucum. Ideo in vitiario primitus cum exit vitis, tota resicari solet, ut firmiore sarmento e 3 terra exeat atque in pariendis colibus vires habeat maiores. Eiuncidum enim sarmentum propter infirmitatem sterile neque ex se potest eicere vitem, quam vocant minorem flagellum, maiorem et iam unde uvae nascuntur palmam. Prior littera una mutata declinata a venti flatu, similiter ac flabellum flagellum. Posterior, quod ea vitis immittitur ad uvas pariendas, dicta primo videtur a 4 pariendo parilema; exin mutatis litteris, ut in multis, dici coepta palma. Ex altera parte parit capreolum. Is est coliculus viteus intortus, ut cincinnus. Hi sunt enim vitis quibus teneat id quo serpit ad locum capiendum, a quo capiendo capreolus dictus. Omne pabulum, primum ocinum farraginem viciam, novissime faenum, secari. Ocinum dictum a graeco verbo ὠκέως, quod valet cito, similiter quod ocimum in horto. Hoc amplius dictum ocinum, quod citat alvom bubus et ideo iis datur, ut purgentur. Id

5 est ex fabali segete viride sectum, antequam genat siliquas.   Farrago contra ex segete ubi sata admixta hordeum et vicia et legumina pabuli causa viride, aut quod ferro caesa, ferrago, dicta, aut inde, quod primum in farracia segete seri coepta.   Eo equi et iumenta cetera verno tempore purgantur ac saginantur.   Vicia dicta a vinciendo, quod item capreolos habet, ut vitis, quibus, cum susum versus serpit, ad scapum lupini aliumve quem ut haereat, id solet vincire.   Si prata inrigua habebis, simulac faenum sustuleris, inrigare.   In poma, quae insita erunt, siccitatibus aquam addi cotidie vesperi.   A quo; quod indigent potu, poma dicta esse possunt.

The Farmer's Calendar: The Fourth Period

xxxii.   *Scrofa continued: During the fourth period, between the solstice and the rising of* Canicula, *the Dog Star, the harvest was usually gathered: ploughing should be finished, if the ground was being " broken up " it should be worked over a second time to break up the clods.   Legumes were to be sown, old vines hoed a second time, young ones a third if there were any clods still left.*

xxxii. Quarto intervallo inter solstitium et caniculam plerique messem faciunt, quod frumentum dicunt quindecim diebus esse in vaginis, quindecim florere, quindecim exarescere, cum sit maturum. Arationes absolvi, quae eo fructuosiores fiunt, quo caldiore terra aratur.   Si proscideris, offringi oportet, id est iterare, ut frangantur glaebae; prima enim aratione grandes glaebae ex terra scinduntur. 2 Serendum viciam, lentem, cicerculam, ervilam ceteraque, quae alii legumina, alii, ut Gallicani quidam, legarica appellant, utraque dicta a legendo, quod ea non secantur, sed vellendo leguntur.   Vineas veteres iterum occare, novellas etiam tertio, si sunt etiam tum glaebae.

The Farmer's Calendar: The Fifth Period

xxxiii.   *Scrofa continued: In the fifth period, between the rising of the Dog Star and the autumnal equinox straw was to be cut, stacks built, ploughed land harrowed, leaf-fodder gathered, and irrigated meadows mowed a second time.*

xxxiii. Quinto intervallo inter caniculam et aequinoctium autumnale oportet stramenta desecari et acervos constitui, arata offringi, frondem caedi, prata inrigua iterum secari.

The Farmer's Calendar: The Sixth Period

xxxiv, 1-2. *Scrofa went on to say that in the sixth period, from the autumnal equinox, sowing began. No sowing was to be done after the winter solstice, nor should it begin before the equinox. Beans were best sown when the* Vergiliae *set, but the vintage fell between the autumnal equinox and the setting of the* Vergiliae. *Pruning and layering of vines and planting of fruit trees should begin.*

xxxiv. Sexto intervallo ab aequinoctio autumnali incipere scribunt oportere serere usque ad diem nonagensimum unum. Post brumam, nisi quae necessaria causa coegerit, non serere, quod tantum intersit, ut ante brumam sata quae septimo die, post brumam sata quadragesimo die vix existant. Neque ante aequinoctium incipi oportere putant, quod, si minus idoneae tempestates sint consecutae, putescere semina soleant. Fabam optime seri in
2 vergiliarum occasu; uvas autem legere et vindemiam facere inter aequinoctium autumnale et vergiliarum occasum; dein vites putare incipere et propagare et serere poma. Haec aliquot regionibus, ubi maturius frigora fiunt asperiora, melius verno tempore.

The Farmer's Calendar: The Seventh Period

xxxv, 1-2. *Scrofa went on to say that during the seventh period which was between the setting of the* Vergiliae *and the winter solstice, lilies and crocus were to be planted: slips might be cut from a rooted rose-tree: violet beds were not profitable. From the beginning of the west wind to the rising of Arcturus wild thyme should be transplanted: ditching and pruning were also done: the elm tree might be planted.*

xxxv. Septimo intervallo inter vergiliarum occasum et brumam haec fieri oportere dicunt: serere lilium et crocum. Quae iam egit radicem rosa, ea conciditur radicitus in virgulas palmares et obruitur, haec eadem postea transfertur facta viviradix. Violaria in fundo facere non est utile, ideo quod necesse est terra adruenda pulvinos fieri, quos irrigationes et pluviae tempestates abluunt et

2 agrum faciunt macriorem. A favonio usque ad arcturi exortum recte serpillum e seminario transferri, quod dictum ab eo, quod serpit. Fossas novas fodere, veteres tergere, vineas arbustumque putare, dum in XV diebus ante et post brumam, ut pleraque, ne facias. Nec non tum aliquid recte seritur, ut ulmi.

The Farmer's Calendar: The Eighth Period

xxxvi, 1-5. *Scrofa concluded his account with the injunction that during the eighth period which lasted from the solstice to the coming of the west wind, grain lands should be drained or harrowed: vineyards and orchards pruned. Various tasks should be done indoors when work could not be carried out on the land. A written copy of the calendar should be kept posted up on the farm.*

xxxvi. Octavo intervallo inter brumam et favonium haec fieri oportet. De segetibus, siqua est aqua, deduci; sin siccitates sunt et terra teneritudinem habet, sarire. Vineas arbustaque putare. Cum in agris opus fieri non potest, quae sub tecto possunt tunc conficienda antelucano tempore hiberno. Quae dixi scripta et proposita habere in villa oportet, maxime ut vilicus norit.

The Moon and the Sixfold Division of the Year
(*The text is omitted*)

xxxvii, 1-5. *Scrofa continued: The days of the moon should also be observed. In one period it waxed from the new to the full, and then waned again to the new until it reached the* intermenstruum, *the day between two months when it was said to be ending and beginning. Some farm operations were better done when the moon was waxing than when it was waning, and* vice versa, *such as the cutting of corn and leaf-fodder. Stolo then talked also about a six-fold division of times connected with the sun and moon for bringing produce to perfection. These were (1) preparation, (2) sowing, (3) cultivation, (4) harvesting, (5) storage, (6) bringing out of store. The first stage included trenching, digging, making of furrows, or ploughing and digging. Sometimes after a first ploughing, a second was necessary before sowing. The preparation of meadows involved fencing to keep out the cattle and irrigating if they were to be watered.*

71

Of Manuring

(the text is omitted)

xxxviii, 1-3. *Stolo continued: The best manure, according to Cassius, was that of birds, and the best of all was the dung of pigeons. But, in Stolo's opinion, aviaries of field-fares and blackbirds produced the best, because it could be fed to cattle and swine. Cassius further recommended human manure, and after that the dung of goats, sheep, and asses: horse dung was good for meadows and produced a good crop of grass. A manure heap should be made near the farmyard to save the labour of carrying.*

Methods of Propagation

(the text is omitted)

xxxix, 1-3. *Stolo went on with the discussion, saying that planting, which was the second step, must be done with careful regard to the season. Different plants blossomed and grew at different seasons, with the result that they were sown, grafted, and harvested at different times. There were four methods of propagation, from natural seeds, by means of rooted slips, shoots from trees, and lastly grafting.*

Methods of Propagation: By Seeds, Slips, Shoots, and Grafting

xl, 1-6. *Stolo continued, saying that there were two kinds of seed; the first were the primordial seeds described by philosophers which were invisible, but were in the air, and produced spontaneous growth when once they reached the earth: the second were those which the farmer planted: it was necessary that they should be fresh, clean, and unmixed. Age might alter the nature of some seeds so that they did not produce their own kind. Spring, autumn, and the rising of the Dog Star were the best seasons for planting although these were also influenced by the nature of the soil. The third method of planting was by means of shoots placed in the earth: these should be taken from the tree before it budded or blossomed. They should be torn not broken off from the tree. Olive cuttings were to be taken from a tender branch and sharpened at both ends. In the fourth method a shoot from one tree was grafted on to another: only certain trees would take grafts from other species. When you grafted the same species, the graft should be of a superior quality: another method of grafting was to run a small branch of one tree into a branch of another beside it.*

xl. Primum semen, quod est principium genendi, id duplex, unum quod latet nostrum sensum, alterum quod apertum. Latet, si sunt semina in aere, ut ait physicos Anaxagoras, et si aqua, quae influit in agrum, inferre solet, ut scribit Theophrastus. Illud quod apparet ad agricolas, id videndum diligenter. Quaedam enim ad genendum propensa usque adeo parva, ut sint obscura, ut cupressi. Non enim galbuli qui nascuntur, id est tamquam pilae
2 parvae corticiae, id semen, sed in iis intus. Primigenia semina dedit natura, reliqua invenit experientia coloni. Prima quae sine colono, priusquam sata, nata; secunda quae ex iis collecta neque, priusquam sata, nata. Prima semina videre oportet ne vetustate sint exsucta aut ne sint admixta aut ne propter similitudinem sint adulterina. Semen vetus tantum valet in quibusdam rebus, ut naturam commutet. Nam ex semine brassicae vetere sato nasci
3 aiunt rapa et contra ex raporum brassicam. Secunda semina videre oportet ne, unde tollas, nimium cito aut tarde tollas. Tempus enim idoneum, quod scribit Theophrastus, vere et autumno et caniculae exortu, neque omnibis locis ac generibus idem. In sicco et macro loco et argilloso vernum tempus idoneum, quo minus habet umoris: in terra bona ac pingui autumno, quod vere multus umor, quam sationem quidam metiuntur fere diebus XXX.
4 Tertium genus seminis, quod ex arbore per surclos defertur in terram, si in humum demittitur, in quibusdam est videndum ut eo tempore sit deplantatum, quo oportet (id fit tum, antequam gemmare aut florere quid incipit); et quae de arbore transferas ut ea deplantes potius quam defringas, quod plantae solum stabilius, quo latius aut radices facilius mittit. Ea celeriter, antequam sucus exarescat, in terram demittunt. In oleagineis seminibus videndum ut sit de tenero ramo ex utraque parte aequabiliter praecisum, quas alii
5 clavolas, alii taleas appellant ac faciunt circiter pedales. Quartum genus seminis, quod transit ex arbore in aliam, videndum qua ex arbore in quam transferatur et quo tempore et quem ad modum obligetur. Non enim pirum recipit quercus, neque enim si malus pirum. Hoc secuntur multi, qui haruspices audiunt multum, a quibus proditum, in singulis arboribus quot genera insita sint, uno ictu tot fulmina fieri illut quod fulmen concepit. Si in pirum silvaticam inserueris pirum quamvis bonam, non fore tam iucundam,
6 quam si in eam quae silvestris non sit. In quamcumque arborem

inseras, si eiusdem generis est, dumtaxat ut sit utraque malus, ita inserere oportet referentem ad fructum, meliore genere ut sit surculus, quam est quo veniat arbor. Est altera species ex arbore in arborem inserendi nuper animadversa in arboribus propinquis. Ex arbore, qua vult habere surculum, in eam quam inserere vult ramulum traducit et in eius ramo praeciso ac diffisso implicat, eum locum qui contingit, ex utraque parti quod intro est falce extenuatum, ita ut ex una parti quod caelum visurum est corticem cum cortice exaequatum habeat. Eius ramuli, quem inseret, cacumen ut derectum sit ad caelum curat. Postero anno, cum comprendit, unde propagatum est, ab altera arbore praecidit.

Grafts and Shoots

xli, 1-6. Careful attention should be given to the proper time for grafting. The fig, having a soft wood, should be grafted in the hot season to avoid moisture which is harmful to it. If moisture is needed for harder plants a vessel could be fastened above the graft from which water dripped slowly. The shoot should be sharpened to a point and the graft smeared with clay and tied with bark. The vine shoot should be dried for three days beforehand, but drier plants may be grafted at once. Cuttings were more practical for the fig because of the smallness of the seed, although seeds in the dried fruit might be shipped or imported. The olive was more easily grown from cuttings because the seed was a nut which only germinated slowly.

xli. Quo tempore quaeque transferas, haec in primis videnda, quae prius verno tempore inserebantur, nunc etiam solstitiali, ut ficus, quod densa materia non est et ideo sequitur caldorem. A quo fit ut in locis frigidis ficeta fieri non possint. Aqua recenti insito 2 inimica; tenellum enim cito facit putre. Itquae caniculae signo commodissime existimatur ea inseri. Quae autem natura minus sunt mollia, vas aliquod supra alligant, unde stillet lente aqua, ne prius exarescat surculus, quam colescat. Cuius surculi corticem integrum servandum et eum sic exacuendum, ut non denudes medullam. Ne extrinsecus imbres noceant aut nimius calor, argilla 3 oblinendum ac libro obligandum. Itaque vitem triduo antequam inserant desecant, ut qui in ea nimius est umor defluat, antequam inseratur; aut in quam inserunt, in ea paulo infra, quam insitum est, incidunt, qua umor adventicius effluere possit. Contra in fico et malo punica, et siquae etiam horum natura aridiora, continuo. In aliis translationibus videndum ut quod transferat cacumen habeat gemmam, ut in ficis.

4 De his primis quattuor generibus seminum quaedam quod tardiora, surculis potius utendum, ut in ficetis faciunt. Fici enim semen naturale intus in ea fico quam edimus, quae sunt minuta grana; e quibus parvis quod enasci coliculi vix queunt—omnia enim minuta et arida ad crescendum tarda, ea quae laxiora, et fecundiora, ut femina quam mas et pro portione in virgultis item; itaque ficus, malus punica et vitis propter femineam mollitiam ad crescendum prona, contra palma et cupressus et olea in crescendo

5 tarda: in hoc enim umidiora quam aridiora—quare [ex terra] potius in seminariis surculos de ficeto quam grana de fico expedit obruere, praeter si aliter nequeas, ut siquando quis trans mare semina mittere aut inde petere vult. Tum enim resticulam per ficos, quas edimus, maturas perserunt et eas, cum inaruerunt, complicant ac quo

6 volunt mittunt, ubi obrutae in seminario pariant. Sic genera ficorum, Chiae ac Chalcidicae et Lydiae et Africanae, item cetera transmarina in Italiam perlata. Simili de causa, oleae semen cum sit nuculeus, quod ex eo tardius enascebatur colis quam ex aliis, ideo potius in seminariis taleas, quas dixi, serimus.

(the text is omitted)

xlii-xlix. [*These paragraphs contain directions given by Stolo for sowing various fodder crops, and observations on the growth of plants: in xlviii there is a detailed description of the parts of a grain plant with etymological explanations: then follow directions for the hay harvest: the grass must be cut with the sickle and tossed with a fork. When it is dry, it must be made into bundles and taken to the barn. The meadows must then be sickled again.*]

Methods of Harvesting Grain

l, 1-3. *Stolo described three ways of harvesting grain. In Umbria the stalk was cut close to the ground with the hook and the ears then cut off the stalk: in Picenum the ears were cut off at the top with a small iron saw, leaving the straw standing. Near Rome the stalk was cut in the middle and ears and straw were taken to the threshing floor. On easy land one man should reap a* iugerum *in a day.*

1. Messis proprio nomine dicitur in iis quae metimur, maxime in frumento, et ab eo esse vocabulo declinata. Frumenti tria genera sunt messionis, unum, ut in Umbria, ubi falce secundum terram succidunt stramentum et manipulum, ut quemque subsicuerunt, ponunt in terra. Ubi eos fecerunt multos, iterum eos percensent ac de singulis secant inter spicas et stramentum. Spicas coiciunt in corbem atque in aream mittunt, stramenta relincunt

2 in segete, unde tollantur in acervum. Altero modo metunt, ut in Piceno, ubi ligneum habent incurvum bacillum, in quo sit extremo serrula ferrea. Haec cum comprendit fascem spicarum, desecat et stramenta stantia in segeti relinquit, ut postea subsecentur. Tertio modo metitur, ut sub urbe Roma et locis plerisque, ut stramentum medium subsicent, quod manu sinistra summum prendunt; a quo medio messem dictam puto. Infra manum stramentum cum terra

3 haeret, postea subsecatur; contra quod cum spica stramentum haeret, corbibus in aream defertur. Ibi discedit in aperto loco palam; a quo potest nominata esse palea. Alii stramentum ab stando, ut stamen, dictum putant; alii ab stratu, quod id substernatur pecori. Cum est matura seges, metendum, cum in ea in iugerum fere una opera propemodum in facili agro satis esse dicatur. Messas spicas corbibus in aream deferre debent.

The Threshing-Floor

li, 1-2. Stolo said that the threshing-floor should be on raised ground to catch the wind, round in shape, higher at the centre so as to drain off rainwater. The floor should be of beaten earth, and coated with amurca which is poison to vermin and weeds. Some floors were paved. The Bagienni had roofed floors because of rain-storms in their country. In a hot climate there should be a shelter nearby where the workers could go during the heat of the day.

li. Aream esse oportet in agro sublimiori loco, quam perflare possit ventus; hanc esse modicam pro magnitudine segetis, potissimum rutundam et mediam paulo extumidam, ut, si pluerit, non consistat aqua et quam brevissimo itinere extra aream defluere possit; omne porro brevissimum in rutundo e medio ad extremum. Solida terra pavita, maxime si est argilla, ne, aestu peminosa si sit, in rimis eius grana oblitescant et recipiant aquam et ostia aperiant

muribus ac formicis. Itaque amurca solent perfundere, ea enim
2 herbarum et formicarum et talparum venenum. Quidam aream
ut habeant soldam, muniunt lapide aut etiam faciunt pavimentum.
Non nulli etiam tegunt areas, ut in Bagiennis, quod ibi saepe
id temporis anni oriuntur nimbi. Ubi ea retecta et loca calda,
prope aream faciundum umbracula, quo succedant homines in
aestu tempore meridiano.

Methods of Threshing

lii, 1-2. *Stolo then went on to say that the best ears were to be picked
out for seed. Threshing was done either by driving a yoke of oxen and a
sledge over the grain or a plostellum punicum, a cart moving on a series
of toothed rollers. Others drove cattle over the grain so that their hoofs
separated it from the husks. After it had been threshed, the grain had to
be tossed with forks in the wind or separated from the chaff in winnowing
baskets.*

lii. Quae seges grandissima atque optima fuerit, seorsum in
aream secerni oportet spicas, ut semen optimum habeat; e spicis
in area excuti grana. Quod fit apud alios iumentis iunctis ac
tribulo. Id fit e tabula lapidibus aut ferro asperata, quae cum
imposito auriga aut pondere grandi trahitur iumentis iunctis,
discutit e spica grana; aut ex axibus dentatis cum orbiculis, quod
vocant plostellum poenicum; in eo quis sedeat atque agitet quae
trahant iumenta, ut in Hispania citeriore et aliis locis faciunt.
2 Apud alios exteritur grege iumentorum inacto et ibi agitato perticis,
quod ungulis e spica exteruntur grana. Iis tritis oportet e terra
subiectari vallis aut ventilabris, cum ventus spirat lenis. Ita fit ut
quod levissimum est in eo atque appellatur acus ac palea evannatur
foras extra aream ac frumentum, quod est ponderosum, purum
veniat ad corbem.

The Gleaning

liii. *Stolo said that the gleaning should be let, gathered, or pastured.
Expense must be watched.*

liii. Messi facta spicilegium venire oportet aut domi legere
stipulam aut, si sunt spicae rarae et operae carae, compasci. Summa
enim spectanda, ne in ea re sumptus fructum superet.

77

The Vintage

liv, 1-3. *Stolo went on to say that those grapes which ripened earlier than others should be picked first. The grapes were selected, some for wine, some to be preserved in jars, others for the table. When the grapes had been trodden, the stalks and skin were put under the press. The mass which then remained might be cut around and pressed again. Wine called lora was made for the farmhands from pressed skins and added water.*

liv. In vinetis uva cum erit matura, vindemiam ita fieri oportet, ut videas, a quo genere uvarum et a quo loco vineti incipias legere. Nam et praecox et miscella, quam vocant nigram, multo ante coquitur, quo prior legenda, et quae pars arbusti ac 2 vineae magis aprica, prius debet descendere de vite. In vindemia diligentis uva non solum legitur sed etiam eligitur; legitur ad bibendum, eligitur ad edendum. Itaque lecta defertur in forum vinarium, unde in dolium inane veniat; electa in secretam corbulam, unde in ollulas addatur et in dolia plena vinaciorum contrudatur, alia quae in piscinam in amphoram picatam descendat, alia quae in aream in carnarium escendat. Quae calcatae uvae erunt, earum scopi cum folliculis subiciendi sub prelum, ut, siquid reliqui 3 habeant musti, exprimatur in eundem lacum. Cum desiit sub prelo fluere, quidam circumcidunt extrema et rursus premunt et, rursus cum expressum, circumsicium appellant ac seorsum quod expressum est servant, quod resipit ferrum. Expressi acinorum folliculi in dolia coiciuntur, eoque aqua additur; ea vocatur lora, quod lota acina, ac pro vino operariis datur hieme.

The Olive Harvest

lv, 1-7. *Stolo gave instructions for the olive harvest. Olives should be picked not shaken down, to avoid bruising: bare fingers were better than pincers. Those out of reach might be beaten down with a reed. Some olives were eaten, others yielded oil. The latter were allowed to mellow in piles and then put in the press (trapeta) which was an olive mill fitted with hard stones. Amurca, the pulp which flowed out of the press and blackened the ground, was valuable as a fertiliser for trees and as a herbicide.*

lv. De oliveto oleam, quam manu tangere possis e terra ac scalis, legere oportet potius quam quatere, quod ea quae vapulavit

macescit nec dat tantum olei. Quae manu stricta, melior ea quae digitis nudis, quam illa quae cum digitabulis, durities enim eorum
2 quod non solum stringit bacam, sed etiam ramos glubit ac relinquit ad gelicidium retectos. Quae manu tangi non poterunt, ita quati debent, ut harundine potius quam pertica feriantur; gravior enim
3 plaga medicum quaerit. Qui quatiet, ne adversam caedat; saepe enim ita percussa olea secum defert de ramulo plantam, quo facto fructum amittunt posteri anni. Nec haec non minima causa, quod oliveta dicant alternis annis non ferre fructus aut non aeque magnos.
4 Olea ut uva per idem bivium redit in villam, alia ad cibum, alia ut eliquescat ac non solum corpus intus unguat sed etiam extrinsecus.
5 Itaque dominum et in balneas et gymnasium sequitur. Haec, de qua fit oleum, congeri solet acervatim in dies singulos in tabulata, ut ibi mediocriter fracescat, ac primus quisque acervos demittatur per serias ac vasa olearia ad trapetas, quae res molae oleariae ex
6 duro et aspero lapide. Olea lecta si nimium diu fuit in acervis, caldore fracescit et oleum foetidum fit. Itaque si nequeas mature
7 conficere, in acervis iactando ventilare oportet. Ex olea fructus duplex: oleum, quod omnibus notum, et amurca, cuius utilitatem quod ignorant plerique, licet videre e torculis oleariis fluere in agros ac non solum denigrare terram, sed multitudine facere sterilem; cum is umor modicus cum ad multas res tum ad agri culturam pertineat vehementer, quod circum arborum radices infundi solet, maxime ad oleam, et ubicumque in agro herba nocet.

(the text is omitted)

lvi-lxix. [*Chapters lvi-lxi give directions through Stolo for storing the farm products: hay under cover (lvi), wheat in granaries (lvii), grapes in jars (lviii), fruit in fruit houses (lix), olives in brine, salt or oil (lx),* amurca *in jars (lxi). In chapters lxii-lxix, 1, instructions are given for bringing the various products out of store: in lxix, 2-3, the discussion was interrupted by the sacristan's freedman running up to the friends in distress saying that he was late because his master had been stabbed by mistake in the crowd. The surgeon had not been able to save him. The friends parted company, sadder for the misfortunes of life than surprised at such an occurrence in Rome.*]

LIBER SECUNDUS

Praefatio

1-6. *Our ancestors used to think that the Romans who lived in the country were hardier than those who lived in towns. Nowadays one gymnasium apiece is hardly enough, and their villas must ring with Greek names. Fathers of families do not work on the land now, with the result that grain and wine are imported from overseas. Although shepherds founded Rome and taught agriculture to their children, their descendants despite the laws have made pastures out of grain land.*

The owner of a farm should have knowledge of both agriculture and stockbreeding, and, in addition, of the products of the home farm. I offer this book on stockbreeding to you, Turranius Niger, because you enjoy this pursuit. I shall briefly describe stockbreeding, recalling conversations which I once had with cattle owners in Epirus.

Viri magni nostri maiores non sine causa praeponebant rusticos Romanos urbanis. Ut ruri enim qui in villa vivunt ignaviores, quam qui in agro versantur in aliquo opere faciendo, sic qui in oppido sederent, quam qui rura colerent, desidiosiores putabant. Itaque annum ita diviserunt, ut nonis modo diebus urbanas res 2 usurparent, reliquis septem ut rura colerent. Quod dum servaverunt institutum, utrumque sunt consecuti, ut et cultura agros fecundissimos haberent et ipsi valetudine firmiores essent, ac ne Graecorum urbana desiderarent gymnasia. Quae nunc vix satis singula sunt, nec putant se habere villam, si non multis vocabulis retinniat Graecis, quom vocent particulatim loca, procoetona, palaestram, 3 apodyterion, peristylon, ornithona, peripteron, oporothecen. Igitur quod nunc intra murum fere patres familiae correpserunt relictis falce et aratro et manus movere maluerunt in theatro ac circo, quam in segetibus ac vinetis, frumentum locamus qui nobis advehat, qui saturi fiamus ex Africa et Sardinia, et navibus vindemiam condimus ex insula Coa et Chia.

4 Itaque in qua terra culturam agri docuerunt pastores progeniem suam, qui condiderunt urbem, ibi contra progenies eorum propter avaritiam contra leges ex segetibus fecit prata, ignorantes non idem esse agri culturam et pastionem. Alius enim opilio et arator, nec, si possunt in agro pasci armenta, armentarius non aliut ac bubulcus. Armentum enim id quod in agro natum non creat, sed tollit dentibus;

contra bos domitus causa fit ut commodius nascatur frumentum
5 in segete et pabulum in novali. Alia, inquam, ratio ac scientia
coloni, alia pastoris: coloni ea quae agri cultura factum ut nascerentur e terra, contra pastoris ea quae nata ex pecore. Quarum
quoniam societas inter se magna, propterea quod pabulum in fundo
compascere quam vendere plerumque magis expedit domino fundi
et stercoratio ad fructus terrestres aptissima et maxume ad id pecus
appositum, qui habet praedium, habere utramque debet disciplinam,
et agri culturae et pecoris pascendi, et etiam villaticae pastionis.
Ex ea enim quoque fructus tolli possunt non mediocres ex orni-
6 thonibus ac leporariis et piscinis. E quis quoniam de agri cultura
librum Fundaniae uxori propter eius fundum feci, tibi, Niger
Turrani noster, qui vehementer delectaris pecore, propterea quod
te empturientem in campos Macros ad mercatum adducunt crebro
pedes, quo facilius sumptibus multa poscentibus ministres, quod
eo facilius faciam, quod et ipse pecuarias habui grandes, in Apulia
oviarias et in Reatino equarias, de re pecuaria breviter ac summatim
percurram ex sermonibus nostris collatis cum iis qui pecuarias
habuerunt in Epiro magnas, tum cum piratico bello inter Delum et
Siciliam Graeciae classibus praeessem. Incipiam hinc...

The Origin and History of Stockbreeding
(the text is omitted)

i, 1-10. *Once I discussed with my friends the historical aspect of
stockbreeding, reminding them that men and flocks had always existed and
that the rearing of animals had been inherited from the primitive ages described
by Dicaearchus. At first men lived on wild fruits like the arbutus, then
passed to the pastoral stage when they tamed wild animals. The sheep was the
first to be domesticated because of its amenable nature, and its products.
Thirdly, man learnt to till the soil, but the rearing of animals continued
alongside agriculture. The most distinguished men in Greek and Latin
literature were shepherds. The legend of the Golden Fleece and the sheep
which Hercules brought from Africa came to mind. Domestic animals
were included in the signs of the zodiac; many geographical names were
taken from animals; Italy was called after* vituli *(bullocks);* Faustulus,
*the foster-father of Romulus and Remus, was a shepherd. The birthday
of Rome was on the festival of the shepherds, the* Parilia; *the oldest coins
were marked with figures of cattle; the Roman people were purified by*

ARBUTUS UNEDO, L., *the arbutus, or strawberry-tree, a common bush of the Italian countryside.*

the sacrifice of a boar, a ram, and a bull (the suovetaurilia), *and many Roman names were derived from those of domestic animals.*

The Science of Stockbreeding

i, 11-16. *Scrofa began by stating that stockbreeding for profit was a science: that even the word for money* (pecunia) *was derived from cattle* (pecus). *In building up a flock the age of purchased animals was important. Young and old cattle were cheaper, but the young would be more profitable. It was important to be able to recognise the good " points " of each kind of animal: the special breeds should be known, such as Arcadian asses and those of Reate.*

11 Relicum est de scientia pastorali, de qua est dicendum, quod Scrofa noster, cui haec aetas defert rerum rusticarum omnium palmam, quo melius potest, dicet. Cum convertissent in eum ora omnes, Scrofa, Igitur, inquit, est scientia pecoris parandi ac pascendi, ut fructus quam possint maximi capiantur ex eo, a quibus ipsa pecunia nominata est; nam omnis pecuniae pecus fundamentum.

12 Ea partes habet novem, discretas ter ternas, ut sit una de minoribus pecudibus, cuius genera tria, oves capra sus, altera de pecore maiore, in quo sunt item ad tres species natura discreti, boves asini equi. Tertia pars est in pecuaria quae non parantur, ut ex iis capiatur fructus, sed propter eam aut ex ea sunt, muli canes pastores. Harum una quaeque in se generales partes habet minimum novenas,

quarum in pecore parando necessariae quattuor, alterae in pascendo totidem; praeterea communis una. Ita fiunt omnium partes minimum octoginta et una, et quidem necessariae nec parvae.

13 Primum ut bonum pares pecus, unum scire oportet, qua aetate quamque pecudem parare habereque expediat. Itaque in bubulo pecore minoris emitur annicula et supra decem annorum, quod a bima aut trima fructum ferre incipit neque longius post decimum

14 annum procedit. Nam prima aetas omnis pecoris et extrema sterilis. E quattuor primis altera pars est cognitio formae unius cuiusque pecudis, qualis sit. Magni enim interest, cuius modi quaeque sit ad fructum. Ita potius bovem emunt cornibus nigrantibus quam albis, capram amplam quam parvam, sues procero corpore, capitibus ut sint parvis. Tertia pars est, quo sit seminio quaerendum. Hoc nomine enim asini Arcadici in Graecia nobilitati, in Italia Reatini, usque eo ut mea memoria asinus venierit sestertiis milibus sexaginta et unae quadrigae Romae constiterint

15 quadringentis milibus. Quarta pars est de iure in parando, quem ad modum quamque pecudem emi oporteat civili iure. Quod enim alterius fuit, id ut fiat meum, necesse est aliquid intercedere, neque in omnibus satis est stipulatio aut solutio nummorum ad mutationem domini. In emptione alias stipulandum sanum esse,

16 alias e sano pecore, alias neutrum.

Pasturage, Breeding, and Feeding

i. 16-20. In connection with pasturage, locality, time, and manner were to be considered. The same districts were not equally suitable at all seasons. Flocks of sheep were driven from Apulia to Samnium, and mules from the Rosea to the mountains for summer grazing. The kind of food most acceptable to each species should be considered, and with what extra nourishment it should be supplemented. The time for mating the various animals was to be carefully followed. The period of gestation varied between the species. It was said that certain mares in Lusitania conceived from the wind, just as some hens laid "wind" eggs. Newly-born animals should have a clean soft bed. Attention should be paid to feeding: the time of weaning too differs with different species.

Alterae partes quattuor sunt, cum iam emeris, observandae, de pastione, de fetura, de nutricatu, de sanitate. Pascendi primus locus

qui est, eius ratio triplex, in qua regione quamque potissimum
pascas et quando et qui, ut capras in montuosis potius locis fruticibus
quam in herbidis campis, equas contra. Neque eadem loca aestiva
et hiberna idonea omnibus ad pascendum. Itaque greges ovium
longe abiguntur ex Apulia in Samnium aestivatum atque ad publicanum profitentur, ne, si inscriptum pecus paverint, lege censoria
17 committant. Muli e Rosea campestri aestate exiguntur in Burbures
altos montes. Qui potissimum quaeque pecudum pascatur,
habenda ratio, nec solum quod faeno fit satura equa aut bos, cum
sues hoc vitent et quaerant glandem, sed quod hordeum et faba
interdum sit quibusdam obiciendum et dandum bubus lupinum et
lactariis medica et cytisum; praeterea quod ante admissuram diebus
triginta arietibus ac tauris datur plus cibi, ut vires habeant, feminis
bubus demitur, quod macescentes melius concipere dicuntur.
18 Secunda pars est de fetura. Nunc appello feturam a conceptu ad
partum; hi enim praegnationis primi et extremi fines. Quare
primum videndum de admissione, quo quaeque tempore ut ineant
facere oporteat. Nam ut suillo pecori a favonio ad aequinoctium
vernum putant aptum, sic ovillo ab arcturi occasu usque ad aquilae
occasum. Praeterea habenda ratio, quanto antequam incipiat
admissura fieri mares a feminis secretos habeant, quod fere in
omnibus binis mensibus ante faciunt et armentarii et opiliones.
19 Altera pars est, in fetura quae sint observanda, quod alia alio
tempore parere solet. Equa enim ventrem fert duodecim menses,
vacca decem, ovis et capra quinos, sus quattuor. In fetura res
incredibilis est in Hispania, sed est vera, quod in Lusitania ad
oceanum in ea regione, ubi est oppidum Olisipo, monte Tagro
quaedam e vento concipiunt certo tempore equae, ut hic gallinae
quoque solent, quarum ova hypenemia appellant. Sed ex his equis
qui nati pulli, non plus triennium vivunt. Quae nata sunt matura
et corda, ut pure et molliter stent videndum et ne obterantur.
Dicuntur agni cordi, qui post tempus nascuntur ac remanserunt
20 in volvis intimis . . . vocant chorion, a quo cordi appellati. Tertia
res est, in nutricatu quae observari oporteat, in quo quot diebus
matris sugant mammam et id quo tempore et ubi; et si parum
habet lactis mater, ut subiciat sub alterius mammam, qui appellantur
subrumi, id est sub mamma. Antiquo enim vocabulo mamma

rumis, ut opinor. Fere ad quattuor menses a mamma non diiunguntur agni, haedi tres, porci duo. E quis qui iam puri sunt ad sacrificium, ut immolentur, olim appellati sacres, quos appellat Plautus
cum ait "quanti sunt porci sacres?" Sic boves altiles ad sacrificia
publica saginati dicuntur opimi.

The Health of the Cattle
(the text is omitted)

i, 21-23. *Scrofa went on: the fourth division was that of health, an
important matter because a sickly herd could cause the farmer heavy losses.
Sometimes an animal-doctor had to be called in, at others a careful
shepherd could administer treatment. The usual causes of sickness were
heat, cold, overwork, and lack of exercise. In the case of overwork the
animal should be drenched with water, rubbed down with oil and wine,
fed, and kept covered. If this was not effective, blood should be let from
the head. The herdsman should keep a written record of all other diseases.*

The Size of the Herd
(the text is omitted)

i, 24-28. *Scrofa said that finally the optimum size for a herd had to
be considered by the owner. The number in a herd and the number of
herds should be decided on, in relation to the pasture available. If the
weaker ones were weeded out when too many young were born, the rest
flourished better.*

*Atticus asked how this consideration of stock numbers and breeding
applied in the case of mules and herdsmen since mules cannot produce
young. He admitted, however, that perhaps it was apposite for herdsmen
since they frequently had women with them in their summer or winter
quarters and so added to the number of slaves. I [Varro] remarked that
the classifications were not meant to be exact and so considerations of
breeding and foaling could be left out in the case of mules. Scrofa then
remarked that if there had to be a replacement, the gap could be filled
from two sources of revenue, wool from sheep and goats, and milk and
cheese.*

Of Sheep

ii, 1-6. *I next invited my friends from Epirus to take up the discussion. Atticus spoke in answer, about sheep. They should be in good
condition when bought, neither too young nor too old. A good sheep*

85

should be plump, with a soft, thick fleece. The flock should come from a good stock. He described the characteristics of a good ram. In purchasing the buyer and seller might choose what reservations to make but usually the old legal formula was followed. After the price had been agreed, a stipulation regarding soundness was made. Even then the flock had not changed hands until the money had been paid down: either party could sue the other in case of default.

Sed quoniam nos nostrum pensum absolvimus ac limitata est pecuaria quaestio, nunc rursus vos reddite nobis, o Epirotae, de una quaque re, ut videamus, quid pastores a Pergamide Maledove 2 potis sint. Atticus, qui tunc Titus Pomponius, nunc Quintus Caecilius cognomine eodem, Ego opinor, inquit, incipiam primus, quoniam in me videre coniecisse oculos, et dicam de primigenia pecuaria. E feris enim pecudibus primum dicis oves comprehensas ab hominibus ac mansuefactas. Has primum oportet bonas emere, quae ita ab aetate, si neque vetulae sunt neque merae agnae, quod alterae nondum, alterae iam non possunt dare fructum. Sed ea 3 melior aetas, quam sequitur spes, quam ea quam mors. De forma ovem esse oportet corpore amplo, quae lana multa sit et molli, villis altis et densis toto corpore, maxime circum cervicem et collum, ventrem quoque ut habeat pilosum. Itaque quae id non habent, maiores nostri apicas appellabant ac reiciebant. Esse oportet cruribus humilibus; caudis observate ut sint in Italia prolixis, in Syria brevibus. In primis videndum ut boni seminis pecus habeas. 4 Id fere ex duabus rebus potest animadverti, ex forma et progenie: ex forma, si arietes sint fronte lana vestiti bene, tortis cornibus pronis ad rostrum, ravis oculis, lana opertis auribus, ampli, pectore et scapulis et clunibus latis, cauda lata et longa. Animadvertendum quoque lingua ne nigra aut varia sit, quod fere qui eam habent nigros aut varios procreant agnos. Ex progenie autem animadvert- 5 itur, si agnos procreant formosos. In emptionibus iure utimur eo, quo lex praescripsit. In ea enim alii plura, alii pauciora excipiunt; quidam enim pretio facto in singulas oves, ut agni cordi duo pro una ove adnumerentur, et si quoi vetustate dentes absunt, item binae pro singulis ut procedant. De reliquo antiqua fere formula utuntur. Cum emptor dixit "tanti sunt mi emptae?" et ille respondit "sunt" et expromisit nummos emptor stipulatur prisca 6 formula sic, "illasce oves, qua de re agitur, sanas recte esse, uti pecus ovillum, quod recte sanum est extra luscam surdam minam, id est ventre glabro, neque de pecore morboso esse habereque recte licere, haec sic recte fieri spondesne?" Cum id factum est, tamen grex dominum non mutavit, nisi si est adnumeratum; nec non emptor pote ex empto vendito illum damnare, si non tradet, quamvis non solverit nummos, ut ille emptorem simili iudicio, si non reddit pretium.

Of Sheep: Pasturage

ii, 7-12. *Atticus continued, saying that the fold on the home farm should be protected from the wind and face east: the ground should be sloping and kept clean. There should be separate enclosures for the pregnant and the sick. For those that fed away from the farm, materials for making folds must be carried: winter and summer grazing grounds were often far apart. I reminded the others about my flocks that wintered in Apulia but passed the summer on the mountains near Reate. Atticus went on to say that those flocks fed differently during the seasons. They fed in summer at daybreak and sheltered from the heat of noonday under cliffs and trees: and fed again in the evening. Sheep might profitably graze stubble fields. In winter and spring they were driven out to feed the whole day and once to water.*

7 De alteris quattuor rebus deinceps dicam, de pastione, fetura, nutricatu, sanitate. Primum providendum ut totum annum recte pascantur intus et foris; stabula idoneo loco ut sint, ne ventosa, quae spectent magis ad orientem quam ad meridianum tempus. Ubi stent, solum oportet esse eruderatum et proclivum, ut everri facile possit ac fieri purum. Non enim solum ea uligo lanam
8 corrumpit ovium, sed etiam ungulas, ac scabras fieri cogit. Cum aliquot dies steterunt, subicere oportet virgulta alia, quo mollius requiescant purioresque sint; libentius enim ita pascuntur. Faciendum quoque saepta secreta ab aliis, quo incientes secludere possis, item quo corpore aegro. Haec magis ad villaticos greges animad-
9 vertenda. Contra illae in saltibus quae pascuntur et a tectis absunt longe, portant secum crates aut retia, quibus cohortes in solitudine faciant, ceteraque utensilia. Longe enim et late in diversis locis pasci solent, ut multa milia absint saepe hibernae pastiones ab aestivis. Ego vero scio, inquam; nam mihi greges in Apulia hibernabant, qui in Reatinis montibus aestivabant, cum inter haec

bina loca, ut iugum continet sirpiculos, sic calles publicae distantes
10 pastiones. Eaeque ibi, ubi pascuntur in eadem regione, tamen
temporibus distinguntur, aestate quod cum prima luce exeunt
pastum, propterea quod tunc herba ruscida meridianam, quae est
aridior, iucunditate praestat. Sole exorto potum propellunt, ut
11 redintegrantes rursus ad pastum alacriores faciant. Circiter
meridianos aestus, dum defervescant, sub umbriferas rupes et
arbores patulas subigunt, quaad refrigeratur. Aere vespertino
rursus pascunt ad solis occasum. Ita pascere pecus oportet, ut
averso sole agat; caput enim maxime ovis molle est. Ab occasu
parvo intervallo interposito ad bibendum appellunt et rursus
pascunt, quaad contenebravit; iterum enim tum iucunditas in
herba redintegrabit. Haec a vergiliarum exortu ad aequinoctium
12 autumnale maxime observant. Quibus in locis messes sunt factae,
inigere est utile duplici de causa: quod et caduca spica saturantur
et obtritis stramentis et stercoratione faciunt in annum segetes
meliores. Reliquae pastiones hiberno ac verno tempore hoc mutant,
quod pruina iam exalata propellunt in pabulum et pascunt diem
totum ac meridiano tempore semel agere potum satis habent.

Of Sheep: Breeding, Feeding, Health

*ii. 13-20. Atticus continued: the rams should be kept apart for two
months before mating. The best time for mating was from May to the
end of July. Then the lambs were born in autumn when the grass was
sprouting again. The rams should be kept apart after mating and ewes
should not breed under two years. The lambs were born in pens kept
separate, and put near a fire. After a few days the ewes went out to
pasture and suckled the lambs in the evening and the morning. After
about ten days the lambs were tethered to stakes: they were given green
fodder for four months. When they were being weaned they should be
coaxed with dainty food. Lambs were castrated at the fifth month.
"Jacketed" sheep (i.e. those whose fleece is protected by coverings made
of skins, to prevent soiling) received special care in regards to folding and
food; the head shepherd should keep a written record of remedies for
sheep. In Epirus they had one shepherd to a hundred sheep and two to
the "jacketed" sheep.*

13   Quod ad pastiones attinet, haec fere sunt; quod ad feturam,
quae dicam. Arietes, quibus sis usurus ad feturam, bimestri

tempore ante secernendum et largius pabulo explendum. Cum
redierunt ad stabula e pastu, hordeum si est datum, firmiores fiunt
ad laborem sustinendum. Tempus optimum ad admittendum ab
arcturi occasu ad aquilae occasum, quod quae postea concipiuntur,
14 fiunt vegrandes atque imbecillae. Ovis praegnas est diebus CL.
Itaque fit partus exitu autumnali, cum aer est modice temperatus
et primitus oritur herba imbribus primoribus evocata. Quam diu
admissura fit, eadem aqua uti oportet, quod commutatio et lanam
facit variam et corrumpit uterum. Cum omnes conceperunt, rursus
arietes secernendi, iam factis praegnatibus quod sunt molesti,
obsunt. Neque pati oportet minores quam bimas saliri, quod
neque natum ex his idoneum est, neque non ipsae fiunt deteriores;
et non meliores quam trimae admissae. Deterrent ab saliendo, et
fiscellas e iunco aliave qua re quod alligant ad naturam; commodius
servantur, si secretas pascunt.

15 In nutricatu, cum parere coeperunt, inigunt in stabula, eaque
habent ad eam rem seclusa, ibique nata recentia ad ignem prope
ponunt, quaad convaluerunt. Biduum aut triduum retinent, dum
adcognoscant matrem agni et pabulo se saturent. Deinde matres
cum grege pastum prodeunt, retinent agnos, ad quos cum reductae
ad vesperum, aluntur lacte et rursus discernuntur, ne noctu a
matribus conculcentur. Hoc item faciunt mane, antequam matres
16 in pabulum exeant, ut agni satulli fiant lacte. Circiter decem dies
cum praeterierunt, palos offigunt et ad eos alligant libro aut qua alia
re levi distantes, ne toto die cursantes inter se teneri delibent aliquid
membrorum. Si ad matris mammam non accedet, admovere
oportet et labra agni unguere buturo aut adipe suilla et olfacere labra
lacte. Diebus post paucis obicere iis viciam molitam aut herbam
17 teneram, antequam exeunt pastum et cum reverterunt. Et sic
nutricantur, quaad facti sunt quadrimestres. Interea matres
eorum iis temporibus non mulgent quidam; qui id melius, omnino
perpetuo, quod et lanae plus ferunt et agnos plures. Cum depulsi
sunt agni a matribus, diligentia adhibenda est, ne desiderio senescant. Itaque deliniendum in nutricatu pabuli bonitate et a frigore
18 et aestu ne quid laboret curandum. Cum oblivione iam lactis non
desiderat matrem, tum denique compellendum in gregem ovium.
Castrare oportet agnum non minorem quinque mensum, neque
antequam calores aut frigora se fregerunt. Quos arietes summittere

89

volunt, potissimum eligunt ex matribus, quae geminos parere solent. Pleraque similiter faciendum in ovibus pellitis, quae propter lanae bonitatem, ut sunt Tarentinae et Atticae, pellibus integuntur, ne lana inquinetur, quo minus vel infici recte possit vel

19 lavari ac putari. Harum praesepia ac stabula ut sint pura maiorem adhibent diligentiam, quam hirtis. Itaque faciunt lapide strata, urina necubi in stabulo consistat. His quaecumque lubenter vescuntur, ut folia ficulnea et palea, vinacea, furfures, obiciuntur modice, ne parum aut nimium saturentur. Utrumque enim ad corpus alendum inimicum, ut maxime amicum cytisum et medica. Nam et pingues facit facillime et genit lacte.

20 De sanitate sunt multa; sed ea, ut dixi, in libro scripta magister pecoris habet, et quae opus ad medendum, portat secum. Relinquitur de numero, quem faciunt alii maiorem, alii minorem. Nulli enim huius moduli naturales. Illut fere omnes in Epiro facimus, ne minus habeamus in centenas oves hirtas singulos homines, in pellitas binos.

Of Goats

iii, 1-10. *Cossinius talked about goats, recommending that young goats should be assembled for a flock: he described the characteristics of a good breed. Goats were more trouble than sheep: the domestic goat was derived from the wild species. In purchasing no guarantee of soundness could be required as they were unhealthy. The stalls should be warm and paved. The goat fed on bushes and undergrowth and was injurious to all kinds of plants. It was therefore illegal for a tenant to pasture goats on the farm. Mating was timed for the kids to be born in spring. The head goatherds should keep a record of remedies especially those for flesh wounds. The number in a flock should be smaller than for sheep because they scattered when grazing. Herds were kept small in the Ager Gallicus to guard against epidemics. A Roman knight, Gaberius, had lost a herd of a thousand by disease. There should be one buck to ten or more does.*

Cui Cossinius, Quoniam satis balasti, inquit, o Faustule noster, accipe a me cum Homerico Melanthio cordo de capellis, et quem ad modum breviter oporteat dicere, disce. Qui caprinum gregem constituere vult, in eligendo animadvertat oportet primum aetatem, ut eam paret, quae iam ferre possit fructum, et de iis eam

2 potius, quae diutius; novella enim quam vetus utilior. De forma videndum ut sint firmae, magnae, corpus leve ut habeant, crebro pilo, nisi si glabrae sunt (duo enim genera earum); sub rostro duas ut mammulas pensiles habeant, quod eae fecundiores; ubere sint grandiore, ut et lac multum et pingue habeant pro portione. Hircus molliori et potissimum pilo albo ac cervice et collo brevi, gurgulione longiore. Melior fit grex, si non est ex collectis comparatus, sed ex
3 consuetis una. De seminio dico eadem, quae Atticus in ovibus; hoc aliter, ovium semen tardius esse, quo eae sint placidiores; contra caprile mobilius esse, de quarum velocitate in Originum libro Cato scribit haec: "in Sauracti et Fiscello caprae ferae sunt, quae saliunt e saxo pedes plus sexagenos." Oves enim, quas pascimus, ortae sunt ab ovibus feris, sic quas alimus caprae a capris feris ortae,
4 a quis propter Italiam Caprasia insula est nominata. De capris quod meliore semine eae quae bis pariant, ex his potissimum mares solent summitti ad admissuras. Quidam etiam dant operam ut ex insula Melia capras habeant, quod ibi maximi ac pulcherrimi existimantur fieri haedi.
5 De emptione aliter dico atque fit, quod capras sanas sanus nemo promittit; numquam enim sine febri sunt. Itaque stipulantur paucis exceptis verbis, ac Manilius scriptum reliquit sic: "illasce capras hodie recte esse et bibere posse habereque recte licere, haec spondesne?" de quibus admirandum illut, quod etiam Archelaus scribit: non ut reliqua animalia naribus, sed auribus spiritum ducere solere pastores curiosiores aliquot dicunt.
6 De alteris quattuor quod est de pastu, hoc dico: Stabulatur pecus melius, ad hibernos exortos si spectat, quod est alsiosum. Id, ut pleraque, lapide aut testa substerni oportet, caprile quo minus sit uliginosum ac lutulentum. Foris cum est pernoctandum, item in eandem partem caeli quae spectent saepta oportet substerni virgultis, ne oblinantur. Non multo aliter tuendum hoc pecus in pastu atque ovillum, tamen habent sua propria quaedam, quod
7 potius silvestribus saltibus delectantur quam pratis; studiose enim de agrestibus fruticibus pascuntur atque in locis cultis virgulta carpunt. Itaque a carpendo caprae nominatae. Ab hoc in lege locationis fundi excipi solet, ne colonus capra natum in fundo pascat. Harum enim dentes inimici sationi, quas etiam astrologi ita receperunt in caelum, ut extra lembum duodecim signorum excluserint;

8 sunt duo haedi at capra non longe a tauro. Quod ad feturam pertinet, desistente autumno exigunt a grege in campo hircos in caprilia, item ut in arietibus dictum. Quae concepit, post quartum mensem reddit tempore verno. In nutricatu haedi, trimestres cum sunt facti, tum submittuntur et in grege incipiunt esse. Quid dicam de earum sanitate, quae numquam sunt sanae? Nisi tamen illud unum: quaedam scripta habere magistros pecoris, quibus remediis utantur ad morbos quosdam earum ac vulneratum corpus, quod usu venit iis saepe, quod inter se cornibus pugnant atque in
9 spinosis locis pascuntur. Relinquitur de numero, qui in gregibus est minor caprino quam in ovillo, quod caprae lascivae et quae dispargant se; contra oves quae se congregent ac condensent in locum unum. Itaque in agro Gallico greges plures potius faciunt quam magnos, quod in magnis cito existat pestilentia, quae ad
10 perniciem eum perducat. Satis magnum gregem putant esse circiter quinquagenas. Quibus adsentiri putant id quod usu venit Gaberio, equiti Romano. Is enim cum in suburbano mille iugerum haberet et a caprario quodam, qui adduxit capellas ad urbem decem, sibi in dies singulos denarios singulos dare audisset, coegit mille caprarum, sperans se capturum de praedio in dies singulos denarium mille. Tantum enim fefellit, ut brevi omnes amiserit morbo. Contra in Sallentinis et in Casinati ad centenas pascunt. De maribus et feminis idem fere discrimen, ut alii ad denas capras singulos parent hircos, ut ego; alii etiam ad quindecim, ut Menas; non nulli etiam, ut Murrius, ad viginti.

Of Swine

(the text is omitted)

iv, 1-22. *Scrofa, who now began to talk of pigs, explained how his grandfather first received the name "Scrofa" after a victory in battle. For a good herd, young animals with good characteristics and of handsome appearance should be chosen. The usual formula of purchase with stipulations was to be followed. Swine like wet ground and their usual diet of mash, beans, and barley gave the flesh a good flavour. In summer they were to be pastured twice daily with a noontide rest in the shade: in winter after the frost had melted. The best time for mating was in early spring: the young were then born in summer when the land was rich in food. Swine were the earliest victims for sacrifice. The best flitches came from Gaul. Each sow should have a separate sty which should be three feet high. Sows with young needed two or more pounds of barley per day. Usually the sow should bear as many piglets as she has teats, but portents have occurred. The sow of Aeneas at Lavinium bore thirty white pigs: thirty years later the people of Lavinium founded Alba. The pigs are still commemorated in bronze images at Lavinium and the sow's body preserved in brine.*

The sow is allowed to graze ten days after farrowing, and the pigs when grown follow her to pasture. The swineherd should train the animals to come for food at the sound of a horn. Boars are castrated at a year. Ten boars were enough for a hundred sows. A herd was usually a hundred in number.

Of Oxen

(the text is omitted)

v, 1-5. Quintus Lucienus arrived at this point and after greeting the company went away again to pay his dues to the Lares taking Murrius with him. Vaccius was ready to lead the next discussion which was on oxen but I [Varro] broke in with some mythological references to cows: Vaccius added several others.

Of Oxen

v, 6-11. Vaccius first described the four stages of life in oxen. Whoever bought a herd should see that they were still able to bear. He outlined the characteristics of the ideal ox; the males especially should be of good appearance. Gallic oxen were good workers, the Ligurian worthless: those from Epirus were the best in all Greece and Italy. The large Italian oxen were preferred for ritual offerings because of their size and white colour. In the purchase of oxen the stipulations were different for trained and untrained animals, but in both the buyer should be protected from suits for damage. Cattle were best pastured in wooded land.

6 Primum in bubulo genere aetatis gradus dicuntur quattuor, prima vitulorum, secunda iuvencorum, tertia bovum novellorum, quarta vetulorum. Discernuntur in prima vitulus et vitula, in

secunda iuvencus et iuvenca, in tertia et quarta taurus et vacca.
Quae sterilis est vacca, taura appellata; quae praegnas, horda.
Ab eo in fastis dies hordicidia nominatur, quod tum hordae boves
7 immolantur. Qui gregem armentorum emere vult, observare debet
primum, ut sint eae pecudes aetate potius ad fructos ferendos
integrae quam iam expartae; ut sint bene compositae, ut integris
membris, oblongae, amplae, nigrantibus cornibus, latis frontibus,
oculis magnis et nigris, pilosis auribus, compressis malis subsimae,
8 ne gibberae, spina leviter remissa, apertis naribus, labris subnigris,
cervicibus crassis ac longis, a collo palea demissa, corpore bene
costato, latis umeris, bonis clunibus, codam profusam usque ad
calces ut habeant, inferiorem partem frequentibus pilis subcrispam,
cruribus potius minoribus rectis, genibus eminulis distantibus inter
se, pedibus non latis, neque ingredientibus qui displudantur, nec
cuius ungulae divarent, et cuius ungues sint leves et pares, corium
tactu non asperum ac durum, colore potissimum nigro, deinde
robeo, tertio helvo, quarto albo; mollissimus enim hic, ut durissimus
9 primus. De mediis duobus prior quam posterior in eo prior,
utrique plures quam nigri et albi. Neque non praeterea ut mares
seminis boni sint, quorum et forma est spectanda, et qui ex his orti
sunt respondent ad parentum speciem. Et praeterea quibus
regionibus nati sint refert; boni enim generis in Italia plerique
10 Gallici ad opus, contra nugatorii Ligusci; transmarini Epirotici non
solum meliores totius Graeciae, sed etiam quam Italiae. Tametsi
quidam de Italicis, quos propter amplitudinem praestare dicunt,
victimas faciunt atque ad deorum servant supplicia, qui sine
dubio ad res divinas propter dignitatem amplitudinis et coloris
praeponendi. Quod eo magis fit, quod albi in Italia non tam
frequentes quam in Thracia ad μέλανα κόλπον, ubi alio colore pauci.
Eos cum emimus domitos, stipulamur sic: "illosce boves sanos
11 esse noxisque praestari"; cum emimus indomitos, sic: "illosce
iuvencos sanos recte deque pecore sano esse noxisque praestari
spondesne?" Paulo verbosius haec, qui Manili actiones secuntur
lanii, qui ad cultrum bovem emunt; qui ad altaria, hostiae sanitatem
non solent stipulari.

Pascuntur armenta commodissime in nemoribus, ubi virgulta
et frons multa; hieme cum hibernant secundum mare, aestu
abiguntur in montes frondosos.

Of Oxen: Breeding, Health

v, 12-18. *Vaccius continued, saying that before mating cows should have less food and drink, but the bulls should be fattened: he kept two bulls to every seventy cows. Cows were best mated at four years: the best time was from the rising of the Dolphin. Cows in calf should be pastured in watered meadows but some breeders kept them in stall because of harmful insects. When they were near calving fresh fodder should be provided for them to nibble. The calves should be kept separate and suckled twice a day: the calves would soon begin to eat green fodder. The stalls should be paved. Calves should be castrated at two years. The herd should be culled every year. The head herdsman kept a written record of remedies taken from Magos' writings. The usual number of bulls to sixty cows was two, and the size of the herd was usually one hundred.*

12   Propter feturam haec servare soleo.   Ante admissuram mensem unum ne cibo et potione se impleant, quod existimantur facilius macrae concipere.   Contra tauros duobus mensibus ante admissuram herba et, palea ac faeno facio pleniores et a feminis secerno. Habeo tauros totidem, quot Atticus, ad matrices LXX duo, unum anniculum, alterum bimum.   Haec secundum astri exortum facio,
13 quod Graeci vocant lyran, fidem nostri.   Tum denique tauros in gregem redigo.   Mas an femina sit concepta, significat descensu taurus, cum init, quod, si mas est, in dexteriorem partem abit; si femina, in sinisteriorem.   Cur hoc fiat, vos videritis, inquit mihi, qui Aristotelem legitis.   Non minores oportet inire bimas, ut trimae pariant, eo melius, si quadrimae.   Pleraeque pariunt in decem annos, quaedam etiam plures.   Maxime idoneum tempus ad concipiendum a delphini exortu usque ad dies quadraginta aut paulo plus.   Quae enim ita conceperunt, temperatissimo anni tempore pariunt; vaccae enim mensibus decem sunt praegnates.
14 De quibus admirandum scriptum inveni, exemptis testiculis si statim admiseris taurum, concipere.   Eas pasci oportet locis viridibus et aquosis.   Cavere oportet ne aut angustius stent aut feriantur aut concurrant.   Itaque quod eas aestate tabani concitare solent et bestiolae quaedam minutae sub cauda ali, ne concitentur, aliqui solent includere saeptis.   Iis substerni oportet frondem aliudve quid in cubilia, quo mollius conquiescant.   Aestate ad
15 aquam appellendum bis, hieme semel.   Cum parere coeperunt,

95

secundum stabula pabulum servari oportet integrum, quod egredientes degustare possint; fastidiosae enim fiunt. Et providendum, quo recipiunt se, ne frigidus locus sit; algor enim eas et famis macescere cogit.

16 In alimoniis armenticium pecus sic contuendum. Lactantes cum matribus ne cubent; obteruntur enim. Ad eas mane adigi oportet, et cum redierunt e pastu. Cum creverent vituli, levandae matres pabulo viridi obiciendo in praesepiis. Item his, ut fere in omnibus stabulis, lapides substernendi aut quid item, ne ungulae putrescant. Ab aequinoctio autumnali una pascuntur cum matri

17 bus. Castrare non oportet ante bimum, quod difficulter, si aliter feceris, se recipiunt; qui autem postea castrantur, duri et inutiles fiunt. Item ut in reliquis gregibus pecuariis dilectus quotannis habendus et reiculae reiciundae, quod locum occupant earum quae ferre possunt fructus. Siquae amisit vitulum, ei supponere oportet eos, quibus non satis praebent matres. Semestribus vitulis obiciunt furfures triticios et farinam hordeaceam et teneram herbam et

18 ut bibant mane et vesperi curant. De sanitate sunt complura, quae exscripta de Magonis libris armentarium meum crebro ut aliquid legat curo. Numerus de tauris et vaccis sic habendus, ut in sexaginta unus sit anniculus, alter bimus. Quidam habent aut minorem aut maiorem numerum; nam apud Atticum duo tauri in septuaginta matribus sunt. Numerum gregum alius facit alium, quidam centenarium modicum putant esse, ut ego. Atticus centum viginti habet, ut Lucienus.

Of Asses

(the text is omitted)

vi, 1-5. *Murrius who had returned with Lucienus went on to talk about asses: his asses bred at Reate were the best. If anyone wished to start a herd he should choose young, sturdy animals from the best districts. There were two kinds, the wild and the domesticated: the wild was good for breeding. In purchase the usual legal formulae were employed. Asses were best fed on grain and barley-bran. Mating should be before the solstice. The period of gestation was a year. The foals stayed with their mothers for a year; in the third year they were trained for work. Asses were not kept in herds but used for work at the mill, or for hauling*

and ploughing in light soil. Traders employed trains of asses for trans-
porting merchandise in panniers.

Of Horses

vii, 1-16: (8-9 are omitted). *Lucienus then discussed horses.*
In establishing a herd age should be considered. The age of a horse
could be judged from the teeth. Stallions should be comely in appearance.
It was important to choose from good breeds: famous ones were Thessalici
from Thessaly and Roseani *from the Rosea. That colt would prove to*
be a good horse which showed spirit and took the lead in the herd. Pur-
chase was according to the laws of Manilius. Mating should be from
the vernal equinox until the solstice. Mares in foal should not be over-
worked nor kept in cold places. In the stalls the ground should be dry and
poles fixed to keep them separate. Mares should bear every other year.
Colts should be handled while still with their mothers. At three years
the training began when a boy should mount them first: then they could
be exercised gently. Different horses served different purposes, and were
chosen by different standards. The head groom should keep symptoms
of disease and their remedies in written form.

Lucienus: Ego quoque adveniens aperiam carceres, inquit,
et equos emittere incipiam, nec solum mares, quos admissarios
habeo, ut Atticus, singulos in feminas denas. E quis feminas Q.
Modius Equiculus, vir fortissimus, etiam patre militari, iuxta ac
mares habere solebat. Horum equorum et equarum greges qui
habere voluerunt, ut habent aliqui in Peloponneso et in Apulia,
primum spectare oportet aetatem, quam praecipiunt sic. Videmus
2 ne sint minores trimae, maiores decem annorum. Aetas cognoscitur
et equorum et fere omnium qui ungulas indivisas habent et etiam
cornutarum, quod equus triginta mensibus primum dentes medios
dicitur amittere, duo superiores, totidem inferiores. Incipientes
quartum agere annum itidem eiciunt et totidem eiciunt proxumos
eorum quos amiserunt, et incipiunt nasci, quos vocant columellares.
3 Quinto anno incipiente item eodem modo amittere binos, cum
cavos habeat tum renascentes, ei sexto anno impleri, septumo omnes
habere solet renatos et completos. Hoc maiores qui sunt, intellegi
negant posse, praeterquam cum dentes sint facti brocchi et supercilia
cana et sub ea lacunae, ex observatu dicunt eum equom habere
4 annos sedecim. De forma esse oportet magnitudine modica, quod

nec vastos nec minutos decet esse, equas clunibus ac ventribus latis. Equos, ad admissuram quos velis habere, legere oportet amplo corpore, formosos, nulla parte corporis inter se non congruenti.

5 Qualis futurus sit equus, e pullo coniectari potest: si caput habet non magnum nec membris confusis si est, oculis nigris, naribus non angustis, auribus applicatis, iuba crebra, fusca, subcrispa subtenuibus saetis, inplicata in dexteriorem partem cervicis, pectus latum at plenum, umeris latis, ventre modico, lumbis deorsum versus pressis, scapulis latis, spina maxime duplici, si minus, non extanti, coda ampla subcrispa, cruribus rectis aequalibus intro versus potius figuratis, genibus rutundis ne magnis, ungulis duris; toto corpore ut habeat venas, quae animadverti possint, quod qui huiusce 6 modi sit, cum est aeger, ad medendum appositus. De stirpe magni interest qua sint, quod genera sunt multa. Itaque ab hoc nobiles a regionibus dicuntur, in Graecia Thessalici equi a Thessalia, in Italia ab Apulia Apuli, ab Rosea Roseani. Equi boni futuri signa, si cum gregalibus in pabulo contendit in currendo aliave qua re, quo potior sit; si, cum flumen travehundum est gregi, in primis progreditur ac non respectat alios. Emptio equina similis fere ac boum et asinorum, quod eisdem rebus in emptione dominum mutant, ut in Manili actionibus sunt perscripta.

7 Equinum pecus pascendum in pratis potissimum herba, in stabulis ac praesepibus arido faeno; cum pepererunt, hordeo adiecto, bis die data aqua. Horum feturae initium admissionis facere oportet ab aequinoctio verno ad solstitium, ut partus idoneo tempore fiat; duodecimo enim mense die decimo aiunt nasci. Quae post tempus nascuntur, fere vitiosa atque inutilia existunt.

(*paragraphs 8 and 9 omitted*)

10 Cum conceperunt equae, videndum ne aut laborent plusculum aut ne frigidis locis sint, quod algor maxime praegnatibus obest. Itaque in stabulis et umore prohibere oportet humum, clausa habere ostia ac fenestras, et inter singulas a praesepibus intericere longorios, qui eas discernant, ne inter se pugnare possint. Praegnatem neque 11 implere cibo neque esurire oportet. Alternis qui admittant, diuturniores equas, meliores pullos fieri dicunt, itaque ut restibiles segetes esse exuctiores, sic quotannis quae praegnates fiant.

In decem diebus secundum partum cum matribus in pabulum prodigendum, ne ungulas comburat stercus tenellas. Quinque-

mestribus pullis factis, cum redacti sunt in stabulum, obiciendum farinam hordeaciam molitam cum furfuribus, et siquid aliud terra
12 natum libenter edent. Anniculis iam factis dandum hordeum et furfures, usque quaad erunt lactantes. Neque prius biennio confecto a lacte removendum; eosque, cum stent cum matribus, interdum tractandum, ne, cum sint deiuncti, exterreantur; eademque causa ibi frenos suspendendum, ut eculi consuescant et videre
13 eorum faciem et e motu audire crepitus. Cum iam ad manus accedere consuerint, interdum imponere iis puerum bis aut ter pronum in ventrem, postea iam sedentem. Haec facere, cum sit trimus; tum enim maxime crescere ac lacertosum fieri. Sunt qui dicant post annum et sex menses eculum domari posse, sed melius post trimum, a quo tempore farrago dari solet. Haec enim purgatio maxime necessaria equino pecori. Quod diebus decem facere
14 oportet, nec pati alium ullum cibum gustare. Ab undecimo die usque ad quartum decimum dandum hordeum, cottidie adicientem minutatim; quod quarto die feceris, in eo decem diebus proximis manendum. Ab eo tempore mediocriter exercendum et, cum consudarit, perunguendum oleo. Si frigus erit, in equili faciendus ignis.
15　Equi quod alii sunt ad rem militarem idonei, alii ad vecturam, alii ad admissuram, alii ad cursuram, non item sunt spectandi atque habendi. Itaque peritus belli alios eligit atque alit ac docet; aliter quadrigarius ac desultor; neque idem qui vectorios facere vult ad ephippium aut ad raedam, quod qui ad rem militarem, quod ut ibi ad castra habere volunt acres, sic contra in viis habere malunt placidos. Propter quod discrimen maxime institutum ut castrentur equi. Demptis enim testiculis fiunt quietiores, ideo quod semine carent. Ii cantherii appellati, ut in subus maiales, gallis gallinaceis
16 capi. De medicina vel plurima sunt in equis et signa morborum et genera curationum, quae pastorem scripta habere oportet. Itaque ab hoc in Graecia potissimum medici pecorum ἱππιατροί appellati.

Of Mules and Hinnies
(the text is omitted)

viii. 1-6. *A freedman came from Menates to say that the sacrifice was ready but I answered that the third act must be played out. Murrius then discussed mules and hinnies: they were useful for work but were not*

99

profitable for breeding. Heavy and handsome breeding jacks could be selected from the Arcadian or the Reatine herd. In purchase the same code was employed as in the case of horses. The mating was at the same time as for horses. In assembling a herd, age and build should be considered: they should be strong for hauling and of handsome appearance. They drew vehicles on the roads in pairs.

Of Dogs

ix, 1-16. Atticus led the discussion on dogs, saying that a dog was the guardian of the flock. Sheep and goats needed his protection. There were two sorts: the hunting-dog and the watch-dog. He described in detail the points of good sheep-dogs. They should come from good breeds such as the Lacones, *the* Epirotici, *and the* Sallentini. *It was better to buy a bitch, either trained or untrained, from a shepherd. A dog became more attached to the shepherd than the sheep. A dog had been legally purchased when it was delivered to the new owner: the formula was the same as for sheep, but stipulations should be made about the pups. Dogs ate meat and bones together with barley-bread soaked in milk. They were mated in early spring and the puppies weaned after two months. At first they were tied with thin leashes and trained not to gnaw them. Kennels should be kept dry and warm with bedding of leaves or fodder or chaff. A paste of crushed almonds rubbed on the ears and between the toes would keep off insects. Some wore collars studded with iron nails as a protection against wild beasts. One dog to a shepherd was usually enough, but those who drove the transhumanting sheep needed more.*

Relinquitur, inquit Atticus, de quadripedibus quod ad canes attinet, quod pertinet maxime ad nos, qui pecus pascimus lanare. Canes enim ita custos pecoris eius quod eo comite indiget ad se defendendum. In quo genere sunt maxime oves, deinde caprae. Has enim lupus captare solet, cui opponimus canes defensores. In suillo pecore tamen sunt quae se vindicent, verres, maiales, scrofae. Prope enim haec apris, qui in silvis saepe dentibus canes occiderunt.

2 Quid dicam de pecore maiore? cum sciam mulorum gregem, cum pasceretur et eo venisset lupus, ultro mulos circumfluxisse et ungulis caedendo eum occidisse, et tauros solere diversos adsistere clunibus continuatos et cornibus facile propulsare lupos. Quare de canibus quoniam genera duo, unum venaticum et pertinet ad feras bestias silvestres, alterum quod custodiae causa paratur et

pertinet ad pastorem, dicam de eo ad formam artis expositam in novem partes.

3 Primum aetate idonea parandi, quod catuli et vetuli neque sibi neque ovibus sunt praesidio et feris bestiis non numquam praedae. Facie debent esse formosi, magnitudine ampla, oculis nigrantibus aut ravis, naribus congruentibus, labris subnigris aut rubicundis neque resimis superioribus nec pendulis subtus, mento suppresso et ex eo enatis duobus dentibus dextra et sinistra paulo eminulis, 4 superioribus directis potius quam brocchis, acutos quos habeant labro tectos, capitibus et auriculis magnis ac flaccis, crassis cervicibus ac collo, internodis articulorum longis, cruribus rectis et potius varis quam vatiis, pedibus magnis et latis, qui ingredienti ei displodantur, digitis discretis, unguibus duris ac curvis, solo ne ut corneo ne nimium duro, sed ut fermentato ac molli; a feminibus summis corpore suppresso, spina neque eminula neque curva, cauda crassa; latrato gravi, hiatu magno, colore potissimum 5 albo, quod in tenebris facilius agnoscuntur, specie leonina. Praeterea feminas volunt esse mammosas aequalibus papillis. Item videndum ut boni seminii sint; itaque et a regionibus appellantur Lacones, Epirotici, Sallentini. Videndum ne a venatoribus aut laniis canes emas; alteri quod ad pecus sequendum inertes, alteri, si viderint leporem aut cervum, quod eum potius quam oves sequentur. Quare a pastoribus empta melior, quae oves sequi consuevit, aut sine ulla consuetudine quae fuerit. Canis enim facilius quid adsuescit, eaque consuetudo firmior, quae fit ad pastores, quam 6 quae ad pecudes. Publius Aufidius Pontianus Amiterninus cum greges ovium emisset in Umbria ultima, quibus gregibus sine pastoribus canes accessissent, pastores ut deducerent in Metapontinos saltus et Heracleae emporium, inde cum domum redissent qui ad locum deduxerant, e desiderio hominum diebus paucis postea canes sua sponte, cum dierum multorum via interesset, sibi ex agris cibaria praebuerunt atque in Umbriam ad pastores redierunt. Neque eorum quisquam fecerat, quod in agri cultura Saserna praecepit: qui vellet se a cane sectari, ut ranam obiciat coctam. Magni interest ex semine esse canes eodem, quod cognati maxime 7 inter se sunt praesidio. Sequitur quartum de emptione: fit alterius, cum a priore domino secundo traditus est. De sanitate et noxa stipulationes fiunt eaedem, quae in pecore, nisi quod hic

utiliter exceptum est: alii pretium faciunt in singula capita canum, alii ut catuli sequantur matrem, alii ut bini catuli unius canis numerum obtineant, ut solent bini agni ovis, plerique ut accedant canes, qui consuerunt esse una.

8 Cibatus canis propior hominis quam ovis. Pascitur enim eduliis et ossibus, non herbis aut fronde. Diligenter ut habeat cibaria providendum. Fames enim hos ad quaerendum cibum 9 ducet, si non praebebitur, et a pecore abducet; nisi si, ut quidam putant, etiam illuc pervenerint, proverbium ut tollant anticum vel etiam ut μῦθον aperiant de Actaeone atque in dominum adferant 10 dentes. Nec non ita panem hordeacium dandum, ut non potius eum in lacte des intritum, quod eo consueti cibo uti a pecore non cito desciscunt. Morticinae ovis non patiuntur vesci carne, ne ducti sapore minus se abstineant. Dant etiam ius ex ossibus et ea ipsa ossa contusa. Dentes enim facit firmiores et os magis patulum, propterea quod vehementius diducuntur malae, acrioresque fiunt propter medullarum saporem. Cibum capere consuescunt inter- 11 diu, ubi pascuntur, vesperi, ubi stabulantur. Feturae principium admittendi faciunt veris principio; tum enim dicuntur catulire, id est ostendere velle se maritari. Quae tum admissae, pariunt circiter solstitium; praegnates enim solent esse ternos menses. In fetura dandum potius hordeacios quam triticios panes; magis enim eo 12 aluntur et lactis praebent maiorem facultatem. In nutricatu secundum partum, si plures sunt, statim eligere oportet quos habere velis, reliquos abicere. Quam paucissimos reliqueris, tam optimi in alendo fiunt propter copiam lactis. Substernitur eis acus aut quid item aliut, quod molliore cubili facilius educentur. Catuli diebus XX videre incipiunt. Duobus mensibus primis a partu non diiunguntur a matre, sed minutatim desuefiunt. Educunt eos plures in unum locum et inritant ad pugnandum, quo fiunt acriores, 13 neque defatigari patiuntur, quo fiunt segniores. Consue quoque faciunt ut alligari possint primum levibus vinclis; quae si abrodere conantur, ne id consuescant facere, verberibus eos deterrere solent. Pluviis diebus cubilia substernenda fronde aut pabulo duabus de 14 causis: ut ne oblinantur aut perfrigescant. Quidam eos castrant, quod eo minus putant relinquere gregem; quidam non faciunt, quod eos credunt minus acres fieri. Quidam nucibus Graecis in aqua tritis perungunt aures et inter digitos, quod muscae et ricini

15 et pulices soleant, si hoc unguine non sis usus, ea exulcerare. Ne vulnerentur a bestiis, imponuntur his collaria, quae vocantur melium, id est cingulum circum collum ex corio firmo cum clavulis capitatis, quae intra capita insuitur pellis mollis, ne noceat collo duritia ferri; quod, si lupus aliusve quis his vulneratus est, reliquas
16 quoque canes facit, quae id non habent, ut sint in tuto. Numerus canum pro pecoris multitudine solet parari; fere modicum esse putant, ut singuli sequantur singulos opiliones. De quo numero alius alium modum constituit, quod, si sunt regiones, ubi bestiae sint multae, debent esse plures, quod accidit his qui per calles silvestres longinquos solent comitari in aestiva et hiberna. Villatico vero gregi in fundum satis esse duo, et id marem et feminam. Ita enim sunt adsiduiores, quod cum altero item alter fit acrior, et si alteruter aeger est, ne sine cane grex sit.

Of Shepherds

x, 1-11. It was now Cossinius' turn to talk about shepherds. Sturdy types were needed for transhumance. The shepherds must graze their flocks together during the whole day but each sleep with his own flock at night. The head-shepherd should be superior in age and knowledge. Men suitable to be shepherds should be strong, quick, nimble, and able to throw the javelin. There were six methods of legal purchase: the peculium *usually went with the slave: stipulations should be made concerning thefts or damage. The head-shepherd was responsible for all equipment and food supplies.*

For breeding purposes, shepherds at home had a female slave, but the transhumanting shepherds too were often accompanied by women. They both worked and bore children, a reproach to fashionable women who lay for days under mosquito nets. I [Varro] said that in Illyria a pregnant woman often gave birth to her child when at work. Cossinius went on to say that the head-shepherd should keep directions for health of humans and cattle in written form: he had one shepherd to eighty sheep. In the case of mares two men were needed, one of which should have a mare for his own use, in districts where the mares were rounded up.

Cum circumspiceret, nequid praeterisset, Hoc silentium, inquam, vocat alium ad partes. Relicum enim in hoc actu, quot et quod genus sint habendi pastores. Cossinius: Ad maiores pecudes aetate superiores, ad minores etiam pueros, utrosque horum firmiores

qui in callibus versentur, quam eos qui in fundo cotidie ad villam redeant (itaque in saltibus licet videre iuventutem, et eam fere armatam, cum in fundis non modo pueri sed etiam puellae pascant).
2 Qui pascunt, eos cogere oportet in pastione diem totum esse, pascere communiter, contra pernoctare ad suum quemque gregem, esse omnes sub uno magistro pecoris; eum esse maiorem natu potius quam alios et peritiorem quam reliquos, quod ei qui aetate et
3 scientia praestat animo aequiore reliqui parent. Ita tamen oportet aetate praestare, ut ne propter senectutem minus sustinere possit labores. Neque enim senes neque pueri callium difficultatem ac montium arduitatem atque asperitatem facile ferunt, quod patiendum illis, qui greges secuntur, praesertim armenticios ac caprinos, quibus rupes ac silvae ad pabulandum cordi. Formae hominum legendae ut sint firmae ac veloces, mobiles, expeditis membris, qui non solum pecus sequi possint, sed etiam a bestiis ac praedonibus defendere, qui onera extollere in iumenta possint, qui
4 excurrere, qui iaculari. Non omnis apta natio ad pecuariam, quod neque Bastulus neque Turdulus idonei, Galli appositissimi, maxime ad iumenta. In emptionibus dominum legitimum sex fere res perficiunt: si hereditatem iustam adiit; si, ut debuit, mancipio ab eo accepit, a quo iure civili potuit; aut si in iure cessit, qui potuit cedere, et id ubi oportuit; aut si usucepit; aut si e praeda sub corona emit; tumve cum in bonis sectioneve cuius publice veniit.
5 In horum emptione solet accedere peculium aut excipi et stipulatio intercedere, sanum esse, furtis noxisque solutum; aut, si mancipio non datur, dupla promitti, aut, si ita pacti, simpla. Cibus eorum debet esse interdius separatim unius cuiusque gregis, vespertinus in cena, qui sunt sub uno magistro, communis. Magistrum providere oportet ut omnia sequantur instrumenta, quae pecori et pastoribus opus sunt, maxime ad victum hominum et ad medicinam pecudum. Ad quam rem habent iumenta dossuaria domini, alii equas, alii pro iis quid aliut, quod onus dosso ferre possit.
6 Quod ad feturam humanam pertinet pastorum, qui in fundo perpetuo manent, facile est; quod habent conservam in villa, nec hac venus pastoralis longius quid quaerit. Qui autem in saltibus et silvestribus locis pascunt et non villa, sed casis repentinis imbres vitant, iis mulieres adiungere, quae sequantur greges ac cibaria pastoribus expediant eosque assiduiores faciant, utile arbitrati

PLATE III

A—A donkey mill from Pompeii. (Photo: *Fototeca Unione.*)

B—Terracotta plaque from a miller's tomb on the Isola Sacra near Ostia, showing a mill in action. The animal appears to be a mule. The apparatus above the *catillus* is not clear, but there seems to be a hopper feeding grain into the mill. The slave holds a whip. (*Deutsches Archäologisches Institut, Rome.*)

PLATE IV

A reconstruction of an oil-mill from Pompeii. (Photo: *Alinari*.)

7 multi. Sed eas mulieres esse oportet firmas, non turpes, quae in
opere multis regionibus non cedunt viris, ut in Illyrico passim
videre licet, quod vel pascere pecus vel ad focum afferre ligna ac
8 cibum coquere vel ad casas instrumentum servare possunt. De
nutricatu hoc dico, easdem fere et nutrices et matres. Simul aspicit
ad me et, Ut te audii dicere, inquit, cum in Liburniam venisses, te
vidisse matres familias eorum afferre ligna et simul pueros, quos
alerent, alias singulos, alias binos, quae ostenderunt fetas nostras,
quae in conopiis iacent dies aliquot, esse eiuncidas ac contem-
9 nendas. Cui ego, Certe, inquam; nam in Illyrico hoc amplius,
praegnatem saepe, cum venit pariendi tempus, non longe ab opere
discedere ibique enixam puerum referre, quem non peperisse, sed
invenisse putes; nec non etiam hoc, quas virgines ibi appellant,
non numquam annorum viginti, quibus mos eorum non denegavit,
ante nuptias ut succumberent quibus vellent et incomitatis ut vagari
10 liceret et filios habere. Quae ad valitudinem pertinent hominum
ac pecoris et sine medico curari possunt, magistrum scripta habere
oportet. Is enim sine litteris idoneus non est, quod rationes
dominicas pecuarias conficere nequiquam recte potest. De numero
11 pastorum alii angustius, alii laxius constituere solent. Ego in
octogenas hirtas oves singulos pastores constitui, Atticus in centenas.
Greges ovium si magni sunt, quos miliarios faciunt quidam, facilius
de summa hominum detrahere possis, quam de minoribus, ut
sunt et Attici et mei. Septingenarii enim mei; tu, opinor, octin-
genarios habuisti, nec tamen non, ut nos, arietum decumam partem.
Ad equarum gregem quinquagenarium bini homines, utique
uterque horum ut secum habeat equas domitas singulas in his
regionibus, in quibus in stabula solent equas abigere, ut in Apulia
et in Lucanis accidit saepe.

(the text is omitted)

xi, 1-12. [*This chapter is concerned with milk, the making of cheese,
and sheep-shearing. The details are too long to be given here: the reader is
recommended to refer to the Loeb edition on these points. The conversation ended and the company dispersed when a freedman of Vitulus came
with a message asking his guests, Varro and Scrofa, to come early.*]

LIBER TERTIUS

Introduction and Dedication

(*the text is omitted*)

i, 1-10. *My dear Pinnius, there are two ways of life, one in the country and the other in the city. Country life is far the more ancient because land was created by nature and was here to be cultivated long before there were towns: men built the towns. Farming is better for man and so our ancestors were right to encourage citizens to return to the land, to call it "Mother" and "Ceres", and to regard the tillers of the soil as living a wholesome life. The first farmers sowed and grazed the same piece of land, but with the increase in the size of the herds, there came a division between them and some became farmers and some shepherds. The shepherds' occupation may be divided into two kinds, rearing animals on the farm or grazing them in the country. No one has yet thought of the former as a separate branch although it is a kind of rearing of animals. I have divided farming for profit under three heads, the first on agriculture, the second on cattle-raising, and the third on the rearing of animals around the farm. I have chosen to dedicate this, my third book, to you because you have a fine villa which you have adorned even with your own writings. I remember too the conversations you and I had on the ideal country house.*

Of Various Kinds of Villas

ii, 1-6. *During the election of the aediles, Axius and I were waiting for our candidate in the* Villa Publica. *Appius Claudius and four other friends whose names recalled those of birds were there as well. Appius reminded Axius how much he had enjoyed his birds served to him during his recent visit to Reate. This led on to a discussion about villas. Appius remarked on the simplicity and usefulness of the* Villa Publica, *contrasting it with Axius' own villa with its rich decorations. Axius retorted that the* Villa Publica *was decorated at any rate with paintings and sculpture, but his own villa was also a farm, and embellished with farm equipment, while the* Villa Publica *had no farm attached to it. The conclusion was drawn that if a building was outside the city this in itself did not make it a villa.*

Comitiis aediliciis cum sole caldo ego et Q. Axius senator tribulis suffragium tulissemus et candidato, cui studebamus, vellemus esse praesto, cum domum rediret, Axius mihi, Dum diribentur, inquit, suffragia, vis potius villae publicae utamur

umbra, quam privati candidati tabella dimidiata aedificemus nobis ?
Opinor, inquam, non solum, quod dicitur, "malum consilium
consultori est pessimum," sed etiam bonum consilium, qui consulit
2 et qui consulitur, bonum habendum. Itquae imus, venimus in
villam. Ibi Appium Claudium augurem sedentem invenimus in
subselliis, ut consuli, siquid usus poposcisset, esset praesto. Sedebat
ad sinistram ei Cornelius Merula consulari familia ortus et Fircellius
Pavo Reatinus, ad dextram Minucius Pica et M. Petronius Passer.
Ad quem cum accessissemus, Axius Appio subridens, Recipis
3 nos, inquit, in tuum ornithona, ubi sedes inter aves ? Ille, Ego
vero, inquit, te praesertim, quoius aves hospitales etiam nunc ructor,
quas mihi apposuisti paucis ante diebus in Villa Reatina ad lacum
Velini euenti de controversiis Interamnatium et Reatinorum.
Sed non haec, inquit, villa, quam aedificarunt maiores nostri,
4 frugalior ac melior est quam tua illa perpolita in Reatino ? Nuncubi hic vides citrum aut aurum ? Num minium aut armenium ?
Num quod emblema aut lithostrotum ? Quae illic omnia contra.
Et cum haec sit communis universi populi, illa solius tua; haec quo
succedant e campo cives et reliqui homines, illa quo equae et asini;
praeterea cum ad rem publicam administrandam haec sit utilis, ubi
cohortes ad dilectum consuli adductae considant, ubi arma ostendant,
5 ubi censores censu admittant populum. Tua scilicet, inquit
Axius, haec in campo Martio extremo utilis et non deliciis sumptu-
osior quam omnes omnium universae Reatinae ? Tua enim oblita
tabulis pictis nec minus signis; at mea, vestigium ubi sit nullum
Lysippi aut Antiphilu, at crebra sartoris et pastoris. Et cum illa
non sit sine fundo magno et eo polito cultura, tua ista neque agrum
6 habeat ullum nec bovem nec equam. Denique quid tua habet
simile villae illius, quam tuus avos ac proavos habebat ? Nec enim,
ut illa, faenisicia vidit arida in tabulato nec vindemiam in cella
neque in granario messim. Nam quod extra urbem est aedificium,
nihilo magis ideo est villa, quam eorum aedificia, qui habitant extra
portam Flumentanam aut in Aemilianis.

On the Profits from Villas

ii, 7-18. *Appius asked Axius to describe a villa because he intended to
buy a villa from Marcus Seius, but one without a farm attached. Merula*

had described its amenities. Axius then asked Merula how anything could be a villa which had neither the furnishings of a city nor the apparatus of a farm. Merula in answer said that there was no distinction between Axius' simple farm on the bend of the Velinus and his luxurious villa in the Rosea where he bred asses. The conclusion was drawn that if the latter was rightly to be called a villa, then any estate was a villa from which profits from pasturage were drawn, whatever the source, whether sheep, birds, bees, or boars. Appius remarked that there were two kinds of pasture, cattle-raising on the fields, and the rearing of small animals around the home-farm. Seius drew a large revenue from the latter. I added that my aunt drew large profits from fieldfares on her estate. Axius then asked Merula to talk about this kind of husbandry.

7   Appius subridens, Quoniam ego ignoro, inquit, quid sit villa, velim me doceas, ne labar imprudentia, quod volo emere a M. Seio in Ostiensi villam. Quod si ea aedificia villae non sunt, quae asinum tuum, quem mihi quadraginta milibus emptum ostendebas aput te, non habent, metuo ne pro villa emam in litore Seianas aedes.
8 Quod aedificium hic me Lucius Merula impulit ut cuperem habere, cum diceret nullam se accepisse villam, qua magis delectatus esset, cum apud eum dies aliquot fuisset; nec tamen ibi se vidisse tabulam pictam neque signum aheneum aut marmoreum ullum, nihilo magis
9 torcula vasa vindemiatoria aut serias olearias aut trapetas. Axius aspicit Merulam et, Quid igitur, inquit, est ista villa, si nec urbana habet ornamenta neque rustica membra? Quoi ille; Num minus villa tua erit ad angulum Velini, quam neque pictor neque tector vidit umquam, quam in Rosia quae est polita opere tectorio elegan-
10 ter, quam dominus habes communem cum asino? Cum significas-set nutu nihilo minus esse villam eam quae esset simplex rustica, quam eam quae esset utrumque, et ea et urbana, et rogasset, quid ex iis rebus colligeret, Quid? inquit, si propter pastiones tuus fundus in Rosia probandus sit, et quod ibi pascitur pecus ac stabulatur, recte villa appellatur, haec quoque simili de causa debet vocari
11 villa, in qua propter pastiones fructus capiuntur magni. Quid enim refert, utrum propter oves, an propter aves fructus capias? Anne dulcior est fructus apud te ex bubulo pecore, unde apes nascuntur, quam ex apibus, quae ad villam Sei in alvariis opus faciunt? Et num pluris tu e villa illic natos verres lanio vendis,
12 quam hinc apros macellario Seius? Qui minus ego, inquit Axius,

istas habere possum in Reatina villa? Nisi si apud Seium Siculum
fit mel, Corsicum in Reatino; et hic aprum glas cum pascit empticia,
facit pinguem, illic gratuita exilem. Appius: Posse ad te fieri,
inquit, Seianas pastiones non negavit Merula; ego non esse ipse
13 vidi. Duo enim genera cum sint pastionum, unum agreste, in quo
pecuariae sunt, alterum villaticum, in quo sunt gallinae ac columbae
et apes et cetera, quae in villa solent pasci, de quibus et Poenus
Mago et Cassius Dionysius et alii quaedam separatim ac dispersim
in libris reliquerunt, quae Seius legisse videtur et ideo ex iis pastionibus ex una villa maiores fructus capere, quam alii faciunt ex toto
14 fundo. Certe, inquit Merula; nam ibi vidi greges magnos anserum,
gallinarum, columbarum, gruum, pavonum, nec non glirium,
piscium, aprorum, ceterae venationis. Ex quibus rebus scriba
librarius, libertus eius, qui apparuit Varroni et me absente patrono
hospitio accipiebat, in annos singulos plus quinquagena milia
e villa capere dicebat. Axio admiranti, Certe nosti, inquam,
materterae meae fundum, in Sabinis qui est ad quartum vicesimum
15 lapidem via Salaria a Roma. Quidni? inquit, ubi aestate diem
meridie dividere soleam, cum eo Reate ex urbe aut, cum inde
venio hieme, noctu ponere castra. Atque in hac villa qui est
ornithon, ex eo uno quinque milia scio venisse turdorum denariis
ternis, ut sexaginta milia ea pars reddiderit eo anno villae, bis tantum
quam tuus fundus ducentum iugerum Reate reddit. Quid?
16 sexaginta, inquit Axius, sexaginta, sexaginta? derides. Sexaginta,
inquam. Sed ad hunc bolum ut pervenias, opus erit tibi aut
epulum aut triumphus alicuius, ut tunc fuit Scipionis Metelli, aut
collegiorum cenae, quae nunc innumerabiles excandefaciunt
annonam macelli. Reliquis annis omnibus si non hanc expectabis
summam, spero, non tibi decoquet ornithon; neque hoc accidit
his moribus nisi raro ut decipiaris. Quotus quisque enim est
annus, quo non videas epulum aut triumphum aut collegia non
epulari? Sed propter luxuriam, inquit, quodam modo epulum
17 cotidianum est intra ianuas Romae. Nonne item L. Abuccius,
homo, ut scitis, apprime doctus, cuius Luciliano charactere sunt
libelli, dicebat in Albano fundum suum pastionibus semper vinci
a villa? Agrum enim minus decem milia reddere, villam plus
vicena. Idem secundum mare, quo loco vellet, si parasset villam,
se supra centum milia e villa recepturum. Age, non M. Cato

nuper, cum Luculli accepit tutelam, e piscinis eius quadraginta
18 milibus sestertis vendidit pisces? Axius, Merula mi, inquit, recipe
me quaeso discipulum villaticae pastionis. Ille: Quin simulac
promiseris minerval, incipiam, inquit. Ego vero non recuso, vel
hodie vel ex ista pastione crebro. Appius: Credo simulac primum
ex isto villatico pecore mortui erunt anseres aut pavones. Cui ille:
Quid enim interest, utrum morticinas editis volucres an pisces,
quos nisi mortuos estis numquam? Sed oro te, inquit, induce me
in viam disciplinae villaticae pastionis ac vim formamque eius
expone.

The Three Divisions of Home-Feeding
(the text is omitted)

iii, 1-10. *Merula described the three departments of home-feeding,
the aviaries, preserves, ponds for fresh- or salt-water fish. Hens also
were reared and were the first to be kept on the villa. In earlier times
there was only the hen-run and the pigeon-cote for aviaries, but now they
were called* ornithones *and in them were kept birds which were luxuries:
now preserves covered many acres to keep boar and deer and not just the
hare: fish-ponds had now been built out into the sea to house deep-sea fish.*

Aviaries
(the text is omitted)

iv, 1-3. *Axius asked Merula to talk first about the ornithon because
of the large profit to be gained from fieldfares. Merula answered that there
were two kinds, one like Varro's at Casinum made for the owner's pleasure
and the other for profit such as those in the Sabine country. Lucullus
had tried to combine the two types by having a dining-room in his aviary,
but the smell made it impracticable.*

The Aviary for Profit

v, 1-7. *Merula said that a domed building or a peristyle was needed
large enough to house several thousand birds: water should be conducted
into it into narrow channels and the ground kept dry. The entrance
should be winding. The aviary should be in semi-darkness and be faced
around doors and windows with smooth plaster. Perches and rods should*

be fixed inside and water and food put on the ground. The birds should be
fattened for twenty days before they were needed. Those for market were
killed in a separate shed. Some birds were migratory, such as fieldfares,
turtle-doves, and quails. Flocks of these came to rest on the Pontian
islands, both when on passage south and on passage north.

Sed quod te malle arbitror, Axi, dicam de hoc quod fructus
causa faciunt, unde, non ubi, sumuntur pingues turdi. Igitur
testudo, aut peristylum tectum tegulis aut rete, fit magna, in qua
2 milia aliquot turdorum ac merularum includere possint, quidam
cum eo adiciant praeterea aves alias quoque, quae pingues veneunt
care, ut miliariae ac coturnices. In hoc tectum aquam venire
oportet per fistulam et eam potius per canales angustas serpere, quae
facile extergeri possint (si enim late ibi diffusa aqua, et inquin-
atur facilius et bibitur inutilius), et ex eis caduca quae abundat
3 per fistulam exire, ne luto aves laborent. Ostium habere humile
et angustum et potissimum eius generis, quod cocliam appellant,
ut solet esse in cavea, in qua tauri pugnare solent; fenestras raras,
per quas non videantur extrinsecus arbores aut aves, quod earum
aspectus ac desiderium marcescere facit volucres inclusas. Tantum
locum luminis habere oportet, ut aves videre possint, ubi assidant,
ubi cibus, ubi aqua sit. Tectorio tacta esse levi circum ostia ac
4 fenestras, nequa intrare mus aliave quae bestia possit. Circum
huius aedifici parietes intrinsecus multos esse palos, ubi aves
assidere possint, praeterea perticis inclinatis ex humo ad parietem
et in eis traversis gradatim modicis intervallis perticis adnexis ad
speciem cancellorum scenicorum ac theatri. Deorsum in terram
esse aquam, quam bibere possint, cibatui offas positas. Eae
maxime glomerantur ex ficis et farre mixto. Diebus viginti antequam
tollere vult turdos, largius dat cibum, quod plus ponit et farre
subtiliore incipit alere. In hoc tecto caveas, quae caveae tabulata
5 habeant aliquot ad perticarum supplementum. Contra hic aviarius
quae mortuae ibi sunt aves, ut domino numerum reddat, solet
ibidem servare. Cum opus sunt, ex hoc aviario ut sumantur
idoneae, excludantur in minusculum aviarium, quod est coniunctum
cum maiore ostio, lumine illustriore, quod seclusorium appellant.
Ibi cum eum numerum habet exclusum, quem sumere vult, omnes
6 occidit. Hoc ideo in secluso clam, ne reliqui, si videant, despondeant animum atque alieno tempore venditoris moriantur. Non

ut advenae volucres pullos faciunt, in agro ciconiae, in tecto hirundines, sic aut hic aut illic turdi, qui cum sint nomine mares, re vera
feminae quoque sunt. Neque id non secutum ut esset in merulis,
7 quae nomine feminino mares quoque sunt. Praeterea volucres
cum partim advenae sint, ut hirundines et grues, partim vernaculae,
ut gallinae ac columbae, de illo genere sunt turdi adventicio ac
quotannis in Italiam trans mare advolant circiter aequinoctium
autumnale et eodem revolant ad aequinoctium vernum, et alio
tempore turtures ac coturnices immani numero. Hoc ita fieri
apparet in insulis propinquis Pontiis, Palmariae, Pandateriae.
Ibi enim in prima volatura cum veniunt, morantur dies paucos
requiescendi causa itemque faciunt, cum ex Italia trans mare remeant.

The Aviary for Pleasure

v, 8-18. *Appius asked me to describe my aviary which I had built for
pleasure on my estate near Casinum. I told how there was a walk alongside
the river which ran through my estate, and how the aviary was situated off it.
It was enclosed by high walls on the right and left. The shape of it was
like that of a writing-tablet with a rounded top. On each side were rows
of stone columns on the outside and dwarf trees on the inside: these
colonnades were covered with netting and filled with all kinds of birds.
In the open quadrangle were two fish-ponds and between them a path leading
to a domed colonnaded building. Surrounding this was a wood planted
with large trees, and enclosed with a high wall. There was a space five feet
wide between the outer stone columns and the inner ones of fir. The outer
columns were surrounded with fine netting to give a view into the wood:
the interior columns were also enclosed with a net. In between the two
rows of columns there was a kind of bird-theatre built up of perches
attached to the columns. All kinds of birds, especially singing birds, were
kept inside. Below the columns was a stone facing above a platform
which itself rose above a pond. There were shelters for ducks hollowed
out under the platform. On the island in the pond there was a spoked
wheel for a revolving table when desired. A stream ran out into the fish-
ponds from the pond and cold and warm water could be turned on for the
guests from cocks on the table. Under the dome the morning and evening
stars revolved and showed the time. There was also a compass of the eight
winds like that in the Tower of the Winds at Athens.*

*At this point Pantuleius Parra came to say that a man had been caught
cheating over the ballot-boxes. Pavo went off at once to see if it affected his
candidate.*

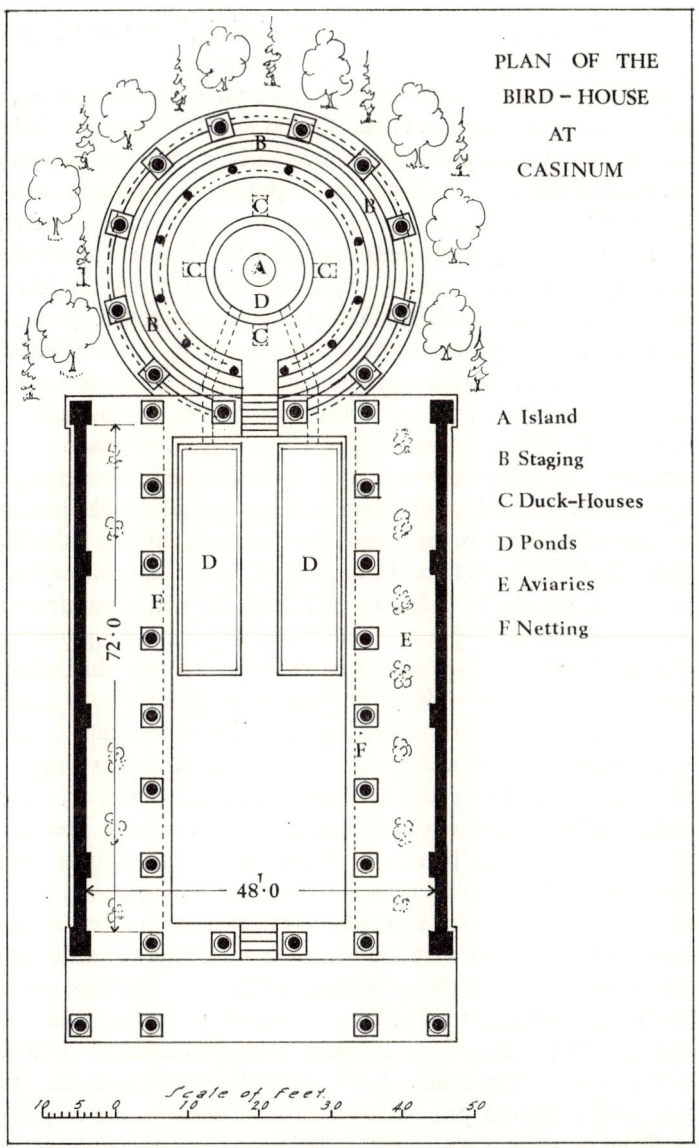

PLAN OF THE
BIRD – HOUSE
AT
CASINUM

A Island

B Staging

C Duck–Houses

D Ponds

E Aviaries

F Netting

Suggested plan of Varro's bird-house at Casinum, after Ll. Storr-Best, Varro on
Farming, *London 1912.   It should be noted that this plan does not correspond
with the reconstruction on p. 115 by Des Anges & Seure.*

8 Appius Axio, Si quinque milia hoc coieceris, inquit, et erit epulum ac triumphus, sexaginta milia quae vis statim in fenus des licebit multum. Tum mihi, tu dic illut alterum genus ornithonis, qui animi causa constitutus a te sub Casino fertur, in quo diceris longe vicisse non modo archetypon inventoris nostri ornithotrophion M. Laeni Strabonis, qui Brundisii hospes noster primus in peristylo habuit exhedra conclusas aves, quas pasceret obiecto rete, sed etiam 9 in Tusculano magna aedificia Luculli. Quoi ego: Cum habeam sub oppido Casino flumen, quod per villam fluat, liquidum et altum marginibus lapideis, latum pedes quinquaginta septem, et e villa in villam pontibus transeatur, longum pedes DCCCCL derectum ab insula, quae est in imo fluvio, ubi confluit altera amnis, 10 ad summum flumen, ubi est museum, circum huius ripas ambulatio sub dio pedes lata denos, ab hac est in agrum versus ornithonis locus ex duabus partibus dextra et sinistra maceriis altis conclusus. Inter quas locus qui est ornithonis deformatus ad tabulae litterariae speciem cum capitulo, forma qua est quadrata, patet in latitudinem pedes XLVIII, in longitudinem pedes LXXII; qua ad capitulum 11 rutundum est, pedes XXVII. Ad haec, ita ut in margine quasi infimo tabulae descripta sit, ambulatio, ab ornithone plumula, in qua media sunt caveae, qua introitus in aream est. In limine, in lateribus dextra et sinistra porticus sunt primoribus columnis lapideis, pro mediis arbusculis humilibus ordinatae, cum a summa macerie ad epistylum tecta porticus sit rete cannabina et ab epistylo ad stylobaten. Hae sunt avibus omnigenus oppletae, quibus 12 cibus ministratur per retem et aqua rivolo tenui affluit. Secundum stylobatis interiorem partem dextra et sinistra ad summam aream quadratam e medio diversae duae non latae oblongae sunt piscinae ad porticus versus. Inter eas piscinas tantummodo accessus semita in tholum, qui est ultra rutundus columnatus, ut est in aede Catuli, si pro parietibus feceris columnas. Extra eas columnas est silva manu sata grandibus arboribus, ut infima perluceat, tota saepta 13 maceriis altis. Intra tholi columnas exteriores lapideas et totidem interiores ex abiete tenues locus est pedes quinque latus. Inter columnas exteriores pro pariete reticuli e nervis sunt, ut prospici in silvam possit et quae ibi sunt videri neque avis ea transire. Intra interiores columnas pro pariete rete aviarium est obiectum. Inter has et exteriores gradatim substructum ut theatridion avium,

Suggested reconstruction of Varro's aviary, after C. des Anges & G. Seure, Revue de Philologie, de Litterature et d'Histoire Ancienne, 3rd série VI, 1932, p. 241.

14 mutuli crebri in omnibus columnis impositi, sedilia avium. Intra
retem aves sunt omnigenus, maxime cantrices, ut lusciniolae ac
merulae, quibus aqua ministratur per canaliculum, cibus obicitur
sub retem. Subter columnarum stylobaten est lapis a falere pedem
et dodrantem alta; ipsum falere ad duo pedes altum a stagno,
latum ad quinque, ut in culcitas et columellas convivae pedibus
circumire possint. Infimo intra falere est stagnum cum margine
pedali et insula in medio parva. Circum falere et navalia sunt
15 excavata anatium stabula. In insula est columella, in qua intus
axis, qui pro mensa sustinet rotam radiatam, ita ut ad extremum,
ubi orbile solet esse, arcuata tabula cavata sit ut tympanum in
latitudinem duo pedes et semipedem, in altitudinem palmum.
Haec ab uno puero, qui ministrat, ita vertitur, ut omnia una ponantur
et ad bibendum et ad edendum et admoveantur ad omnes convivas.
16 Ex suggesto faleris, ubi solent esse peripetasmata, prodeunt anates
in stagnum ac nant, e quo rivus pervenit in duas, quas dixi, piscinas,
ac pisciculi ultro ac citro commetant, cum et aqua calida et frigida
ex orbi ligneo mensaque, quam dixi in primis radiis esse, epitoniis
17 versis ad unum quemque factum sit ut fluat convivam. Intrinsecus sub tholo stella lucifer interdiu, noctu hesperus, ita circumeunt
ad infimum hemisphaerium ac moventur, ut indicent, quot sint
horae. In eodem hemisphaerio medio circum cardinem est orbis
ventorum octo, ut Athenis in horologio, quod fecit Cyrrestes;
ibique eminens radius a cardine ad orbem ita movetur, ut eum
tangat ventum, qui flet, ut intus scire possis.
18 Cum haec loqueremur, clamor fit in campo. Nos athletae
comitiorum cum id fieri non miraremur propter studia suffragatorum
et tamen scire vellemus, quid esset, venit ad nos Panteleius Parra,
narrat ad tabulam, cum diriberent, quendam deprensum tesserulas
coicientem in loculum, eum ad consulem tractum a fautoribus
competitorum. Pavo surgit, quod eius candidati custos dicebatur
deprensus.

The Rearing of Birds
(the text is omitted)

vi, vii, viii. [*In these chapters there are detailed instructions for the
rearing of peafowl* (vi), *pigeons* (vii), *and turtle-doves* (viii). *It is not*

thought practicable to give a summary of these: the reader who wishes for more information could profitably consult the Loeb edition.]

Poultry Farming

ix, 1-16. Merula recommended that to ensure profits from a poultry-farm, consideration must be given to purchase, breed, eggs, chicks, and feeding. There were hens, cocks, and capons: the cocks were castrated by cauterising the leg. He described the characteristics of the best breed of hen and cock. For two hundred birds two large coops were needed, with a run in front. The caretaker should have a room in which were the nesting boxes. He then gave instructions for the care of sitting hens and the eggs, tests for infertility of eggs, and directions for the care, feeding, and rearing of newly hatched chicks.

Axius, Ego quae requiro farturae membra, de gallinis dic sodes, Merula: tum de reliquis siquid idoneum fuerit ratiocinari, licebit. Igitur sunt gallinae quae vocantur generum trium: 2 villaticae et rusticae et Africanae. Gallinae villaticae sunt, quas deinceps rure habent in villis. De his qui ornithoboscion instituere vult, id est adhibita scientia ac cura ut capiant magnos fructus, ut factitaverunt Deliaci, haec quinque maxime animadvertant oportet: de emptione, cuius modi et quam multas parent; de fetura, quem ad modum admittant et pariant; de ovis, quem ad modum incubent et excudant; de pullis, quem ad modum et a quibus educentur; hisce appendix adicitur pars quinta, quem ad modum saginentur. 3 Ex quis tribus generibus proprio nomine vocantur feminae quae sunt villaticae gallinae, mares galli, capi semimares, qui sunt castrati. Gallos castrant, ut sint capi, candenti ferro inurentes ad infima crura, usque dum rumpatur, et quod extat ulcus, oblinunt figlina 4 creta. Qui spectat ut ornithoboscion perfectum habeat, scilicet genera ei tria paranda, maxime villaticas gallinas. E quis in parando eligat oportet fecundas, plerumque rubicunda pluma, nigris pinnis, imparibus digitis, magnis capitibus, crista erecta, 5 amplas; hae enim ad partiones sunt aptiores. Gallos salaces qui animadvertunt, si sunt lacertosi, rubenti crista, rostro brevi pleno acuto, oculis ravis aut nigris, palea rubra subalbicanti, collo vario aut aureolo, feminibus pilosis, cruribus brevibus, unguibus longis, caudis magnis, frequentibus pinnis; item qui elati sunt ac

vociferant saepe, in certamine pertinaces et qui animalia quae nocent gallinis non modo non pertimescant, sed etiam pro gallinis propug-
6 nent. Nec tamen sequendum in seminio legendo Tanagricos et Melicos et Chalcidicos, qui sine dubio sunt pulchri et ad proeliandum inter se maxime idonei, sed ad partus sunt steriliores. Si ducentos alere velis, locus saeptus adtribuendus, in quo duae caveae coniunctae magnae constituendae, quae spectent ad exorientem versus, utraeque in longitudinem circiter decem pedum, latitudine dimidio minores, altitudine paulo humiliores: in utraque fenestra lata tripedalis, et eae pede altiores e viminibus factae raris, ita ut lumen praebeant multum, neque per eas quicquam ire intro possit, quae nocere
7 solent gallinis. Inter duas ostium sit, qua gallinarius, curator earum, ire possit. In caveis crebrae perticae traiectae sint, ut omnes sustinere possint gallinas. Contra singulas perticas in pariete exclusa sint cubilia earum. Ante sit, ut dixi, vestibulum saeptum, in quo diurno tempore esse possint atque in pulvere volutari. Praeterea sit cella grandis, in qua curator habitet, ita ut in parietibus circum omnia plena sint cubilia gallinarum aut exculpta aut adficta
8 firmiter. Motus enim, cum incubat, nocet. In cubilibus, cum parturient, acus substernendum; cum pepererunt, tollere substramen et recens aliut subicere, quod pulices et cetera nasci solent, quae gallinam conquiescere non patiuntur; ob quam rem ova aut inaequabiliter maturescunt aut consenescunt. Quae velis incubet, negant plus XXV oportere ova incubare, quamvis propter fecun-
9 ditatem pepererit plura, optimum esse partum ab aequinoctio verno ad autumnale. Itaque quae ante aut post nata sunt et etiam prima eo tempore, non supponenda; et ea quae subicias, potius vetulis quam pullitris, et quae rostra aut ungues non habeant acutos, quae debent potius in concipiendo occupatae esse quam incubando.
10 Adpositissimae ad partum sunt anniculae aut bimae. Si ova gallinis pavonina subicias, cum iam decem dies fovere coepit, tum denique gallinacia subicere, ut una excudat. Gallinaciis enim pullis bis deni dies opus sunt, pavoninis ter noveni. Eas includere oportet, ut diem et noctem incubent, praeterquam mane et vespere, dum
11 cibus ac potio is detur. Curator oportet circumeat diebus interpositis aliquot ac vertere ova, ut aequabiliter concalefiant. Ova plena sint atque utilia necne, animadverti aiunt posse, si demiseris in aquam, quod inane natet, plenum desidit. Qui ut hoc intellegant

concutiant, errare, quod vitales venas confundant in iis. Idem aiunt, cum ad lumen sustuleris, quod perluceat, id esse inane.

12 Qui haec volunt diutius servare, perfricant sale minuto aut muria tres aut quattuor horas eaque abluta condunt in furfures aut acus. In supponendo ova observant ut sint numero imparia. Ova, quae incubantur, habeantne semen pulli, curator quadriduo post quam incubari coepit intellegere potest. Si contra lumen tenuit et purum unius modi esse animadvertit, putant eiciendum et aliud subiciun-

13 dum. Excusos pullos subducendum ex singulis nidis et subicien-dum ei quae habeat paucos; ab eaque, si reliqua sint ova pauciora, tollenda et subicienda aliis, quae nondum excuderunt et minus habent triginta pullos. Hoc enim gregem maiorem non faciendum. Obiciendum pullis diebus XV primis mane subiecto pulvere, ne rostris noceat terra dura, polentam mixtam cum nasturti semine et aqua aliquanto ante factam intritam, ne tum denique in eorum

14 corpore turgescat; aqua prohibendum. Qua de clunibus coeperint habere pinnas, e capite, e collo eorum crebro eligendi pedes; saepe enim propter eos consenescunt. Circum caveas eorum incenden-dum cornum cervinum, ne quae serpens accedat, quarum bestiarum ex odore solent interire. Prodigendae in solem et in stercilinum,

15 ut volutare possint, quod ita alibiliores fiunt; neque pullos, sed omne ornithoboscion cum aestate, tum utique cum tempestas sit mollis atque apricum; intento supra rete, quod prohibeat eas extra saepta evolare et in eas involare extrinsecus accipitrem aut quid aliut; evitantem caldorem et frigus, quod utrumque iis adversum. Cum iam pinnas habebunt, consuefaciundum ut unam aut duas sectentur gallinas, ceterae ut potius ad pariendum sint expeditae,

16 quam in nutricatu occupatae. Incubare oportet incipere secundum novam lunam, quod fere quae ante, pleraque non succedunt. Diebus fere viginti excudunt. De quibus villaticis quoniam vel nimium dictum, brevitate reliqua compensabo.

Gallinae rusticae sunt in urbe rarae nec fere nisi mansuetae in cavea videntur Romae, similes facie non his gallinis villaticis nostris, sed Africanis.

The Rearing of Birds and Preservation of Other Animals
(the text is omitted)

ix (17-21), x, xi, xii, xiii, xiv, xv. [*In chapters* ix (17-21) *to* xv *there are instructions for the rearing of "wild hens"* (ix), *geese* (x), *ducks* (xi), *the preservation of hare and deer* (xii), *boar* (xiii), *and the rearing of snails* (xiv) *and dormice* (xv). *A summary is not provided but the reader who wishes for more information should consult the Loeb edition.*]

The Bees: Their Natural History

xvi, 1-9. *Appius spoke about the natural history of the bees. He had not been able to afford honeyed wine until he became rich. Bees were produced either from bees or a rotted carcass: they lived a community life based on reason and skill. They had three tasks, food-gathering, home-making, and labour. The cell in the comb had six angles and so was economical of space. They produced a substance welcome to both gods and men. They had a king and government. They would seek only that which was pure. The bees did no damage. They could be gathered together by means of a loud clanging or clapping noise. They followed their king everywhere. Workers themselves, they made a habit of killing off the drones. They used a substance called* erithace *for filling up openings in the hive. They lived under martial law, sending out colonies and obeying orders.*

Appius, Igitur relinquitur, inquit, de pastione villatica tertius actus de piscinis. Quid tertius? inquit Axius. An quia tu solitus es in adulescentia tua domi mulsum non bibere propter parsimoniam, nos mel neclegemus? Appius nobis, Verum dicit,
2 inquit. Nam cum pauper cum duobus fratribus et duabus sororibus essem relictus, quarum alteram sine dote dedi Lucullo, a quo hereditate me cessa primum et primus mulsum domi meae bibere coepi ipse, cum interea nihilo minus paene cotidie in convivio
3 omnibus daretur mulsum. Praeterea meum erat, non tuum, eas novisse volucres, quibus plurimum natura ingeni atque artis tribuit. Itaque eas melius me nosse quam te ut scias, de incredibili earum arte naturali audi. Merula, ut cetera fecit, historicos quae sequi melitturgoe soleant demonstrabit.
4 Primum apes nascuntur partim ex apibus, partim ex bubulo corpore putrefacto. Itaque Archelaus in epigrammate ait eas esse

βοὸς φθιμένης πεπλανημένα τέκνα,

idem

ἵππων μὲν σφῆκες γενεά μόσχων δὲ μέλισσαι.

Apes non sunt solitaria natura, ut aquilae, sed ut homines. Quod si in hoc faciunt etiam graculi, at non idem, quod hic societas operis et aedificiorum, quod illic non est, hic ratio atque ars, ab his opus 5 facere discunt, ab his aedificare, ab his cibaria condere. Tria enim harum: cibus, domus, opus, neque idem quod cera cibus, nec quod mel, nec quod domus. Non in favo sex angulis cella, totidem quot habet ipsa pedes? Quod geometrae hexagonon fieri in orbi rutundo ostendunt, ut plurimum loci includatur. Foris pascuntur, intus opus faciunt, quod dulcissimum quod est, et deis et hominibus est acceptum, quod favus venit in altaria et mel ad principia convivi 6 et in secundam mensam administratur. Haec ut hominum civitates, quod hic est et rex et imperium et societas. Secuntur omnia pura. Itaque nulla harum adsidit in loco inquinato aut eo qui male oleat, neque etiam in eo qui bona olet unguenta. Itaque iis unctus qui accessit, pungunt, non, ut muscae, ligurriunt, quod nemo has videt, ut illas, in carne aut sanguine aut adipe. Ideo modo considunt 7 in eis quorum sapor dulcis. Minime malefica, quod nullius opus vellicans facit deterius, neque ignava, ut non, qui eius conetur disturbare, resistat; neque tamen nescia suae imbecillitatis. Quae cum causa Musarum esse dicuntur volucres, quod et, si quando displicatae sunt, cymbalis et plausibus numero redducunt in locum unum; et ut his dis Helicona atque Olympon adtribuerunt homines, 8 sic his floridos et incultos natura adtribuit montes. Regem suum secuntur, quocumque it, et fessum sublevant, et si nequit volare, succollant, quod eum servare volunt. Neque ipsae sunt inficientes nec non oderunt inertes. Itaque insectantes ab se eiciunt fucos, quod hi neque adiuvant et mel consumunt, quos vocificantes plures persecuntur etiam paucae. Extra ostium alvi opturant omnia, qua 9 venit inter favos spiritus, quam erithacen appellant Graeci. Omnes ut in exercitu vivunt atque alternis dormiunt et opus faciunt pariter et ut colonias mittunt, iique duces conficiunt quaedam ad vocem ut imitatione tubae. Tum id faciunt, cum inter se signa pacis ac belli habeant. Sed, O Merula, Axius noster ne, dum haec audit physica, macescat, quod de fructu nihil dixi, nunc cursu lampada tibe trado.

The Bees: Profits, Apiary, Hive, Varieties

xvi, 10-19. Merula took up the discussion remarking on the large profits which Seius and I made from our bees. I knew two brothers who had a good revenue from one iugerum where they had an apiary and planted it with bee-plants. Apiaries were to be in a quiet, cool place near the villa with a good supply of food and water at hand. Bee-plants should be cropped if necessary, especially thyme. The best situation was in the portico of the villa. Hives were of wood, earthenware, or even fennel stalks: they should be set on brackets on walls in rows up to three. The bee-keeper should examine the hives three times a month in spring and summer, and if necessary remove troublesome kings. The best variety of bee was small and striped. There were wild and domesticated bees.

10   Merula, De fructu, inquit, hoc dico, quod fortasse an tibi satis sit, Axi, in quo auctorem habeo non solum Seium, qui alvaria sua locata habet quotannis quinis milibus pondo mellis, sed etiam hunc Varronem nostrum, quem audivi dicentem duo milites se habuisse in Hispania fratres Veianios ex agro Falisco locupletes, quibus cum a patre relicta esset parva villa et agellus non sane maior iugero uno, hos circum villam totam alvaria fecisse et hortum habuisse ac relicum thymo et cytiso opsevisse et apiastro, quod alii meliphyllon,
11 alii melissophyllon, quidam melittaenam appellant. Hos numquam minus, ut peraeque ducerent, dena milia sestertia ex melle recipere esse solitos, cum dicerent velle expectare, ut suo potius tempore mercatorem admitterent, quam celerius alieno. Dic igitur, inquit, ubi et cuius modi me facere oporteat alvarium, ut magnos
12 capiam fructus. Ille, melittonas ita facere oportet, quos alii melitrophia appellant, eandem rem quidam mellaria. Primum secundum villam potissimum, ubi non resonent imagines (hic enim sonus harum fugae existimatur esse protelum), esse oportet aere temperato, neque aestate fervido neque hieme non aprico, ut spectet potissimum ad hibernos ortus, qui prope se loca habeat ea, ubi
13 pabulum sit frequens et aqua pura. Si pabulum naturale non est, ea oportet dominum serere, quae maxime secuntur apes. Ea sunt rosa, serpyllon, apiastrum, papaver, faba, lens, pisum, ocimum, cyperum, medice, maxime cytisum, quod minus valentibus utilissimum est. Etenim ab aequinoctio verno florere incipit et permanet
14 ad alterum aequinoctium. Sed ut hoc aptissimum ad sanitatem

apium, sic ad mellificium thymum.   Propter hoc Siculum mel fert
palmam, quod ibi thymum bonum frequens est.   Itaque quidam
thymum contundunt in pila et diluunt in aqua tepida; eo consper-
15 gunt omnia seminaria consita apium causa.   Quod ad locum
pertinet, hoc genus potissimum eligendum iuxta villam, non quo
non in villae porticu quoque quidam, quo tutius esset, alvarium
collocarint.   Ubi sint, alii faciunt ex viminibus rutundas, alii e
ligno ac corticibus, alii ex arbore cava, alii fictiles, alii etiam ex
ferulis quadratas longas pedes circiter ternos, latas pedem, sed ita,
ubi parum sunt quae compleant, ut eas conangustent, in vasto
loco inani ne despondeant animum.   Haec omnia vocant a mellis
alimonio alvos, quas ideo videntur medias facere angustissimas, ut
16 figuram imitentur earum.   Vitiles fimo bubulo oblinunt intus et
extra, ne asperitate absterreantur, easque alvos ita collocant in
mutulis parietis, ut ne agitentur neve inter se contingant, cum in
ordinem sint positae.   Sic intervallo interposito alterum et tertium
ordinem infra faciunt et aiunt potius hinc demi oportere, quam addi
quartum.   Media alvo, qua introeant apes, faciunt foramina parva
17 dextra ac sinistra.   Ad extremam, qua mellarii favum eximere
possint, opercula imponunt.   Alvi optimae fiunt corticeae, deter-
rimae fictiles, quod et frigore hieme et aestate calore vehementissime
haec commoventur.   Verno tempore et aestivo fere ter in mense
mellarius inspicere debet fumigans leniter eas et ab spurcitiis
18 purgare alvum et vermiculos eicere.   Praeterea ut animadvertat ne
reguli plures existant; inutiles enim fiunt propter seditiones.   Et
quidam dicunt, tria genera cum sint ducum in apibus, niger ruber
varius, ut Menecrates scribit, duo, niger et varius, qui ita melior,
ut expediat mellario, cum duo sint in eadem alvo, interficere
nigrum, cum sit cum altero rege, esse seditiosum et corrumpere
19 alvom, quod fuget aut cum multitudine fugetur.   De reliquis
apibus optima est parva varia rutunda.   Fur qui vocabitur, ab aliis
fucus, est ater et lato ventre.   Vespa, quae similitudinem habet
apis, neque socia est operis et nocere solet morsu, quam apes a se
secernunt.   Hae differunt inter se, quod ferae et cicures sunt.
Nunc feras dico, quae in silvestribus locis pascitant, cicures, quae
in cultis.   Silvestres minores sunt magnitudine et pilosae, sed
opifices magis.

The Bees: Purchase, Food, Drink

xvi, 20-28. *Merula continued, saying that bees when purchased should show signs of good health: sickly ones looked shaggy. The transfer should be done to a suitable place, preferably in spring near suitable pasturage. The new hive should be smeared with balm and have some full combs inside. They produce* propolis *for protecting the entrance, in front of the hive;* erithace *with which they weld together the ends of the comb: wax from which the cells of the comb are made, and honey, different substances come from different plants and not all from the same: the quality of honey differs in different plants. The best is from thyme. Clear running water is essential, so shallow that they can settle on stones and drink. Food must also be provided at the hive for bad weather: honeyed water, or raisins and figs soaked in wine are recommended.*

In emendo emptorem videre oportet, valeant an sint aegrae.
20 Sanitatis signa, si sunt frequentes in examine et si nitidae et si opus quod faciunt est aequibile ac leve. Minus valentium signa, si sunt pilosae et horridae, ut pulverulentae, nisi opificii eas urget tempus;
21 tum enim propter laborem asperantur ac marcescunt. Si transferendae sunt in alium locum, id facere diligenter oportet et tempora, quibus id potissimum facias, animadvertendum et loca, quo transferas, idonea providendum: tempora, ut verno potius quam hiberno, quod hieme difficulter consuescunt quo translatae manere, itaque fugiunt plerumque. Si e bono loco transtuleris eo, ubi idonea pabulatio non sit, fugitivae fiunt. Nec, si ex alvo in alvum in eodem
22 loco traicias, neglegenter faciendum, sed et in quam transiturae sint apes, ea apiastro perfricanda, quod inlicium hoc illis, et favi melliti intus ponendi a faucibus non longe, ne, cum animadverterint aut inopiam esse ... habuisse dicit. Is ait, cum sint apes morbidae propter primores vernos pastus, qui ex floribus nucis graecae et
23 cornus fiunt, coeliacas fieri atque urina pota reficiendas. Propolim vocant, e quo faciunt ad foramen introitus protectum ante alvum maxime aestate. Quam rem etiam nomine eodem medici utuntur in emplastris, propter quam rem etiam carius in sacra via quam mel venit. Erithacen vocant, quo favos extremos inter se conglutinant, quod est aliut melle et propoli; itaque in hoc vim esse illiciendi. Quocirca examen ubi volunt considere, eum ramum
24 aliamve quam rem oblinunt hoc admixto apiastro. Favus est, quem fingunt multicavatum e cera, cum singula cava sena latera habeant,

quot singulis pedes dedit natura. Neque quae afferunt ad quattuor res faciendas, propolim, erithacen, favum, mel, ex iisdem omnibus rebus carpere dicunt. Simplex, quod e malo punico et asparago cibum carpant solum, ex olea arbore ceram, e fico mel, sed non
25 bonum. Duplex ministerium praeberi, ut e faba, apiastro, cucurbita, brassica ceram et cibum; nec non aliter duplex quod fit e malo et piris silvestribus, cibum et mel; item aliter duplex quod e papavere, ceram et mel. Triplex ministerium quoque fieri, ut ex nuce Graeca et e lapsano cibum, mel, ceram. Item ex aliis floribus ita
26 carpere, ut alia ad singulas res sumant, alia ad plures, nec non etiam aliut discrimen sequantur in carptura aut eas sequatur, ut in melle, quod ex alia re faciant liquidum mel, ut e siserae flore, ex alia contra spissum, ut e rore marino; sic ex alia re, ut e fico mel insuave, e
27 cytiso bonum, e thymo optimum. Cibi pars quod potio et ea iis aqua liquida, unde bibant esse oportet, eamque propinquam, quae praeterfluat aut in aliquem lacum influat, ita ut ne altitudine escendat duo aut tres digitos; in qua aqua iaceant testae aut lapilli, ita ut extent paulum, ubi adsidere et bibere possint. In quo diligenter habenda cura ut aqua sit pura, quod ad mellificium
28 bonum vehementer prodest. Quod non omnis tempestas ad pastum prodire longius patitur, praeparandus his cibus, ne tum melle cogantur solo vivere aut relinquere exinanitas alvos. Igitur ficorum pinguium circiter decem pondo decoquont in aquae congiis sex, quas coctas in offas prope apponunt. Alii aquam mulsam in vasculis prope ut sit curant, in quae addunt lanam puram, per quam sugant, uno tempore ne potu nimium impleantur aut ne incidant in aquam. Singula vasa ponunt ad alvos, haec supplentur. Alii uvam passam et ficum cum pisierunt, affundunt sapam atque ex eo factas offas apponunt ibi, quo foras hieme in pabulum procedere tamen possint.

The Bees: Swarming

xvi, 29-31. *The signs that preceded swarming were when they hung together in masses in front of the hive and a loud humming noise. A few flew out first. The keeper should throw dust and beat brass around them and smear some spot with sweet-smelling herbs where they will settle. They could be induced to enter a hive smeared with the same herbs and remain there.*

29   Cum examen exiturum est, quod fieri solet, cum adnatae prospere
sunt multae ac progeniem ut coloniam emittere volunt, ut olim
crebro Sabini factitaverunt propter multitudinem liberorum, huius
quod duo solent praeire signa, scitur: unum, quod superioribus
diebus, maxime vespertinis, multae ante foramen ut uvae aliae ex
30 aliis pendent conglobatae; alterum, quod, cum iam evolaturae
sunt aut etiam inceperunt, consonant vehementer, proinde ut
milites faciunt, cum castra movent. Quae primum exierunt,
in conspectu volitant reliquas, quae nondum congregatae sunt,
respectantes, dum conveniant. A mellario cum id fecisse sunt
animadversae, iaciundo in eas pulvere et circumtinniendo aere
31 perterritae, quo volunt perducere, non longe inde oblinunt erithace
atque apiastro ceterisque rebus, quibus delectantur. Ubi consederunt, afferunt alvum eisdem inliciis litam intus et prope apposita
fumo leni circumdato cogunt eas intrare. Quae in novam coloniam
cum introierunt, permanent adeo libenter, ut etiam si proximam
posueris illam alvum, unde exierunt, tamen novo domicilio potius
sint contentae.

The Bees: The Honey Yield

xvi, 32-38. *Merula talked next about the yield. When the comb was
covered with a membrane of wax the honey could be taken. A proportion
must be left for the bees. There could be three harvests but some must be
left for winter. Weaker bees should be separated and put under another
king. Fighting could be stopped by sprinkling honey-water over them.
They should be protected from heat and cold, and could be revived if
overcome by rain by being placed in a vessel and by the application of
warmth. They would revive in the sun and return to their work.*

32   Quod ad pastiones pertinere sum ratus quoniam dixi, nunc iam,
quoius causa adhibetur ea cura, de fructu dicam. Eximendorum
favorum signum sumunt ex ipsis uiris alvos habeat nem congerminarit
coniecturam capiunt, si intus faciunt bombum et, cum intro eunt
ac foras, trepidant et si, opercula alvorum cum remoris, favorum
33 foramina obducta videntur membranis, cum sint repleti melle. In
eximendo quidam dicunt oportere ita ut novem partes tollere,
decumam relinquere; quod si omne eximas, fore ut discedant.

Alii hoc plus relincunt, quam dixi. Ut in aratis qui faciunt restibiles segetes, plus tollunt frumenti ex intervallis, sic in alvis, si non quotannis eximas aut non aeque multum, et magis his assiduas
34 habeas apes et magis fructuosas. Eximendorum favorum primum putant esse tempus vergiliarum exortu, secundum aestate acta, antequam totus exoriatur arcturus, tertium post vergiliarum occasum, et ita, si fecunda sit alvos, ut ne plus tertia pars eximatur mellis, reliquum ut hiemationi relinquatur; sin alvus non sit fertilis, ne quid eximatur. Exemptio cum est maior, neque universam neque palam facere oportet, ne deficiant animum. Favi qui eximuntur, siqua pars nihil habet aut habet incunatum, cultello
35 praesicatur. Providendum ne infirmiores a valentioribus opprimantur, eo enim minuitur fructus; itaque imbecilliores secretas subiciunt sub alterum regem. Quae crebrius inter se pugnabunt, aspargi eas oportet aqua mulsa. Quo facto non modo desistunt pugna, sed etiam conferciunt se lingentes, eo magis, si mulso sunt asparsae, quo propter odorem avidius applicant se atque obstupes-
36 cunt potantes. Si ex alvo minus frequentes evadunt ac subsidit aliqua pars, subfumigandum et prope apponendum bene olentium
37 herbarum maxime apiastrum et thymum. Providendum vehementer ne propter aestum aut propter frigus dispereant. Si quando subito imbri in pastu sunt oppressae aut frigore subito, antequam ipsae providerint id fore, quod accidit raro ut decipiantur, et imbris guttis uberibus offensae iacent prostratae, ut efflictae, colligendum eas in vas aliquod et reponendum in tecto loco ac tepido, proximo die quam maxime tempestate bona cinere facto e ficulneis lignis infriandum paulo plus caldo quam tepidiore. Deinde
38 concutiendum leviter ipso vaso, ut manu non tangas, et ponendae in sole. Quae enim sic concaluerunt, restituunt se ac revivescunt, ut solet similiter fieri in muscis aqua necatis. Hoc faciendum secundum alvos, ut reconciliatae ad suum quaeque opus et domicilium redeant.

Fish Ponds

(the text is omitted)

xvii, 1-10. *Pavo returned with the news that they were casting the lots for the tribes and the herald had begun to announce who had been elected aedile by each tribe. Appius went off to congratulate his candidate and so home. Merula did the same. Axius then gave an account of fish-ponds with many details of the keeping of luxury fish both fresh-water and salt-, and saying how profitable it was. As we talked our candidate came into the villa wearing the broad stripe which showed that he had been elected. We escorted him to the Capitol and then went away to our homes.*

PLATE V

Detail of a fourth century mosaic in the Church of Santa Costanza in Rome showing slaves treading grapes. (Photo: *Deutsches Archäologisches Institut, Rome.*)

PLATE VI

A general view of Cassino (ancient Casinum). The hillside with the great Benedictine monastery of Monte Cassino overlooks the wide valley crossed by the river Rapido. Mountains half-surround the valley. Varro's estate was somewhere round here. (Photo: *Alinari*.)

Notes

LIBER PRIMUS

The Place and Occasion of the Conversation

Those who take part have been invited to dinner by Lucius Fundilius the sacristan of the temple of Tellus in Rome. The occasion is the festival of Tellus, the *sementivae feriae,* and the year soon after 48 B.C. They meet in the temple and while waiting for their host (who has been delayed) talk about agriculture. The speakers are:—

Marcus Terentius Varro: the writer of the book.

Gaius Fundanius: Varro's father-in-law: *tribunus plebis;* tribune of the plebs: *curator viarum;* overseer of roads.

Gaius Agrius: A knight: a follower of the Socratic school of philosophers.

Publius Agrasius: a tax collector.

Gaius Licinius Stolo: a descendant of that Gaius Licinius who introduced the *lex Licinia* in 367 B.C.

Gnaeus Tremelius Scrofa, a member of the Commission of Twenty for the allotment of Campanian land; a leading agronomist.

The freedman of Lucius Fundilius.

Varro's five friends all have names connected with farming and agriculture. The first four, Varro, Fundanius, Agrius, and Agrasius begin the conversation: soon after they are joined by Stolo and Scrofa. They continue to while away the time of waiting, but their host, Lucius Fundilius, does not appear. The freedman, his servant, at last arrives in great agitation with the news that his master has been murdered in the crowd: this ends the conversation and the speakers disperse.

I. i. 1-4. *Introduction and Dedication*

I. i. 1. Otium ... scriberem: the use of two different tenses in this conditional sentence is logical; the statement as a whole is contrary to fact, but the *si* clause (*protasis*) refers to past time, " if I had had . . ." (but the fact is I had not) and the main clause (*apodosis*) to the present " I should now be writing . . ." (but the fact is that I am not).

Fundania: she is Varro's wife to whom the work is dedicated: since she has just bought a farm and he is an old man, nearing eighty, it would seem that she is considerably younger than her husband. She is mentioned again in II, *praefatio,* 6, and III. i. 9 (text omitted). Her father, Gaius Fundanius, Varro's father-in-law, takes part in the first of the three dialogues (or conversations) of which the *Res Rusticae* is composed. The *gens Fundania* was a plebeian family of good standing: Gaius was a tribune of the people (*tribunus plebis*) and also held the office of *curator viarum* (overseer of roads) in 72 B.C. He may have been a friend of Cicero since the name appears in one of Cicero's letters to his brother Quintus (*Ad Quintum Ciceronem,* 1.ii.10) and a man of this name was defended by him in a lawsuit. The origin of the name from *fundus* is clear, and can be compared with the English surname " Farmer ". The fact that Fundania has bought a farm shows that Roman women could possess real property, and could work their own estates, an important point in connection with the position of women in Varro's time.

esse properandum: this is the normal impersonal use of the gerundive of an intransitive verb, here in the accusative and infinitive construction after *cogitans,* " there is need of haste ", " I must hurry ".

eo magis: " all the more ", literally " by that the more ": *eo* is in the ablative of " measure of difference ".

senex: supply *est bulla.* Varro is fond of introducing proverbial sayings, perhaps deliberately, as being a characteristic of a countryman.

annus octogesimus: Varro was seventy-nine at the time he was writing, *i.e.* in his eightieth year.

sarcinas conligam: the term is a military one, describing a soldier's preparations before setting out on the march.

antequam proficiscar: this verb is to be regarded either as (1) in the future, in accordance with the normal usage of *antequam* and other conjunctions expressing " time " which are followed by an indicative; the verb is in the future because the main clause *admonet . . . conligam* implies future time: (the English idiom requires a present tense: " before I set out . . ."), or (2) in the present subjunctive in a subordinate clause in indirect speech, depending on *admonet* and regarded as part of what is said. A further consideration is that the future simple in this construction is very rare; *proficiscar* therefore may be in the subjunctive whether it is to be regarded as in indirect speech or not (see Woodcock 227).

2. quem = *et eum.*

quare . . . experiar: this part of the sentence contains a looseness of syntax which is grammatically inconsistent: *id* is difficult but seems to represent the whole idea already stated.

colendo: instrumental ablative of the gerund, " by . . .".

curare: an infinitive expressing purpose and going closely with *id mihi.*

me . . . roges: " and since you ask me to consider it as mine to take care of," *i.e.* " since you ask me to give it my attention ".

quare . . . mortem: this sentence is composed of various clauses: (*quoniam . . .*) (*quem [et eum] cum velis . . .*) *meque* [*-que* repeats *cum* from the previous clause] *. . . roges, experiar* (main verb): the rest of the sentence is in the form of a purpose clause (*ut . . . moneam*) which contains a temporal clause (*quoad vivam*) and an indirect question (*quid . . . oporteat*).

quoad vivam: " as long as I live ". *vivam* is in either (1) the present subjunctive because it is in a dependent clause in indirect speech after *roges,* or (2) in the future indicative in a statement where the present is used in English: see *proficiscar* in 1 above and note.

oporteat: the subjunctive is in an indirect question, dependent on *moneam: ut* is merely repeated from the previous clause: a frequent form of redundancy in Varro.

3. neque . . . perisset: the subjunctives in *viveret* and *prodessent* are in clauses in indirect speech, implied in *patiar,* but *prodessent* probably also expresses purpose " prophecies which were to benefit . . .".

dum viveret: *dum,* meaning " while ", is normally associated with the present

indicative even in indirect speech, but the subjunctive is possible as here.   The
present indicative, however, appears in the second half of the sentence below,
ne dum vivo quidem.   Varro's constant use of variety in construction regardless
of style or proximity in the context probably explains this inconsistency.
ignotissimis: " completely unknown to her ".   For long ages after her
death, the Sibyl's prophecies continued to benefit men unknown to her.
post: is here an adverb.
cum desideramus: supply *scire*:   " when we desire (to know) ".
sit: the subjunctive is in an indirect question.
ex aliquo portento: a portent (*portentum*) was a startling event of some kind,
such as for instance a temple struck by lightning, which might to a Roman
show the anger of the gods.   This would have to be appeased by special
ritual learned by recourse to the Sibylline books (see note below).
quod prosit: the subjunctive expresses purpose: " such as to benefit " but
is also in indirect speech, the sentence *neque facere* is carefully balanced, but
interrupted by a parenthesis (*ad ... portento*): *me* contrasts with *Sibyllam*
and is governed by *patiar* which is repeated.   " I shall not allow the Sibyl . . .
and not (allow) myself to write something which can benefit . . ."
Sibyllam: *Sibylla* is the name for a prophetess: many such were known in
different cult centres and were usually connected with the worship of Apollo.
In Rome a collection of prophetic books (*libri Sibyllini*) associated with the Sibyl
of Cumae, a Greek colony on the Bay of Naples, were in the possession of
the state and carefully preserved in the temple of Apollo by a special college
of priests, in Varro's time composed of fifteen, known as the *quindecimviri
sacris faciundis.*   They were consulted only at the command of the Senate and
in times of emergency.   The story was told how a priestess offered nine books
said to contain oracles concerning the welfare of Rome to Tarquinius Priscus,
the king.   When he refused she burnt three and on his refusal to buy six she
burnt three more.   When she asked him to buy the three that were left at the
same price, the king became alarmed and purchased them.   These books were
destroyed when the Capitol was burnt in 83 B.C. but were replaced by a new
collection: both these could have been known to Varro.

   4. revertare: shorter form of the second person singular of the present
subjunctive of *reverti* which is deponent in the present forms but active
in the perfect, expressing purpose.
quaeres: the verb is in the future because it is dependent on *scribam*, the main
verb which is also in the future: (see note on *proficiscar* above, I. i. 1).
oporteat: the subjunctive is in an indirect question dependent on *quaeres.*
dei facientes adiuvant: " the gods help those who are active in work ".
Here we have another proverbial expression signified by the words *ut aiunt.*
Ennius: (239-169 B.C.) was regarded by the Romans as one of the greatest
of their poets.   He wrote both epic and drama, and was often called " the
father of Roman poetry ".   His most important work was the *Annales,* an
epic poem in eighteen books in which was described the early history of Rome;
in addition he wrote some twenty tragedies and several books of satire.   Only
a few scattered fragments of his poetry survive, most of them from the *Annales.*
In Varro's time they existed in their entirety and were widely read and known.
Musas: The Muses were the nine goddesses of poetry in all its aspects,
including literature, music, the dance, astronomy, philosophy and all intellectual activities.   Their chief sanctuary was in a glen on Mount Helicon in

Boeotia in Northern Greece where games were held every fourth year in their honour. Remains of a temple and a theatre and statues of the Muses still exist there: in their literary form the ancient poets looked upon them as personifications of intellectual and inspirational thought: an invocation to the Muses is found in the *Iliad*, II, 484-93 to introduce the catalogue of the ships, and in the first line of the Odyssey the Muse of Epic is addressed. English poets have repeatedly imitated the same imagery.

duodecim deos Consentes: Varro will call on twelve Councillor-gods to assist his writing, but these are not those twelve well known to all in Rome whose statues, after his time (2nd-3rd centuries A.D.) were erected in the *Porticus Deorum Consentium* at the south-west corner of the Forum, of which much still remains. The names of the twelve Olympians who formed the council (under Jupiter) are preserved in two hexameters of Ennius:

> "*Juno, Vesta, Minerva, Ceres, Diana, Venus, Mars,*
> *Mercurius, Iovi', Neptunus, Vulcanus, Apollo.*"

Varro will call not on these but on twelve different gods who preside over country life.

i. 5-7. *An Invocation to the Gods of Country Life*

5-7. Varro opens his work by calling to inspire him the powers that watch over the many aspects of nature and of the agricultural round: we see passing before us the annual cycle of the country festivals which have to be scrupulously observed for the good of the land and of its produce. Scenes of tillage on the farm and the Italian countryside with its rustic cults and lore are described. The deities in Varro's catalogue are essentially Italian and go back in origin to primitive times. They are grouped in pairs as befit their various functions.

5. Iovem et Tellurem: these are both objects of *invocabo* carried on from the previous paragraph: so also are the names of the deities which follow, as far as *Floram* after which another verb is introduced for variety, *adveneror*.

Iovem: Jupiter is in origin the old Italian sky-god: he gives the rain and it is clearly in this connection that he is placed first in Varro's catalogue and grouped with Tellus: as the father Jupiter sends the rain into the lap of Tellus (Mother Earth). Apart from his predominant position in the Roman state religion, he was worshipped as a rustic god at the two wine festivals of the *Vinalia*, one on April 23rd when wine-skins containing last year's vintage were broached, the other and more important on August 19th when the priest in Rome, the *flamen dialis*, who had charge of his worship cut the first grapes as a sign that the vintage might begin. In another aspect, as the power who gives the warmth to ripen the corn he received an offering of a sacred meal (*daps*) of wine and roasted meat at the beginning of sowing.

Tellurem: Tellus was an ancient Italian goddess, one of the many forms of the earth-mother worshipped throughout the Mediterranean region. Her responsibilities were for vegetation, sowing, and harvest. Particular rites (*feriae*), of variable date in the last days of January—decided each year by the *pontifices*, high priests, in accordance with weather conditions—were observed in her honour and were connected with the sowing of the corn. In Rome the festival was called the *feriae sementivae*, the name of which from *semen*, " seed ", clearly shows the sphere of activity to which the cult relates, and in country districts the *Paganalia*, the feast, that is, of the *pagani*, the farm labourers.

For an account of the details of the rites which reads like that of an eye-witness see Ovid, *Fasti* I, 657-74. The *Fordicidia* (also spelt *Hordicidia*) celebrated in Rome on April 15th was dedicated to Tellus; a pregnant cow was slaughtered on the Capitol. The unborn calf was burnt by the Vestal Virgins and the ashes kept for ritual use at the *Parilia*, the shepherds' festival of April 21st and also that of the corn-goddess Ceres with whom Tellus was often associated. This is one of the oldest forms of ritual in the Roman religious calendar and points to the great antiquity of the worship of Tellus. Her temple in Rome is the setting for the first of the three coversations in this book: (see I, ii, 1 below).

parentes magni: these words come in meaning in the predicative position after *dicuntur*, that is they are closely associated with the verb.

Solem et Lunam: *Sol* and *Luna* were the deities of the sun and moon, the observation of whose movements was vital to the countryman of ancient times: their cults were said to have been introduced into Rome by the Sabine King Titus Tatius and appropriately enough on the Quirinal Hill in Rome, which was settled in early times by the Sabines, there was an ancient cult of Sol, associated with a festival on August 9th. Also in Rome there was a temple sacred to Luna in the Aventine Hill of which little is known beyond the date of the festival on March 31st.

Cum quaedam ... conduntur: " at sowing and harvest ".

Cererem et Liberum: *Ceres* and *Liber* taken together ensure the yield of the farm in food and wine: Ceres presides over the corn and the harvest, Liber over the vine and its produce. Ceres, often associated with Tellus, was originally an old Italian earth-goddess thought to control the growing power of the soil. Her own festival, the *Cerealia*, on April 19th (a time when the corn is well grown in Italy) followed closely on the *Fordicidia* which was celebrated in honour of Tellus on the 15th, and both goddesses were revered at the *feriae sementivae*. There was a temple of Ceres at the foot of the Aventine Hill near the docks said to have been founded as early as 493 B.C. in consequence of a famine. Liber was another old Italian god of fertility and especially of wine. He was associated with Ceres in the Aventine cult, and his feast, the *Liberalia*, appears in the ancient calendars on March 17th. He seems to have been worshipped only in the country. Vergil describes the ritual in *Georgic II*, 385-396.

6. Robigum et Floram: the first protects from mildew, the second ensures that growth is healthy and seasonal. The name Robigus is identical with *robigo*, the red rust or mildew which attacks cereals when the ear is forming: it is a form of fungus and is more common on the continent than in Britain. At the festival of *Robigalia* on April 25th a red dog (probably regarded as the spirit of mildew) and a sheep were offered together with prayers for the safety of the harvest. The ritual is described by Ovid, *Fasti* IV, 911-942. The festival of Flora (*Floralia*) followed on May 3rd. Flora was probably an old Italian deity with a character similar to that of the earth-mother which has already been recognised in Tellus and Ceres. In Rome Flora had a *flamen Floralis* and a temple built in 238 B.C.

neque non: the use of two negatives to make a strong affirmative is a characteristic of Varro's style: he employs this frequently and often for no apparent emphasis though here the implication may be that the crops and trees will come to growth at exactly the proper times.

Minervam et Venerem: the fifth pair is the only one composed of two
female deities. Minerva as the Italian patroness of craftsmen and skills of all
kinds perhaps was thought to preside over the making of farm implements, but
Varro regards her as the custodian of the olive groves. This may be due to the
influence of the Greeks by whom she was often identified with Athena who in
Greek mythology was said to have introduced the cultivation of the olive into
Greece. Venus in her beginnings was not a goddess of any great importance
nor the great goddess of love as she became later, but the root of her name
seems to mean " charm ", " delight ", and might well relate to well-tilled
ground, especially a vegetable plot or a herb garden. Then since she presided
over gardens, she might well have been regarded as the tutelary spirit of the
vineyard, though otherwise her ritual connection with the vine is not clearly
defined; she was however revered at the festival of the *Vinalia* on August
19th as Varro states, although it seems primarily to have been dedicated to
Jupiter (see above on I. i. 5).

quo nomine: this refers to Venus to whom the *Vinalia* was dedicated: " in
her honour . . .".

Lympham and Bonum Eventum: the first (Lympha) watches over streams
and fountains, ensuring a plentiful supply of fresh water, the latter (Bonus
Eventus) gives prosperity to all the undertakings of the farm. There is no
known cult of Lympha, but she is thought to have been an old native divinity
of water. The name came to be easily identified with the Greek *nympha*, a
water-spirit (more often found in the plural *nymphae*). *Lympha* is often used
for water itself with the result that the goddess seems to be a personification,
or an indwelling spirit. The last named of all, Bonus Eventus, is of small
importance. The Elder Pliny (*Naturalis Historia*, XXXIV. xix. 77) describes a
statue of this deity in Rome holding a *patera* in his right hand and in his left an
ear of corn and a poppy: this clearly shows his connection with the harvest.

frustratio . . . non cultura: " not farming, but frustration ".

7. sermones: " conversations ". These expressed in literary form in a
particular setting with several speakers in discussion and centred on a particular
theme are also called " dialogues ": see chapter III.

oporteat: subjunctive in an indirect question dependent on *animadvertere
poteris:* so too *repetas* in the following line dependent on *indicabo*.

et nostris: there were many Roman as well as Greek writers on agriculture.

i. 8-11.   *Writers on Agriculture*
(*the text is omitted.*)

8-11. Most of the works in Varro's list are entirely lost or exist only in
fragments and little or sometimes nothing is known about the authors. They
are all Greek with the exception of Mago who wrote in Punic. Noteworthy
however are Attalus III, also named Philometor, King of Pergamum, born
about 170 B.C. who after losing his reason cultivated in his palace garden medicinal plants and others which produce drugs and poisons, and Amphilochus of
Athens, who wrote in the late fourth and early third centuries B.C., a treatise
on fodder and seems to have been among the first to recommend the use of
manure (see below xxiii. 1. and note). Many are to be dated to the third
and second centuries B.C., the age when the Greeks were taking a scientific
interest in agriculture. Among those whose works have survived, Hesiod, the

earliest, is thought to have been roughly contemporary with Homer, and in the eighth century B.C. had his home at Ascra on the slopes of Mount Helicon. In the *Works and Days*, he teaches that man must work for his living, and together with moral precepts to that end gives a description of a year's work on the farm.   Xenophon, born 430 B.C., is better known for his historical works, especially the *Anabasis*, but in the *Oeconomicus* he gives in dialogue form a charming account of life on his estate at Scillus near Olympia.   Aristotle, who lived from 384-322 B.C. was the founder of scientific methods, the influence of which has shaped modern science.   He was the first to classify plants and animals and in so doing became the founder of biological studies.   Among his writings perhaps that of greatest use for the agriculturist was the *Historia Animalium* in which is elucidated his scientific system of observation and classification. Theophrastus lived from 369-245 B.C. as far as is known and was Aristotle's most devoted pupil.   Making plants his life-long study he incorporated material gathered from many parts of the then known world and this combined with close observation was published in his two most important books, the *Enquiry into Plants* and the *Aetiology of Plants*.   He built up the first system of scientific botany and his work was of the utmost usefulness for methods of cultivating both trees and plants.   Of far greater reputation in the agricultural world than all the rest was Mago of Carthage who wrote in Punic, a form of Hebrew, a comprehensive treatise in twenty-eight books including material on some subjects from other writers.   Of such value was his work that after the destruction of Carthage in 146 B.C. a Latin translation was made at the order of the Senate by a Committee supervised by Decimus Silanus.   The date of its publication is not known but it was probably not much later than a few years after 146 B.C. Later the Latin version was abridged and translated into Greek in twenty books by Cassius Dionysius of Utica: these were dedicated to the Sextilius who was praetor in Africa, in about 88 B.C.   A subsequent abridgement was made by Diophanes of Bithynia in six books and dedicated to King Deiotarus who was a contemporary of Julius Caesar and died in 40 B.C.   Mago's dates are uncertain, but the Romans regarded his books as essential for farming: he is cited several times by Varro and later Roman agronomists in terms that show the high regard in which he was held.   It becomes clear from this chapter that there was an extensive *corpus* of writers on agriculture belonging to the fourth, third, and second centuries B.C.: it is perhaps permitted to surmise that Varro possessed copies of all or most of those he mentions and that the manuscripts were grouped together in a section of his extensive library if these had escaped Antony's depredations in 43 B.C. (see Varro's Life).

ii. 1-2.   *The Setting and Occasion of the Conversation on Agriculture*

1. Sementivis feriis: " on (the day of) the Festival of Sowing ", locatival ablative expressing " time when ", or " point of time ".   The date cannot be given exactly because it was a movable feast (*conceptivae*) (see note on I. i. 5). The festival was a holiday for townsmen as well as countrymen: it is fitting therefore that the conversation with which Book I opens should be set in Rome in the temple of Tellus (whose festival it was) where Varro and certain friends gather at their leisure, to meet the sacristan who has invited them to dinner. **in aedem Telluris:** the temple of Tellus in Rome was on the Esquiline Hill (*mons Esquilinus*) in the quarter known as the Carinae: a fashionable part of the city in Varro's time where wealthy citizens had their homes. The original temple was built in fulfilment of a vow by Publius Sempronius Sophus when an earthquake occurred during a battle with the Picentes in 268 B.C. It was located on the site formerly occupied by the house of Spurius Cassius which was said to have been demolished in 495 B.C. The temple adjoined the house of Cicero's brother, Quintus, and was restored by him about 54 B.C. The date of its dedication was December 13th. It is thought that since the worship of Tellus is very ancient, that there had been a much earlier cult centre on the site. Nothing now remains of the temple but it probably stood somewhere between the present Via del Colosseo and the Via dei Serpenti.

veneram: the subject is Varro who explains the situation.

ab aeditumo ... ab aedituo: " by the *aeditumus* (the sacristan) as we have been taught to call him by our fathers, or as we are corrected by our present-day men of culture, by the *aedituus*". *aeditumus* of which another form is *aeditimus* derived from *aedes* is used for the sacristan or keeper of a temple. In Varro's time the form *aedituus* was in general use in cultured society. This remark made by Varro is one of many which reflect his interest in language, and more especially in the history and derivation of words: (see Varro's Life).

offendi ibi C. Fundanium ... Agrasium: *Fundanius, Agrius,* and *Agrasius* are authentic Roman names and evidently chosen because they contain puns on the words *fundus* and *ager*: (see note on Fundanius above, i. 1). They might be regarded as the equivalents of the English surnames " Farmer ", " Fielding ", " Fieldman ". Varro's interest in language is reflected in his fondness for playing on words. Most of the other personages who take part in the three conversations have names connected with some aspect of agriculture.

pictam Italiam: they are studying a map of Italy. This would probably be in the form of a long painted strip stretching across the wall from one end to the other. Although the Romans had learnt a certain amount of scientific geography from the Greeks, surveying and cartography were not yet understood fully enough to enable them to draw maps as we know them now.

Quid vos hic ? : a colloquial expression: " What are you doing here ? "

otiosos: supply *vos*: " on holiday ", **solebant:** supply *adducere* from *adduxerunt.*

num: this adverb here only signifies a question, and does not mean as it does normally " surely not ".

2. eadem causa quae te: = *eadem causa (adduxit) quae te (adduxit).*

morere: second person singular, shorter form, of the present subjunctive (deponent) of *morari*, governed by the impersonal verb *oportet*. The subjunctive expressing necessity is a semi-dependent jussive, although the subordinating conjunction *ut* is normally omitted; the infinitive is more usual with *oportet* in this meaning.

dum ... revertatur: the subjunctive is used in a final sense, because of the note of expectancy, though the meaning is simply " until he returns ".

ab aedile: an aedile (*aedilis*) was one of several minor magistrates elected annually whose duties included in the first place the care of public buildings, as the name (from *aedes*) implies. In addition the maintenance of public order in the city, the distribution of the corn-supply, the charge of the water-supply and of the market including the inspection of weights and measures were all within their sphere of office. The aediles also supervised religious matters and cult practices and in this connection conducted the games (*ludi*) which were held on the

occasion of the greater religious festivals. As a consequence of this duty
it became customary for the aediles to bear personally the expenses of the
games as a vote-catching measure, for although the aedileship was not a distinguished office, the holding of it could mark the beginning of a political career.
It is clear from Varro that an aedile had the supervision (*procuratio*) of the
temple of Tellus, nor is it surprising that this particular *aedilis* should send for
the sacristan of the temple on the day of the festival.

nos ... rogaret: " he left (a man) to ask us to wait for him ".   Some such
pronoun as *quemdam* is to be supplied, as the antecedent to *qui*.

qui ... rogaret: the relative clause expresses purpose.

voltis ... usurpemus: this polite form of command is comparable with the
English " would you like to ... ? " The subjunctive is in semi-dependence on
voltis: see the note on *oportet* in the previous sentence.   The verb *usurpare*,
compounded from *usus* and *rapere* means literally " take to one's use ", hence
" exercise ", " practise ".

dum ille venit: this is the normal use of the present indicative in a *dum* clause
with the meaning " until ".

Romanus ... vincit: a reference to the strategy of Fabius Maximus Cunctator,
the Roman leader during the Second Punic War.   By avoiding pitched battles
with Hannibal and so drawing him down into the heel of Italy and further
and further away from his base, he wore down his strength, and contributed
greatly to the eventual victory over Carthage.   His delaying tactics, successful
as they were, earned for him the title " Cunctator " (" Delayer ").

voltis ... venit: it seems as if this sentence is spoken by Varro.   Although
the text does not indicate a change of speaker, Agrius appears to answer a
suggestion made by someone else, with the word *sane*, which follows.

cogitans ... esse: " remarking that the longest part of the journey is said
to be the (arrival at) the gate ", a proverbial saying in the accusative and
infinitive construction after *cogitans*: the implication is perhaps that when the
gate is in sight the distance looks shorter than it is in reality, or that the pre-
liminaries before the traveller may pass through the gate are tedious at the
end of a journey.   Most Roman towns were surrounded by walls pierced with
gates.   Agrius means to say that " now that they have all arrived (*i.e.* passed
the gate) they can take a rest ".   The usual meaning of *cogitare* is " think "
but it is clear that Agrius speaks his thoughts aloud to the others.

ii. 3-5.   *The Geographical Situation of Italy*

3. multas terras: among the speakers Varro had travelled widely in Spain,
Greece, and Asia Minor in the course of his political career: (see Varro's Life).

ecquam: supply *terram*.

nullam ... culta: " I think there is none which is so intensively cultivated as
a whole ".

3-4. Primum ... interior: this long sentence is made up of three concessive clauses, followed by the main verb, which is itself followed by two
causal clauses.   The framework of the sentence is *cum ... divisus sit ... et
quoniam ... est ...., et (quoniam) ... sunt) ...., dicendum (est) ... quod est,
... quod ... (est) ...*; the subjunctive with *cum* meaning " although " and
" since " and the indicative with *quoniam* follow the normal usage: the main
verb in the form of an impersonal gerundive with *est* to be supplied governs the

accusative and infinitive construction in the phrase *utique . . . Asiam*: the verb in both the *quod* clauses is in the indicative because plain facts are stated.

3. Ab Eratosthene: the Greek scholar Eratosthenes, lived from 275 to 194 B.C. After several years spent in study at Athens, in 247 B.C. he was made head of the library at Alexandria by Ptolemy Euergetes in succession to Apollonius Rhodius. Such an appointment showed that he was acknowledged as the most versatile scholar of his time. Although he wrote books on textual and literary criticism, chronology, mathematics, astronomy and geography, his fame rests chiefly on the last named subject. As the first systematic geographer, he calculated almost exactly the circumference of the earth, and the magnitude and distance from the earth of the sun and moon. He also divided the earth into two parts which he called Europa (included in which was North Africa) and Asia by means of an imaginary line from the Pillars of Hercules (now the Straits of Gibraltar) to the island of Rhodes and at the latter point drew two main axes of latitude and longitude. In the northern part were the Mediterranean countries (including North Africa) which enjoyed a temperate climate, but farther north was the frozen land of the Arctic: in the southern part was a torrid zone where the heat forbade cultivation.

4. maxime secundum naturam: " according to a very natural division ". Eratosthenes was guided most by nature in the formulation of his geographical system.

quae . . . fructuosiora: the change to the neuter plural is surprising coming after *pars* in the previous clause but is after Varro's style in which agreements are not always observed; perhaps a general statement is made with *loca* in mind " (places) which are more healthful are for that reason more fruitful ". The clause stated in full is: *et (quoniam ea) salubriora (sunt) illa fructuosiora (sunt)*.

sex mensibus continuis: in this expression which suggests a length of time the accusative case might have been expected: it is not the passing months which are thought of, so much as the point of time regarded as an entity.

posse: an impersonal use: " to be possible ".

5. Fundanius: supply some word of speaking such as *inquit*: he breaks in with an exclamation.

Em: this gives a colloquial touch: it is an interjection frequent in the comedies of Plautus and Terence.

ubi . . . natum: " where (in that part of the world) could you think that anything could germinate or grow once it germinated ? " The sentence is in the form of a repudiating deliberative question sometimes called " rhetorical ", with the implication that the answer if given would be in the negative. Surprise, indignation, incredulity may all be expressed in the form of a question to which an answer is not expected. It is styled rhetorical because it is common in speeches. The subjunctive is normally used, and is closely allied to the potential use. When it comes in Indirect Speech a rhetorical question is put in the accusative and infinitive: this shows that it is really a statement of fact accompanied with some kind of emotion. (See Woodcock, sections 175, 267.)

quicquam: this pronoun is usually associated with a negative: here the negative is virtually implied in the rhetorical question: that is, if an answer were given it would be " nowhere ".

illud Pacuvi: supply *verbum*; Pacuvius, the Roman writer of tragedies and nephew of Ennius (see note above on I. i. 4), lived from 220-130 B.C. The titles of some twelve plays are known but his works exist now only in fragments quoted by other writers.

sit: the present subjunctive is usual in conditional clauses which imply probability: in this passage however the imperfect would have been expected because the supposition is an impossibility, but *sit* makes the supposition graphic especially since it is followed by the present infinitive in the accusative and infinitive construction governed by *illud* which is equal to a verb of speaking.

tamen: "even". The underlying thought is that even in Italy where there are not extremes of heat and cold Fundanius finds his midday rest essential.

si diffinderem ... possum: in a conditional sentence where a supposition is made which is contrary to fact in present time, the imperfect subjunctive is normally employed in both the "if" clause (*protasis*) and the main clause (*apodosis*) but an exception usually occurs with *posse* where the statement remains true and is not influenced by the supposition: (see Woodcock, section 200). In Varro's sentence on the other hand *vivere non possum* is certainly contingent on *si diffinderem* and might without violating grammar have been expressed in the subjunctive.

meridie: it is the custom in hot countries to take a rest (called in Italy the *siesta*). This taking of a *siesta* is the regular practice in Italy throughout the year; it is clear that it has not changed since the times of the Romans. Work goes on longer in consequence into the evening than in cooler climates.

ii. 6-8. *Praise of Italian Agriculture*

6. Illic: this refers to the arctic regions which have just been described.

quicquam: a negative with which this pronoun is usually associated is implied in the expected answer to the question.

Illic ... possit: a repudiating deliberative, or rhetorical question, is put (see note on ii. 5 above).

Quod far conferam Campano?: "what spelt could I compare to that grown in Campania?".

far: also called *ador* is often translated as "spelt", but is to be regarded as emmer: it was a primitive kind of hulled wheat which was the basic diet of the people of antiquity. It was known in the time of Homer and from the earliest ages to the late Empire was the staple food of the Romans. Traces of it have been found in the earliest *strata* of the Roman forum. The plant was very resistant and could be grown even in poor soils: each spikelet had, as the botanical name *triticum dicocchum* implies, two grains with a closely adhering involucre which formed the husk. To be prepared for food the grains had first to be parched so as to loosen the husk and then pounded twice, first to remove the husks and secondly to break up the grain. Porridge was made from the pounded grain. The antiquity of *far* is illustrated by its use in ritual and in the most ancient form of marriage, the *confarreatio*, in which a cake of *far* figured in the ceremony.

conferam: the subjunctive is deliberative, used when the subject wonders what to do.

Campano: "from *Campania*". Campania, still called by the Roman name, is one of the most abundantly fertile regions not only of Italy but of the Mediterranean, and with every good reason is called "the Garden of Italy". It lies on the western sea-board and extends from the borders of Latium south-eastwards beyond Naples to the Sarno river and the gulf of Sorrento. It is gifted as few other lands are. In the first place the geographical situation of the whole region is ideal for agriculture exactly as Cato in the *De Agricultura* I. i. 3 requires a landed estate to be: " Let it be at the foot of a mountain and face the south ". So is Campania; and behind are the mounting ridges of the Apennines: in consequence it enjoys to the full the unstinted Italian sunshine and is kept free from northern and easterly winds by the sheltering heights. These also provide perennial streams of water which are fed by melting snow on the higher summits even in the hot dry summers. To these natural advantages are added the special properties of the soil which contains rich beds of lava from volcanoes now long extinct, and *alluvium* from the rivers small and great which flowing in gentle meanders spread evenly their rich deposits. The most import-ant are the *Volturnus*, the *Clanius*, and the *Sarno*. The soil is light, powdery and yet absorbent: water which lies beneath can be tapped by means of primitive wells worked by a cord attached to a long pole and weighted with a large stone. An important feature for the farmer is the ease with which it can be worked. Varro I. xx. 4 tells us that the ploughing is done not with heavy oxen but with cows or donkeys. Such was the fertility of this land combined with the favourable climate that it produced at least four crops a year of cereals and vegetables. It was also famous for its wines, olives, and fruit. To see Campania today is to understand the saying " every grain of soil is cultivated ". The vines are trained on screens of poplars and slung on nets from tree to tree: between the rows grow orchard trees such as apricots, plums, cherries, and almonds. Below all these are grown many kinds of vegetables. There is not a corner left uncultivated. The *far* grain which was grown in Campania was thought to be the best according to the geographer Strabo (V, iv, 3) and was called *zea*. " A proof of the fertility of the country is that it produces the best grain—I mean that from which porridge is made, which is superior not only to every kind of rice, but also to almost every kind of cereal. It is said that in one year some of the land is seeded twice with *zea*, the third time with millet and others still the fourth time with vegetables."

Apulo: " from Apulia ", a district (now called Puglia) in south-eastern Italy extending from the Gargano peninsula (*Mons Garganus*) to Calabria. The soil is dry but fertile especially in the north and along the coast.

Falerno: " from the *ager Falernus* ", a district of northern Campania lying between the *Mons Massicus* (now the Monte Massico) and the course of the river *Volturnus*, (the modern Volturno). The soil of volcanic origin was particularly fertile and produced what was considered in Varro's time some of the best wine in Italy. " Falernian " is mentioned in several ancient writers, especially by Horace, *Carm:* i, 20, and in several other passages.

Venafro: " from Venafrum " a town in Campania situated on the slopes of Monte Santa Croce to the north overlooking the valley of the *Volturnus* and set in some of the finest landscape in the district. It was noted for olives in Roman times and still produces olive oil of best quality which is greatly prized even today.

consita: supply *est*: this verb is frequently omitted especially with a perfect passive participle as here: compare *cooperta* (*est*) in the following line.

ut ... videatur: supply *esse*: " with the result that . . ." or " in consequence ".

7. **Phrygia:** the region in north-west Asia Minor in which Troy was situated: Homer calls it " vine-clad ".

quam haec: " than this (land) ", *i.e.* Italy.

Argos: a city in south-eastern Greece in the southern part of the Argive plain, three miles from the sea. Homer describes the state as " rich in corn ". Supply *appellat* from the previous line with *idem poeta*.

iugerum: the *iugerum* was the general standard of land measurement originally based on the amount which a yoke of oxen could plough in a day: the word is therefore probably derived from *iugum*, " yoke ". It was a rectangular area, based on the *actus quadrati*, a square of one hundred and twenty Roman feet, two of which joined end to end, formed a *iugerum*: the measurement was therefore twenty-eight thousand, eight hundred square feet, which is equivalent to five-eighths of an acre.

denos et quinos denos: the distributive numerals are used to indicate the yield of each *iugerum*.

quot quaedem: supply *ferunt* from *fert* in the previous line. Varro frequently leaves words to be understood, or to be repeated from a previous clause or sentence. The most obvious omissions will not be commented on further in these notes but those difficult to determine will be pointed out.

culleos: the Romans measured liquids by the *hemina* equal to half a pint, twelve of which made a *congius*: and by the *congius*, equal to three quarts, eight of which made an *amphora*. One *amphora* was equal approximately to six gallons and twenty of these made one *culleus*. Since the *culleus* was therefore equal to one hundred and twenty gallons, a yield as large as fifteen *cullei* to the *iugerum* would amount to some eighteen hundred gallons. Three hundred *amphorae* equal the same amount and ten *cullei* equal twelve hundred. The *amphora*, equal in volume to one cubic Roman foot, was also called the *quadrantal* and was used for determining the burden, or ton, of ships. The word *amphora* is also used for a vessel for holding liquids, very often wine, as well as for the measure: derived from the Greek it describes an earthenware pitcher with two " ears " or handles by which it can be carried.

M. Cato . . . Originum: Marcus Porcius Cato, often called the Censor, Cato Major, or the Elder, lived from 234-149 B.C. He was born at Tusculum, some 15 miles from Rome but lived when a boy on his father's farm at Reate in the Sabine country where Varro was later to be born and brought up. Cato came to own extensive country estates worked by slave labour. His early years were spent in military service in the course of which he was distinguished in the Second Punic War. He also held several political offices in which his personal qualities of severity, strict honesty, and narrow-minded patriotism became evident. He stood for the old Roman simplicity and hard living. He was a man of wide learning and wrote on a variety of subjects. He is considered to be the father of Latin prose, being the first Roman to have left prose writings of any importance. His works exist now only in fragments (with one exception) and this is true of the seven books of the *Origines* written in his old age, in which he treated of the origins of the Italian tribes and the history of Italy to 169 B.C. It is clear that in the time of Varro it existed in its entirety and we may be permitted to assume that there was a copy in his library. Also an agricultural expert, Cato wrote the earliest book on agriculture in Latin which we possess, the *De Agri Cultura*: this alone among his works has been preserved entire. It is more a collection of precepts for farmers than a systematic handbook and shows a roughness in language. Cato's own simplicity and severity characterise it throughout.

ager Gallicus . . . Picentium: *cis*, " this side ", implies from the point of view of a person in Rome as does *ultra*, " beyond ". The *ager Gallicus* was the narrow strip of the Adriatic coast which extended south-eastwards from Ariminum (modern Rimini) to the district of Ancona on the borders of the territory of the *Piceni*; it lay between the rivers Axis near Ancona and the Rubicon which flowed into the Adriatic between Ariminum and Ravenna to the north. The *ager* was so-called because in early times it was settled by the *Senones*, the last Gallic tribe to invade Italy: they penetrated south beyond the Alps attracted no doubt by the fertility of the soil and the prospect of an outlet to the sea. In 285 B.C. the *ager* was conquered by Manius Curius Dentatus and in 268 a Roman colony was founded at Ariminum. The land came under the dominion of the Romans but the name *Gallicus* remained to recall its earlier history. A further settlement of Roman colonists took place in 232 B.C. when the land was parcelled out (*viritim . . . datus est*) to needy citizens of Rome in accordance with the *lex Flaminia*: for this reason, as Varro states, it was also called *Romanus*.

viritim: literally, " man by man ".

cullea: this is a neuter plural form of *culleus* (see note above) quoted from Cato. The passage *ager . . . fiunt* appears to be taken word for word from his book: if this is so, then *cullea* may be an old form which had fallen out of use by Varro's time.

in agro Faventino: in the district of Faventia. This city, now called Faenza, was situated on a plain at the confluence of the rivers now called Lamone and Marzeno in the region of Romagna of modern times, half-way between Bologna and Rimini and thirty-two kilometres from Ravenna. It is surrounded by fertile and well watered fields which are given over to intensive cultivation and was colonised by Sulla. Its reputation for prosperous farming remains as it was in Varro's day. The modern city is famous for its ceramic industry which produces majolica to which it has given its own name, *faïence*.

trecenas amphoras: the amount is equal to fifteen cullei and so bears out the statement that some districts in Italy did yield eighteen hundred gallons to the *iugerum*.

simul aspicit me: Fundanius turns to Varro when he mentions what his (Varro's) particular friend used to say.

fabrum: this genitive plural form is the usual one with *faber*.

tuos: an older form of the nominative masculine singular, = *tuus*.

Faventiae: locative case.

8. Duo: this neuter plural implies a meaning such as " conditions ".

Italici homines: " the men of Italy ", referring to Italian farmers in general.

spectasse: this is the shorter form of *spectavisse*.

Quorum . . . deducendus: " If either of these two conditions is lacking, and if, in spite of that, any man wishes to farm, he is demented and should be put in charge of his relatives and friends."

adque . . . deducendus: a proverbial saying is quoted. *adgnati* were blood relations on the father's side, such as son, brother, grandson, uncle, cousin, etc.: *gentiles* were members of the same *gens*, who had the same name (*nomen*). If a person were insane, in Roman law his nearest relations took charge of him and his property.

si . . . refici: " if he sees that it will not be repaid ". No sane man will put money into cultivation if he sees that there will be no return for his outlay.

nec . . . dispereant: " nor should he (make an outlay) even if he can raise crops if he sees that they will be destroyed by the unhealthiness of the land ".

eos fore: the grammatical construction is difficult of explanation: it is probable however that *eos* is an anticipated accusative: another but less likely explanation is to regard *fore* as shortened from *futuros esse*.

ut ... dispereant: the clause is consecutive.

ii. 9-10.   *Gaius Licinius Stolo and Gnaeus Tremelius Scrofa*

9. qui ... possint: this generic subjunctive carries the meaning " such as can..." or " who will be able to...".

Nam: is used to introduce an explanation of the statement already made.

Stolonem ... Scrofam: the two friends who now join the group make up the six who take part in the conversation.   They too have been invited to dinner, but come rather late.   Their last names (*cognomina*) are chosen to give an added flavour to the agricultural interest of the meeting in line with Varro's fondness for making puns.   They add colour to the proceedings although they are in reality Roman family names.   Stolo means a " sucker " but does not carry anything of the modern implication: in Varro's estimation it pays a compliment to the holder's good farming.   He looks after his orchards, vineyards, and olives so well that no unwanted shoots are allowed to grow out from the trees.   He even digs them out from the roots before they have a chance to sprout.   Scrofa is the word for " a breeding sow " or any kind of pig, but was not considered a term of abuse by the Romans.   (*Porcella*, for instance, was a girl's name and has been found on sepulchral inscriptions.)

unum: refers to Stolo: it is answered by *alterum collegam* at the beginning of 10 below.

cuius maiores ... civem R.) in 367 B.C. the tribunes Gaius Licinius Stolo and Lucius Sextius proposed laws aimed against the growing wealth of the patricians and in support of the interests of the plebeians, in an attempt to limit the growth of large estates which was developing even in the early republic (see Chapter II. *Latifundia*).   It is interesting to see plebeian sympathies being actively expressed in the fourth century more than two hundred and fifty years before the time of the Gracchi, the great champions of democracy.

cuius ... tulerunt: *legem ferre* is the technical term for introducing a bill before the Senate.

nam Stolonis illa lex: " for that (well-known) law was a Stolo's ", *i.e.* it was a Stolo who proposed it.

qui ... cognomen: " who, as a result of his industry in farming has shown how appropriate is the family name of the Stolones ".   By Varro's time every Roman had three names: to take Gaius Licinius Stolo as an example, the first, Gaius, the *praenomen* was his own individual name, the second, Licinius, the *nomen,* was that of the *gens* or clan, to which his family belonged, and the third, the *cognomen,* Stolo, was the name of his family, which was a part of the *gens.*

quod ... e solo: " because he used to dig from the roots around the trees whatever was growing up out of the soil".   Stolo used to dig out the suckers (or unwanted shoots) as soon as they showed, so as to prevent their taking goodness from the trees: in consequence they did little harm to the main growth and sturdy plants resulted.

nascerentur: this is a generic subjunctive in a descriptive sense.

appellabant: the meaning requires a present tense, but the verb is attracted into the imperfect by analogy with the rest of the sentence which is in the past tense. The subject is people in general, " they ".

Eiusdem gentis ... eduxit: this important change in constitutional procedure took place in 145 B.C. and is attributed to Gaius Licinius Crassus, a tribune of the plebs, by Cicero as well as by Varro, although Plutarch ascribes it to Gaius Gracchus, some twenty years later. In Rome political speeches were delivered from the *rostra*, an open-air platform situated at the east end of the forum, and so called from the beaks (*rostra*) of ships captured in the war against Antium (modern Anzio) in 338 B.C. which were fixed on the side facing the forum. It was situated on the south side of the *comitium* and in front of the *curia Hostilia*, which was the original meeting-place of the Senate. The *comitium* was the place of assembly in the form of an open space at the east end of the forum and served as the political centre of Rome until the second century B.C. until, that is, the change effected by Gaius Licinius Crassus. A recent interpretation of this obscure passage has been given by Miss Lily Ross Taylor in her book *Roman Voting Assemblies* (Michigan, 1966), pp. 23-4. Since the expression *leges audire* is a technical one in legislative procedure for the actual passing of laws, she suggests that his innovation consisted in moving the act of voting to the forum side of the *rostra* where there was room for the masses to be assembled together. Before then, the tribes who voted in the process of approving a bill had probably been called into the *comitium* one after another and the magistrate in charge then faced the *comitium*. Crassus appears to have summoned them all into the forum, inviting them to gather on " their own acres ", in order to exercise their rights. Cicero in the *De Amicitia*, XXV. 96, gives an account of the incident saying that it took place in connection with a proposal made by Crassus that vacancies in the priestly offices should in future be filled by the votes of the people instead of as hitherto, by co-option by the patricians. In his speech in support of the proposed new law he faced the people in the forum and seems to have described it as their own ground. Seven *iugera* was the amount of land allotted to citizens after the expulsion of the kings and is used proverbially. Whatever may be the complete understanding of this reference there is no doubt about Crassus' democratic leanings and his attempt to give effective constitutional powers to the *plebs* in face of patrician privileges. Cicero states in scornful words that the bill was abortive, but the action of Crassus marked a stage in the giving of power to the *plebs*.

post reges exactos: " after the kings had been driven out ", " after the expulsion of the kings". A perfect passive participle is frequently used in Latin where the corresponding English idiom requires an abstract noun. The date is generally regarded as approximately 510 B.C.

ad leges accipiendas: " for the hearing of the laws ", " to hear the laws ". An example of the normal use of *ad* with the gerundive expressing purpose; another occurs in the following sentence, *ad agros dividendos*, " for parcelling the land ", or " to parcel the land ".

primus ... eduxit: " was the first to . . .".

10. **viginti virum qui fuit:** " who was (one) of the Commission of Twenty . . ." (see Varro's Life). *virum* is in the genitive plural, a form frequently found in this and other words as an alternative to that ending in -orum: compare *fabrum* in 7 above.

iucundiore spectaculo: ablative of description or quality.

spectatum: the supine in -um expressing purpose, used only with verbs of motion in this sense.

cum veniant: *cum* with the subjunctive is used here in a concessive sense.

apud Lucullum: the name Lucullus was proverbial then, as it still is, for luxurious living. During campaigns in the East, Lucullus had charge of the war against Mithridates, and after the king's flight he spent the winter of 71 B.C. in restoring law and order to Asia. In the course of his tour of duty he accumulated vast wealth, and after celebrating a triumph in 63 B.C. forsook a military career for a life of ease and luxurious elegance and became a connoisseur of art. He owned estates near Naples where he built villas extending into the sea, near Tusculum in the hills some fifteen miles from Rome, with observatories, banqueting halls, and cloisters, and another in Rome called the *horti Luculli*—the gardens of Lucullus, which was considered the most splendid and costly of any of his time. He collected paintings, as Varro states, and owned a library which was open to all who wished to study his fine manuscripts. His gardens were laid out about 60 B.C. and occupied what is now the most fashionable part of Rome. They covered the slope of the hill near the modern Via due Macelli and the Capo le Case close to the English quarter at the Spanish steps (Piazza di Spagna), and extended towards the Pincian Hill along the line of the Via Porta Pinciana. Nothing remains of them to be seen now. Both Varro and Cicero were friends of Lucullus and must have been frequent visitors to the gardens.

summa sacra via: the *sacra via* was the most ancient street in Rome running through the heart of the primitive centre. It began at the top of the Velia near the temple of the Lares, a ridge that extended from the north side of the Palatine Hill towards the Oppian Hill (*Mons Oppius*) near where later the arch of Titus was built and still stands, and led down to the east end of the forum to the vicinity of the temple of Vesta and the *Regia*. It was probably called *sacra* because beside it were the most sacred places of ancient Rome, those of the Lares and Vesta, as well as the house of the Vestal Virgins and of the Pontifex Maximus, the *Regia*. The road still exists, running across the forum, and is still paved with stones of Augustan date, not much later than Varro's own time. The beginning of the road at the top of the slope of the Velia was known as the *summa sacra via*, perhaps because of the high ground. The road became lined with shops of all kinds, and in Varro's time perhaps on both sides. Clearly the produce sold here was of the best as Varro's reference suggests. Fruit brought its weight figuratively in gold because of its high quality. It is to be noted too that this was a fashionable quarter where were the best shops, where prices would be highest and the merchandise of the choicest.

Huiusce ... imago: "the top of the *sacra via* where fruit is sold for its weight in gold is a picture of this man's (Scrofa's) orchard".

Huiusce: = *huius* + *-ce:* the suffix *-ce* carries a demonstrative meaning and survives in ordinary usage in such words as *illic*.

ii. 11-14. *Stockbreeding and Agriculture*

11. Illi ... nos: Scrofa and Stolo now arrive.

Nam: introduces an explanation of the question just asked. It carries the unexpressed thought " (I am asking because) we do not see Lucius Fundilius ".

Bono animo este: " Keep calm. " The ablative is one of description or quality. The sentence is probably colloquial.

este: the imperative plural of *esse*.
ovom: an old form of *ovum*.
illut: = *illud*.
Nam ... primum: *nam* has the same significance as in the sentence above: " (I have just told you to keep calm) because not only has that egg (not) been lowered which in the games in the Circus marks the finish of the last lap for the four-horse chariots (*i.e.* in the races) but we have not even seen (yet) that egg which is usually served first in the procession (of dishes) at dinner."
non modo when followed by *ne ... quidem* as here, is equal in meaning to *non modo non*. This construction is possible because *non* negatives both *modo* and the rest of the first half of the sentence, as far as *quadrigis*. The negative sense is carried right through and is not limited to *modo*. Speaking metaphorically Agrius assures Stolo and Scrofa that it is not even time for dinner yet because, as it were, the races are still being run: no one, he implies, would think of leaving the Circus for dinner before they are over, but would stay until the end: dinner on the other hand is not even ready because the first course including the customary eggs has not been taken in.
ovom ... quadrigis: The Circus Maximus in Rome ran along the valley (*vallis Murcia*) between the Palatine and Aventine Hills and was a long, narrow enclosure some six hundred metres in length and one hundred and fifty wide. The spectators might number as many as one hundred and fifty thousand. It was used for chariot-racing, sometimes with four-horse chariots, and other kinds of shows and games. A long low wall extended along the centre called the *spina* which divided the arena for the races. On this, among statues of the gods, the goals (*metae*) grouped in three at each end, and other objects, stood seven large egg-shaped objects of wood erected on a high stand. In the course of a race the chariots circled the arena several times, sometimes as many as seven. As each lap was finished one of these " eggs " was taken down to show the spectators how many laps remained to be run. The last " egg " to be taken down showed the finish of the race and at the same time recorded the number of laps which had been run. The Circus can be seen today, but now only the form and outlines remain except for a few fragments of marble seats, later than the time of Varro, at one end.
ne illud ovom ... primum: the main meal of the day among the Romans was *cena*, " dinner ", taken in the late afternoon after the day's work was done. It was a protracted occasion and often went on into the night. There were three parts: first the *gustatio*, to be compared with our *hors d'oevres*, or in Italian *antipasto*, which included eggs, shell-fish, and salad; the main dishes (*ferculae*) were served next; roast meats and poultry of several kinds. The meal ended with the *mensae secundae*, sweets, fruits, and wine. Dinners were sometimes on a grand scale which might account for Varro's words *cenali pompa* suggesting a banquet which was like a procession of all kinds of extravagant and luxurious dishes one following another.
12. dum videatis ... aeditumnus: two differing constructions with *dum* appear here with little or no distinction in the meaning of the present tense. Varieties of this kind used in close juxtaposition are characteristic of Varro's style.
summam: " end " or " aim ".
rudem: the *rudis* was a wooden sword presented to a gladiator at the end of a distinguished career as a sign of his discharge. Since many gladiators had

only the status of slaves this trophy was greatly prized. Many on retirement became *lanistae,* trainers in gladiatorial schools. Stolo has already earned his sword, as it were, in honourable retirement, and is recognised as a master of his craft, and Scrofa has lately followed suit. Both are equally experienced and both are experts equally able to speak with authority and to teach their craft to others.

serantur, inducantur: subjunctives in subordinate clauses in indirect speech (in this instance an indirect question) depending on *discernendum (est),* the impersonal gerundive expressing obligation.

quae serantur: " the things which are sown ", *i.e.* crops, this is used as a general expression for " arable farming ".

quae inducantur: " the things which are led . . .", *i.e.* " taken to pasture ". This is a general expression for stockbreeding.

13. Poenice . . . Latine: the Carthaginian Mago, wrote in his native Punic language (see summary of I. i. 8-11 and notes): this same paragraph contains the names of some fifty Greek writers on agriculture: among those who wrote in Latin Varro was probably thinking especially of Marcus Porcius Cato (see note on I. ii. 7) where some account of him is given. There were few writers in Latin before Cato on this subject, the Sasernae were later than Cato but not esteemed very highly by Varro: (see note on I. ii. 22-8 [text omitted]). The writer who was regarded as the greatest and who exercised (through the translations of his writings into Latin and Greek) an inestimable influence on Roman agriculture was Mago (see note on I. i. 10).

oportuerit: this impersonal verb is in the perfect subjunctive because it stands in a subordinate clause in indirect speech introduced by *video,* but it may at the same time carry a potential meaning: " should . . .". Scrofa means to say that many writers had counted the rearing of sheep and cattle as a part of agriculture, but had gone too far from their main theme in doing so.

eos . . . rem: " I consider that they are not to be imitated at every point, and that certain ones acted all the better who confined (their subject) within narrower limits, leaving out (those) parts which have no bearing on this matter." Stolo agrees with Scrofa that the subject of agriculture should be kept within certain bounds.

tota pastio: " the whole subject of grazing ". This includes the breeding and rearing of horses, cattle, sheep, and goats.

14. Quocirca . . . pecoris: " For this reason the headmen who are put in charge of both occupations are distinguished by different names in that one is called the bailif (*vilicus*) the other the chief herdsman (*magister pecoris.*)"

agri colendi causa: the *vilicus* had to oversee all the activities of the farm connected with the growing, harvesting, and selling of crops. These had to be lifted or hauled at least twice: from the field or garden, vineyard or olive grove to the barns and store rooms, and to the presses in the case of grapes and olives, and a second time when the produce was taken to market.

ab eo: " by him ", the *vilicus* who overlooked all the work of haulage; " under his supervision ".

cum veneunt: the generalising use of *cum,* meaning " whenever they are sold ", is probably to be recognised.

villa: this passage, as far as *velaturam facere* is another instance of Varro's love of etymology. His derivation from *vehere,* to haul, is not now generally accepted. The word is more likely to be a form of the diminutive of *vicus,* *viculus,* meaning a collection of houses and then a country estate. Varro's statement however about the rustic pronunciation of *via* and *villa* is not doubted, and has in fact been found in certain Italic dialects.

ii. 15-18. *Does Stockbreeding belong to Agriculture?*

15. aliut ... succentiva: "stockbreeding is one thing and agriculture is another, but each is related (to the other) just as the right hand pipe is different from the left yet it is linked to it in a certain sense because the one plays the treble notes of the same melody and the other plays the accompaniment". *aliut = aliud.* It is important to notice that *adfinis* is in the singular as is proved in the phrase *ita ... coniuncta* which follows; it is regarded as qualifying both *pastio* and *agricultura* separately.

tibia: the wind instrument known as the *tibia* or "pipe" was similar to the modern clarinet or oboe. It was made of reed, wood, or bone, open at the lower end and pierced with holes along the side. It might be played singly or in pairs. Double pipes were often played at theatrical performances, on other special occasions, and at festivals. In the Phrygian double pipe which perhaps Varro describes here, the left pipe was longer and in consequence deeper in tone. The accompaniment was played on it and could therefore be termed *inferior* both because it extended lower than the right and because it was subordinate in the playing. In the comparison drawn by Varro, it could be said that agriculture played the accompaniment to stockbreeding because stockbreeding was the first to be practised.

ut ... sit: the limiting use of *ut* with the subjunctive, equal to "although" in a concessive sense is to be recognised here.

16. licet ... adicias: "You may add (to what you have already said)." The subjunctive is jussive, dependent on *licet* without a subordinating conjunction.

auctore ... Dicaearcho: this is an ablative absolute "on the authority of ...". Dicaearchus was one of the most learned Greeks of the Alexandrian age. His works were still widely read even in the time of Varro and exercised a strong influence on him and the circle of his learned friends, including Cicero. Born in the fourth century B.C. in Messene (modern Messina) in Sicily (his *floruit* was about 320 B.C.) he spent the greater part of his life in Sparta and the Pelopponese in southern Greece. He followed the teaching of Aristotle and the peripatetic school of philosophers. Most of his works were dedicated to Theophrastus, the first scientific botanist (see note on I. i. 8-11, the text of which is omitted) whose friend he was. His encyclopaedic knowledge covered history, biography, philosophy, music, and geography: in the latter subject he was especially distinguished for his descriptions of the world with maps and for his discovery of a method of measuring mountains. His most important work in three books was called "The Life of Greece", in which he attempted to trace the evolution of mankind starting from a golden age when the earth produced crops and fruits untilled and when men were without greed. He was a stylist greatly admired by Cicero. He is mentioned again by Varro in II. i. 3-5 (text here omitted) where a short account is given of his description of a primitive golden age. Both Lucretius in *De Rerum Natura* V, 925-87, and Vergil in *Eclogue IV*, 18-47, contain passages which are perhaps inspired by Dicaearchus.

qui ... doceat: " who has sketched for us what the life of Greece was like from earliest times in such a way that he teaches us that there was a time in the primitive ages ..."; there is a certain redundancy in this sentence. *ut* introduces a clause expressing consequence or result.

fuerit: the perfect subjunctive in an indirect question in primary sequence dependent on *ostendit* which is in the present tense: this word suggests the drawing of a picture or a sketch.

ab initio: " from earliest times (of which we have any knowledge) ".

ab iis ... culturam: the accusative and infinitive construction depends on *doceat* " (saying that) agriculture was taken up by them at a secondary or lower stage in time ".

quocirca ... foraminibus: " For this reason (agriculture) plays the accompaniment to the pastoral (life) because it is secondary just as the left pipe is secondary (lower) in relation to the stops of the right."

17. tibicen: in apposition to *Tu* in the vocative case, said in a jocular tone: " You, as a piper ", " You with your piping".

domino, servis: datives of disadvantage with the verb of " taking away " (*adimis*).

domini dant ut pascant: " their masters grant that they may graze", " give them the right to graze ". This construction with a final clause is frequent with *dare*. Slaves were allowed to have their own animals and to graze them on the master's land.

leges ... pascat: " You are removing the laws for settlers among which we write (*i.e.* it is written), ' let not a settler graze the off-spring of the she-goat on land planted with young trees'." This law was directed especially against tenant-farmers.

ne ... pascat: *ne* with the present subjunctive in the third person regularly expresses general prohibition. Varro perhaps quotes the actual wording of the law. This could extend to any kind of plantation, whether of vines, olives, or orchard trees. The very fact that laws exist to restrict grazing proves for Agrius that it is generally recognised as a part of agriculture.

capra: the ablative is one of source or origin regularly associated with *natus*, " born from ".

quas ... tauro: " a (race) which astrology has removed to the sky, not far from the bull". *quas* is loosely put in the plural to suggest that the whole race has been removed. Among the twelve signs of the Zodiac are *Capricornus* " the Goat " and *Taurus* " the Bull ". Varro seems to suggest that the constellation is representative of all goats. The thought is that the goat is such a harmful animal that it has been placed in the sky where it can do no harm, near the bull, another grazing animal. This kind of jocularity is of Varro's style.

18. istuc = *istud.*

Cui ... " pecus quoddam ": " Fundanius answers him: ' Be careful, Agrius, that your reference is not wide of the mark since the actual wording of the law is " a certain kind of flock " ' ." Fundanius corrects Agrius because he has not repeated the law exactly. Agrius substitutes for the correct words *pecus quoddam*, the goat (*capra natus*). Fundanius indulges in a legal quibble, but goes on to agree that the goat is harmful to plants although the law includes any beast which might be harmful. Fundanius says in so many words, " Agrius, you must be exact—the law does not say ' goat ', although I agree that it must mean the goat".

149

etiam: " even ", *i.e.* " to be exact ".

veneno: this is probably to be explained as a predicative dative. The goats are poison (*i.e.* act as poison) to plants.

sata: " things sown "; the neuter plural of the perfect passive participle from *serere, sevi, satum,* " sow " is used to mean "crops " in general.

carpendo: the ablative of the gerundive has an instrumental sense, " by nibbling ".

non minimum: " not least ", " especially ".

ii. 19-21. *The Goat and the Conclusion of the Question whether Stockbreeding belongs to Agriculture*

19. diversa de causa: = *de diversa causa,* " for differing reasons ". The singular is used with the implication, " for one reason and then for another".

institutum: supply *est,* " the custom arose ", " it became the custom ".

ut adduceretur: *ut* with the subjunctive here and with *sacrificaretur* below expresses consequence. The use of the imperfect tense is frequentative, in that it describes an action which customarily happened, " used to be . . .".

alii: this form of the genitive is archaic: the classical form is *alius.*

ad alii: suppy *aram* from the previous clause.

ab eodem odio: " because of ", or " as a result of, the same hatred ". The adverbial phrase illustrates the meaning of *ab* " starting from . . ." or " originating from ". The two deities in common loathed the goat, but the ritual connected with them was quite different in each case.

cum alter . . . vellet: " because the one was unwilling to see a goat (at all) while the other wished to see it die ". *cum* with the subjunctives *nollet, vellet* is causal. The feminine form *altera* would have been expected because it refers to Minerva, (see below in the following sentence) but Varro is speaking in general terms. He makes the distinction between the two deities clear in the next sentence.

Sic factum: supply *est.* " So it came about that . . .".

ut immolarentur: *ut* with the subjunctive here expresses result or consequence.

Libero patri: *Liber pater* was an old Italian god of fertility, especially associated with wine: in mythology he was said to have discovered the vine and its cultivation (see note on I. i. 5 above).

ut . . . poenas: " so that they could pay the penalty with their lives ". They would undergo capital punishment for the harm they did to the vines: *caput* is often used to mean " person " or " life ".

contra sterilem: " on the other hand (it came about) that they used to sacrifice no goat at all to Minerva (lit. ' nothing of the race of the goat ') on account of the olive, because they say that any plant which it has nibbled becomes barren".

Minervae: see note below on *Athenis* (20).

immolarent: the change to the active voice in a general sense. " They sacrifice " (*i.e.* people in general) after *immolarentur* in the passive above, is remarkable. It is an instance of Varro's unpolished style.

laeserit: the perfect subjunctive is used in a subordinate clause in primary sequence in indirect speech.

19-20. eius ... possit: all this latter part of the sentence is in the accusative and infinitive construction after *dicunt*.

20. hoc nomine ... inigi: " for this reason (Minerva's hatred of the goat) it is not driven on to the Acropolis in Athens".

Athenis in arcem: *Athenis* is in the locative case.   In the heart of the city of Athens there rises a conspicuous long rocky plateau with precipitous sides: this is the *arx*, *i.e.* the defensive part of the city, the Acropolis.   By Varro's time it was given over to the great cults of the Athenians especially to the worship of Athena, the guardian goddess of Athens, whom the Romans called Minerva.   On the level stretch of rock was built the temple to Athena, called the Parthenon, in the fifth century B.C. after the victory over the Persians. Near by were smaller shrines—one called the Pandroseum was an open walled court in which grew an ancient olive tree.   This was regarded with great reverence because it was sacred to Athena and was said to be the one which she caused to spring out of the earth when she gave the gift of the olive tree to mankind.   This tree had to be carefully protected and for this reason no goat was allowed to come near it.   Another shrine on the Acropolis was sacred to Artemis Brauronia, whose main cult was in Brauron in Beotia.   She was a goddess of the wild and wild animals and of the hunt, and received a goat as an annual sacrifice.   The ritual connected with her shrine which was near the Pandroseum is probably the one to which Varro refers; it was permitted to take a goat on to the Acropolis for this one purpose in spite of the taboo connected with the worship of Athena.

dicitur: although in a subordinate clause in indirect speech, *dicitur* is in the indicative and not the subjunctive as might be expected, because the statement made is not part of what people say, but gives an explanation inserted by the speaker, Fundanius, about the olive tree, as a parenthesis; *i.e.* he makes a remark which is outside the construction of the sentence.

Ne ullae ... possunt: " Neither are any cattle (in addition to the goat)" I concluded, "an essential part of agriculture except those which by their work are able to cause the land to become better cultivated, such as those which are able to plough under the yoke."

possunt is to be supplied with *adiuvare: quo* is regularly used as here as a conjunction with a comparative adjective or adverb and subjunctive to express purpose, *iunctae* implies " yoked together " referring to teams of oxen.

21. cum ... ministrent: this use of *cum* is causal: " since ...".

stercus: the Romans knew well the value to soil of animal manure and how a grazing herd could enrich arable land.

venalium greges: '' gangs of slaves ''; *venalis* may be applied to anything which can be sold and therefore is used as a synonym of *servus*.   The Romans regarded the slave-labourer on the farm as being on a par with the farm beasts. *grex* is primarily applied to herds of animals, but it is also used for crowds of people, not usually, however, in a good sense.

Sic ... agri culturam esse: " By the same train of reasoning we could assert that slave-gangs are (a branch of) agriculture."   The expression is a condensed one: *agri culturae partem esse* might have been expected.

si habendum statuerimus: " if we decide that we must keep one ".   The gerundive with *gregem* understood from *greges* and *esse* expresses purpose. *statuerimus* is in the future perfect in the idiomatic use in a subordinate clause dependent on a future main verb (*dicemus*). The corresponding idiom in English requires the present tense.

Sed ... sequendum: " But the mistake arises from this, that cattle can be (kept) on the land and can therefore bring returns on that land (an argument) which you must not follow." It is a false assumption to think that since cattle are on the land they are a part of agriculture. *error*: the mistake of including cattle in agriculture.

Nam ... artifices: Varro adds that if cattle are a part of agriculture then anything else on the land must be included, even workmen kept for other occupations such as spinning, etc. This is clearly an absurdity.

institutos histonas: " permanent weaving-sheds where looms could be set up". Varro has just spoken of people in mentioning *textores*; now he switches suddenly to the places where people work but the people, weavers, are implied. The argument is that such artisans as spinners and weavers might be employed on the farm but they are not in fact a department of agriculture because it could be operated without them.

Diiungamus: " let us exclude . . .", the jussive subjunctive in the first person plural.

siquis ... aliud: " if any one wishes (to exclude) anything else", " whatever else anyone wishes (to exclude)". The argument as to whether stockbreeding is part of agriculture is now concluded. Scrofa sums it up quite clearly: stockbreeding is to be separated from it. Earlier he stated his opinion that writers who included stockbreeding had wandered too far from their subject (I. ii. 12-13); he has not spoken since in the discussion but now comes in with a definite conclusion.

ii. 22-28.   *The Books of the Sasernae*
(the text is omitted)

The Sasernae who belonged to the *gens* Hostilia were father and son: they wrote on agriculture about the beginning of the first century B.C. midway in time between Cato and Varro. Their works now lost were known not only to Varro, but also to Pliny the Elder and Columella. Stolo and Scrofa, the two grand old men of great experience in agriculture, knew them but did not think very highly of them.

iii.   *Agriculture both an Art and a Science*

quae ... discretum: take in this order, *quoniam discretum (est) cuius modi sint (ea) quae diiungenda essent a cultura:* " since we have decided the nature of those subjects which were to be excluded from agriculture". *sint* is in the subjunctive in an indirect question after *discretum (est); essent* is in the subjunctive in a subordinate clause in indirect speech; the imperfect tense, however, seems to violate the rules of sequence since the main verb is in the perfect. Varro sometimes departs from generally accepted grammatical rules.

de iis ... docete: " tell us in regard to those things what the knowledge (of them) is in agriculture ". These words put the question to the company as to whether agriculture is an art or a science.

scientia: derived from *scire*, " know ", here has the root meaning of " knowledge ": a developed meaning is " science " which is the application of knowledge.

in colendo: " in cultivating ", *i.e.* " in agriculture ".

ars id an quid aliud (*sit*): " whether it is an art or something different ". The neuter pronouns *id, quid, aliud* refer back loosely to *scientia*: the feminine forms might have been expected. *quid* with an indefinite meaning after *an* is not usually found, but is commonly so used after *num*.

a quibus . . . ad metas: " from what starting-point does it race down (the course) to the goal ". The metaphor is taken from the race-course (see note on the Circus Maximus on I. ii. 11 above). The horses and chariots waiting to take part in a race were each stationed in a separate wooden stall called a *carcer*. These stood at the narrow rounded end of the Circus on each side of the main entrance, and were twelve in number. Each was closed by double swing doors. At a given signal the doors were pulled open by a rope worked by a slave to enable the chariots to be driven out to the starting-point which was in a line with the *metae*, " goals ", the three wooden conical pillars on the *spina*. Three other *metae* also stood at the further end of the *spina*: thus one group marked the starting-point, the other the turning-point of each lap of the race: see I. iv. 1 where the two are mentioned.

Tu . . . dicere debes: Scrofa is the grand old man of agriculture (see I. ii. 12 above). Stola defers to him as his superior in knowledge and experience.

necessaria ac magna: " of the utmost importance ".

eaque . . . facienda: it is unusual to find a clause dependent on an abstract noun; in English we must supply " to know "; *eaque* is used idiomatically to give an added statement about a thing already mentioned.

in quoque agro: " in each kind of soil ". *quoque* is from *quisque*.

facienda: these are the necessary activities on the farm such as ploughing, digging, hoeing, planting, harvesting, etc.

quo: a neuter relative representing in general the statement which has gone before, regarded as a whole and introducing a final clause with the subjunctive (*reddat*).

iv. 1-5. *The Farmer's Aims*

1. Eius principia . . . Ennius scribit: " Its elements are the same as those which Ennius says are the elements of the universe ": *principia* is to be repeated with *mundi*; *principia* is the term used by philosophers for the primal material out of which the universe was thought to be made.

Ennius: see note on I. i. 4. Among his other writings Ennius translated into Latin certain works of Epicharmus, the Greek Pythagorean philsopher and writer of comedy who lived, as far as is known, from 540-443 B.C. The translation was called the *Epicharmus* and was much used by the grammarians.

Haec . . . oritur: " These must be learnt ", *i.e.* " you must gain some knowledge of these (elements) before you sow the seed, (an operation) which marks the beginning of all production."

iacias: the subjunctive of the " ideal " second person when a general statement is made: *quod* may be understood as taking up *iacias semina*, with the meaning that sowing is the beginning of the farmer's work, but a more significant meaning is to be recognised if *quod* is taken to refer to knowledge of the elements: they are the underlying cause of successful agriculture and the farmer must first know them before sowing his land. Varro often uses *quod* in the neuter with general reference to an antecedent of a different gender and number.

ad duas metas: the metaphor is carried on from the previous paragraph

(I. iii. 1).   The pursuit of agriculture is like a race in the Circus which has two goals just as there are two groups of *metae*, one at each end of the *spina* which stretches down the middle of the arena.   The first lap is profit and second, pleasure.   It is a gainful occupation and this must be its first aim; pleasure is linked to it but is, as it were, the second lap and is the secondary aim (see I. ii. 11 and iv. 3).

dirigere: this verb is usually transitive and therefore should have an object: perhaps *se* is to be supplied, or a word for chariot.

2. Nec non: the two negatives cancel each other out, to make a positive and emphatic statement: " it is a fact that . . .".

nec . . . pretium: the argument is that cultivation improves land, as for instance when it is planted with ordered rows of orchards, olive groves, or vineyards: its attractive appearance causes it to sell more easily and raises the price.   In this way pleasure is combined with profit.   The good appearance, too, speaks for good cultivation.

Nemo . . . turpis: this sentence seems confused but the gist of it is that a buyer would prefer to pay rather more for an attractive piece of land than one which is just as valuable but has not a good appearance.   The two negatives, *nemo* and *non* cancel out and the statement " there is no one who does not prefer . . ." becomes " anyone prefers to buy for profit the same thing at a higher price because it is more attractive than if (the piece of land) (though) productive is unattractive". *eadem* is difficult but combines the two possibilities, a more attractive object or an unattractive piece of land: the sudden change from the neuter in *formosius* to the masculine *fructuosus turpis* is also difficult but is in keeping with Varro's loosely constructed style: it is necessary to supply *fundus* or *ager*.

pluris: genitive of price.

3. contra . . . patitur: " on the other hand in an unhealthy district misfortune, although on rich land, does not allow the farmer to make a profit". The reference is probably to a malarial district: this used to be true for instance of the coastal district of the Roman Campagna.   The soil might be very good but the district unhealthy and dangerous for the farm workers.

ubi ratio . . . habetur: " where the reckoning is with death ".   No good is to be gained in having land in a district where fatal diseases may be endemic.

alea: a game played with dice, hence a game of pure chance, and in this case with high stakes, the labourer's life.

4. Nec . . . non: another double negative making an emphatic affirmative, " certainly ".

in naturae: repeat *potestate*.

ut . . . possimus: the limiting use of *ut*: " Yet after all there is much in us whereby with that same care we can make lighter those things which are more serious ", *i.e.* " it is well within our power to take measures to improve on the natural disadvantages of the land".

Etenim . . . fenestris: this sentence enlarges on the general statement just made and gives examples: there may be marshy land which gives off a bad odour, or the orientation may make it too hot, or the wind may do it harm; the farmer can improve on these conditions by making some outlay and by studying the general arrangement of his buildings.   The first *si* clause governs the indicative (*est*) but in the following two clauses with which *si* is to be supplied, the subjunctive appears, *aut (si)*, . . . *sit*, *aut (si)* . . . *flet*.   There seems to be little or no difference in the meaning of the two moods: perhaps this is only to be regarded as another example of Varro's looseness of style, and use of variety without any particular significance.

permagni: this is a genitive of value, in the normal usage with *interest*.

ubi ... fenestris: " where the farm-buildings are situated, how large they are, in which direction they face with colonnades, doors, and windows".

ille: a strong demonstrative use: " that well-known ", " that great . . .".

5. Hippocrates medicus: Pliny *N.H.* VII. 123 is our source for the incident. Hippocrates foretold that a plague was spreading from Illyria and sent his pupils round the cities likely to be affected, to give their help. In return for this service the Greeks gave him the same honours as were given to Hercules. He lived from 460-375 B.C. as far as is known and was regarded as the founder of scientific medicine.

Corcyrae: locative case; Corcyra, now Corfu, is an island off the western coast of Greece; for this incident see Varro's Life.

immisso ... permutata: ablatives absolute: Varro let the north, *i.e.* a cool, wind blow in: by cutting new windows he shut off the hot disease-bearing winds, and changed the orientation of doors. He understood the need for the admission of a cool breeze into a dwelling house and seems to have understood how heat generates disease. Corfu faces an almost land-locked bay and is surrounded in part by shallow lagoons. The heat can be stifling in summer, but each day in the afternoon a cool breeze springs up about noon with a resultant fall in temperature. Varro must have taken full advantage of this climatic condition and so saved his troops.

familiam: his personal slaves.

vi. 1-6. *The Soil*

1. primum: indicates that the first of the four main divisions of agriculture is to be discussed, the soil.

videndum haec quattuor: " these (the following) four points must be considered". The impersonal use of the gerundive expressing necessity with an accusative (*haec*) is archaic (see Woodcock 204, note 11). It is found in early Latin literature, but was soon replaced by the personal construction, which here would have been *videnda haec*. In Varro the construction is probably not so much an archaism, as an example of the lack of polish in his style. There are many more examples which will not be commented on unless necessary. The four points are listed in the form of the indirect questions which immediately follow: in the last three supply *sit*.

Formae ... naturali: there are two aspects of the first point to be considered: these are set out in a balanced statement with pairs of corresponding words: *una . . ., altera . . ., prior, quod alius . . . alius, posterior, quod alius . . . alius.*

de naturali: supply *forma*.

2. Igitur: in polished prose *igitur* is not placed at the beginning of a sentence: there are many niceties of this kind which Varro does not observe.

a specie: " in regard to appearance " (*i.e.* natural topography). For this use of *a* compare the expressions *a dextra, a tergo, a fronte;* " in the direction of . . .", " from the point of view of appearance ".

campestre, collinum, montanum: these three words in themselves give a general description of the natural conformation of Italy. Almost the whole of Italy comprises plain, hills, and slopes rising to high mountains: the Apennines run down the length of the country like a backbone on each side of which are lower hills which sink down to a coastal plain. These are set in a north-east to south-west direction, but even north of the plain of Lombardy the same conformation appears where the Alps extend from east to west and afford the same distinctions of terrain as they slope down to the level valley of the Po. In Apulia [see below (3)], the district which stretches to the heel of Italy, the coastal plain broadens out below the promontory of the Gargano. Such an extent of flat land is without shelter and at the mercy of the hot sun in the summer months. In Italy low lying plains become completely parched and lose green vegetation from June to September. This is especially true also of the land around Rome, now called the Roman Campagna, where grain is harvested at the end of May, whereas it is only just showing green blades in the upland country. This difference can strikingly be seen during a journey from Rome into the higher reaches of the valley of the Sangro. The conformation of the country, looked at in broad outline, results in three belts of climate and three resultant belts of vegetation.

quartum: the fourth type of conformation is found when two or even three occur on the same farm. Supply *cum . . . sit:* the *ut* clause is to be regarded as one of consequence: " there is a fourth type which results . . . ".

ut . . . licet videre: " as it is allowed to see ", " as may be seen ". The indicative is normal with *ut* meaning " as ".

cultura: " method of cultivation ".

simplicia cum sunt: a plain statement is made as the indicative shows: *cum* here might be explained as frequentative, since a condition is described which occurs many times in many places.

3. Itaque . . . salubriora: the balance of the sentence is not well maintained: " and so, where there are wide stretches of lowland, there the heat is greater, and in this way in Apulia the region is hotter and more humid, and where the region is mountainous, as on Vesuvius, (because) it is lighter, and for this reason healthier ". *quod* seems superfluous and makes an awkward break in the construction of the sentence.

eo: " by this " (neuter singular), " in this way ", " hence ".

in Vesuvio: the mountain Vesuvius which dominates the bay of Naples: the volcanic soil was especially good for the vine.

deorsum: " on the lowlands ", " on the plains ".

susum: " in the highlands ", " in the mountains ".

Verno tempore . . . coguntur: "in the spring the very same crops are planted earlier in the lowlands than in the uplands and they are harvested earlier here on the lowlands than there on the uplands."

nec non: the two negatives make a strong affirmative.

4. Quaedam . . . prolixiora: " certain things ": a general statement in the neuter, although trees are described which are always feminine. Varro does not always follow a strict pattern in the agreement of adjectives.

abietes: *abies* is the silver fir (*abies alba*) which is straight and tall in growth and is found at its best in the Alps and the Apennines. The wood is light and was prized for building ships, especially triremes.

sappini: appears to be a general term for pine and fir trees.

populi: the poplar tree frequently planted today in Italy to form a wind-break no doubt has served the same purpose all through history.   The black poplar is one of several trees on which the vine can be trained.

tepidiora: supply *loca sunt*.

salices: the *salix alba* or common willow was considered amongst the most useful of trees.   The twigs were used for basket-making and larger branches for vine-supports, and the bark for leather-tanning.   The plant continues to shoot however hard it is cut back.

fertiliora: a neuter agreeing, in a vague, general way, with the statement introduced by *ut*, " things are more fertile ", but the agreement presents a difficulty when it is seen that the adjective qualifies the lists of trees which follow.

arbutus: the arbutus, or wild strawberry tree, is a common woodland tree in Italy: it can be seen, for instance, in the woods at Castel Fusano on the coast near Ostia; the small white bell-shaped flowers and crimson fruit grow together on the tree.   The berry which resembles a small strawberry is edible but has very little flavour.

quercus: various kinds of oak grow in Italy: Varro is perhaps referring to the *genus* in general.   The name *quercus* is often employed in this way.

nuces graecae: these are almonds.

mariscae fici: the mariscan variety of fig was large and inferior.   The fig (*ficus*) was one of the most important fruit-bearing trees for man (see below I. xli. 1).

4. in altis contra: *in altis* (*collibus*): stated in full this implies " on the higher hills the growth is more akin to that of the mountains".

5. discrimina: the differences are determined by climate: crops do best in the heart of the plains and come rapidly to harvest.   Vines grow best on sunny slopes, forest trees flourish in the cold atmosphere of the mountains such as the pine and firs just mentioned.   Thus there are three belts of vegetation according to altitude and the corresponding climate.

putatio: the vines should be pruned three times yearly as well as the trees if these are used as supports.   Orchard trees too need pruning.   Pruning was one of the labourer's heavier tasks.

pabulum: the flocks were taken, and still are, from the plains to mountain pastures in the summer.   See chapter III on Transhumance.

6. eo magis ... eo deterior: the second *eo* is redundant, but has an emphatic effect.

Haec ... momentum: "these considerations and others like them share severally their importance for the cultivation of the three types of conformation of the land".   *momentum* is a shortened form of *movimentum* and came to mean " that which tips the balance " or " which turns the scales ".

vii. 1-3.   *The* Quincunx *Method of Planting*

1. Cato: see note on I. ii. 7 above: the reference is from the *De Agri Cultura*, I. i. 3.

sit et spectet: subjunctives in a subordinate clause in indirect speech.

videtur: supply *mihi*, the phrase is then equal to " I think ".

2. De ... modica: " in regard to the cultivation of the natural confor-mation I assert that it follows that those things which become more attractive in appearance, bring also a greater yield, as for example if those who possess orchards if they are planted (by them) in *quincunx* formation in regular rows and moderate spaces between." Scrofa introduces the second consideration in connection with the conformation of the farm, that is the effect of tillage on landscape. *sequi* is used impersonally and is equal to *sequitur* in direct speech: *fiant* is a subjunctive in indirect speech after *dico*: (with *quae* supply *ea*): the clause (*ea*) . . . *venustiora* comes in meaning after *sequi ut.*

ut . . . sint: a result clause with the subjunctive.

ut . . . modica: the *ut* clause and the main verb *sata sunt* do not connect, but a link can be made if an agent is supplied with *sata*: the real relationship however remains difficult grammatically although the sense is clear, " those who possess orchards should plant them in *quincunces*".

ex arvo . . . consito: " from a piece of land equally large (but) badly planted". Put more simply this is " on a given amount of land laid out by the old-fashioned method".

in quincuncem: " in *quincunx* formation ": planted in alternate rows so that no tree directly faces another. The root of *quincunx* is *quisque -uncia*, five-twelfths of a whole, whether of money, land, weight, etc. It is also used for five spots on a dice which would appear as

In relation to trees planted in rows of this kind it indicates that if any five trees be taken from three rows they would appear to be set out like the *quincunx* in the game of dice. The word came to be used for the planting of trees to make a pattern in alternate rows :

These could also be described as standing in oblique rows which appear if lines are drawn across the rows from tree to tree. Hillsides covered with these ordered rows of olives are to be seen even today for the method has never varied all over Italy in districts good for the olive. In the neighbourhood of Rome, especially in the lower Sabine country, the landscape is beautified by this method of cultivation and is a good instance of how man has added to nature. In this way every tree had its own piece of ground all around it and was not shadowed from the sun and had free play of air. See Vergil's description, *Georgic II*, 273-87, in which he compares the *quincunx* with an army drawn up in battle order.

minus multum: an awkward expression for " less ".

    3. ut . . . possis: this use of *ut* with the subjunctive which appears peculiarly Varronian, meaning " for example ", seems to be an extension of the use of *ut* to express result.

cum . . . possis: *cum* meaning " although ".

fregeris: the idiomatic future perfect used of an action which must happen before another in the future or virtual future (*possis*).

vix . . . possis: " you can hardly pack into a *modius* and a half when you have cracked them". Cracked nuts take up more room than whole ones.

    4. luna: the moon was thought to have had some powers of ripening crops: some country people even today in England still plant their seeds according to the phase of the moon.

Quas ... duo: " which two things two others follow ".
altera illa: carry the meaning " the other and that one " that is two, similar to the English " the one and the other " which amount to a plural: each is regarded separately but can be combined for a plural agreement.
musti, olei: partitive genitives with *plus*.
preti pluris: genitive of value, closely akin to the descriptive genitive.

viii. 1-7. *Methods of Training the Vine*

1. Contra vineam: a terse expression which in full would be: " (arguing) against the vineyard (as a means of profit-making) ...".
inquam: Varro now leads the discussion which is on vines. It appears that he begins speaking at the beginning of the paragraph *Contra vineam*.
sunt qui putent: the subjunctive is generic " there are those who are of the opinion that ...".
sumptu: describes the expense or cost of upkeep of a vineyard which must have been considerable in spite of the measures of economy suggested. Later on (I. xviii. 1, text omitted) Varro quotes Cato (xi. 1) as stipulating that sixteen slaves are needed for a vineyard of a hundred *iugera*. Although slaves were not paid they had to be fed, housed, and clothed all the year round. Varro also mentions (I. xvii. 2, 3) that at the time of the vintage it was necessary to hire freemen for the extra labour; this naturally added to the cost. The vineyard in itself called for expense; it had to be fenced, the vines had to be propped, and although the props were usually grown and made on the estate, they used up land and labour. Despite these costs, however, there is no doubt that vine growing was profitable in Varro's time.
Refert ... eius: this impersonal verb may normally have as a subject, an interrogative, " quod ... sit ". The second *quod* means " because " and since it states a fact, is associated with the indicative (*sunt*).
humiles ... Hispania: the vine is by nature a climbing plant which needs to be raised on some kind of prop if it is to leave the ground. Without props it creeps on the ground: (*humilis* from *humus* has here the root meaning) but if some kind of prop is provided it will climb. *ridica* is a general term for a prop of any kind but is often used for a vine-prop. Varro can speak from his personal experience of the viticulture of Spain gained during the years he spent there under Pompey's command (see Varro's Life). It is clear from this and other references that he observed closely the local and typical agricultural methods of the countries in which he was stationed during his military career.
iugatae: any vines grown on props were called *iugatae*, " yoked ". When lines of vertical poles were joined above by horizontal rods the effect was comparable with that of a yoke formed from two vertical shafts and a cross-bar. Each plant was tied to a central pole and spread on each side along the *juga* (the cross-beams). This method of training on props was the most usual in Italy and is still widely prevalent today.
nomina duo: the two words indicate the two different kinds of props, the verticals (*pedamenta*) and the horizontals (*iuga*).
quibus ... iuga: supply *ea* both with *pedamenta* and *iuga*.
ab eo: " from this formation ": the phrase takes up in general terms the preceding statement.

2. iugorum: these are the horizontal supports of which there are four varieties, each characteristic of a particular region of Italy.

fere: "in general use".

pertica ... Mediolanensi: the four place names are those of four famous vine-growing districts: for *Falerno* (see I. ii. 6 note) *agro* is to be supplied with *ut in Falerno*. *Arpano*: this is the district around *Arpi*, a town in Apulia (now Puglia) in the south-east of Italy. *in Brundisino*: "in the district of Brundisium", the coastal district on the "heel" of Italy: *in Mediolanensi*: "in the district of Mediolanum" (now Milan). This was situated on the plains of Lombardy where vines are still largely grown on trees at the present day.

in agro Canusino: Canusium (now Canosa) was in Apulia (Puglia) on the river Ofanto in a rich agricultural district especially good for viticulture.

derecta: in this method the horizontals are placed along the rows and not across them so that the vineyard is filled with long, separate rows joined along the top and free standing.

compluviata: in the form of a *compluvium* where the rows are joined by transverse cross-bars to make a trellis, or a kind of pergola; this form of trellis is in general use in Italy today. The *compluvium* was the opening in the roof of the main room (*atrium*) of a Roman house which was sloped inwards from the corners and allowed rainwater to collect in a basin (*impluvium*) in the floor below. Trellis of such a nature allowed the vine more room to spread, giving the foliage and ripening fruit more space and shade and made pruning and gathering easier. It conserved warmth and freshness in heat, and guarded against wind: as a result it was a very productive method.

Haec ... sumptum: if the materials out of which the different kinds of props are made, such as wood, reeds, or rushes, can be produced on the farm, the cost is not great.

ubi ... non valde: if the materials can be obtained from a nearby estate, the cost is not too great.

3. Primum genus ... aliquam: in order to produce props on the home farm there must be a plantation of willow, a reed thicket, or a rush bed. The horizontal props in order to avoid injury to the vine branches had to be of some soft, pliant, material: willows and reeds were ideal for the purpose: rope was made out of rushes. Such plantations might take up valuable land but for all three marshy, or at any rate, moist places were best suited and it may be that the plots used in this way were unsuitable for other purposes. Rope was also made from *cannabis* (hemp), *linum* (flax), and *spartum* (esparto grass), all of which could be grown on the estate (I. xxiii. 6).

quartum ... opulos: trees on which it was intended to train vines were planted and grown in rows and kept pruned for the purpose. In addition to maples, characteristic especially of the district around Milan and the plain of Lombardy, vines were commonly trained on elms or on black poplars. The latter are to be seen today for instance over a wide extent of Campania where the effect is like a forest of poplars, tall and gnarled, sprouting into long, slender branches above: below the vines are slung on wires or nets from tree to tree in lines, sometimes spreading fantastically as the branches seem to strive to reach along from one support to another.

Canusini ... in harundulatione in ficis: the reading of *harundulatione* is questionable. If correct it could be taken to mean a trellis-work of reeds attached to fig-trees. These latter would act as supports and the reed trellis would be slung between them. *Canusini:* the people of Canusium, a town in Apulia.

4. Pedamentum: in contrast with the horizontal supports the uprights had to be of stout wood which stood firm. Oak and juniper were said to be the best, but any hard wood could be used such as chestnut, oak, willow, olive, or ash. Posts could be made by the slaves during wet weather or at other slack times.

palus e pertica: the first kind of upright prop was a manufactured one: the second was a branch roughly shaped and undoubtedly peeled, because the bark if not removed would harbour insects. Economy could be practised in the use of the *palus* because it could be reversed when the end which was in the ground had rotted: it would become shorter after each reversal and eventually would have to be discarded.

tertium . . . possit: the third kind of prop made of a bundle of reeds was only used when the others were unobtainable. The bundles were placed in terracotta pipes with open ends which were then fixed in the soil. By means of these superfluous rain water was drained off but a certain amount of moisture was retained which was necessary to prevent the reeds from becoming brittle and so breaking off.

Quartum: the natural prop of the vine is a tree: this is not confined to the elm, as is often thought, but could be any suitable tree, such as chestnut, fig, olive, poplar, and others. The trees were planted about twenty feet apart and pruned in stages with three feet between: in Campania the trees might reach a height of sixty feet. Three to six vines were planted around each tree so that tree and vine were married. The vines might be trained in circles as well as festooned in a kind of trellis from one to another, but they would not be allowed to grow higher than was convenient for the gathering of the grapes and for pruning.

5. Vineae . . . hominis: this is for the ease of the labourers at pruning time and the vintage: supply *est*; although a plain statement is made, the underlying meaning is " ought to be ".

intervalla . . . possint: supply *sunt*: as with (*est*) there is an unexpressed implication, " ought to be . . .".

pedamentorum: a difficult genitive perhaps as nearly possessive as any other: " the spaces between ". The soil of the vineyard had to be kept broken up, loose, and clear of weeds: this was best done by means of oxen and the plough.

Ea . . . vinum: " the least expensive vineyard is that one which provides wine for the pitcher without props". This is a straightforward statement but some losses must be expected in this kind of vineyard.

acratophoro: dative of advantage: a Greek word for a pitcher for unmixed wine that comes straight from the vine, the word is only found here and in Cicero (*De Fin.* III. 4. 15).

cubilia: the vines run along the ground: the earth is their bed.

Asia: Varro must have observed these during his service in Asia (see Varro's Life). In this context *Asia* is roughly equal to Asia Minor.

oppleris: future perfect, shortened from *oppleveris:* although the main verb *fit* is in the present, it has a future significance and is used as a general term; therefore the subordinate verb must be in the future perfect since it will happen before the statement made in *fit*.

Pandateria: one of the Ponzian islands which lie off the west coast near Terracina (see also III. v. 7).

6. ea . . . uvam: " that (branch of the) vine which shows that it is bearing grapes". In this type of vineyard which is without props, the branches on which bunches of grapes are seen to be forming are raised on forked sticks.

cestum: is literally, "an embroidered girdle "; the grapes which ripen when the branches are raised on forks learn to hang down before they are ripe and do not have to wait to be hung up on a string in the fruit-store: this seems to be one of Varro's semi-jocular statements.

Ibi . . . possit: a good example of an economy which can be practised. The forked sticks which had undoubtedly been shaped by the slaves of the farm, were not thrown out but carefully preserved under cover until the next year.

vindemiatoris: a word with the ending -*ator* generally is used for a person who does the action as a profession. The use of the word here can be taken to mean a hired worker, one of those whom Varro says must be employed at the time of the vintage. (I. xvii. 2-3.)

Haec . . . refert: " this variation in culture, therefore, is due chiefly (to the fact) that it is of importance of what kind the soil is". The nature of the soil causes the difference in the method.

non . . . solem: " it does not seek water as (it does) in the drinking-cup, but sun". The Romans frequently diluted wine with water just as the Italians do nowadays.

xi. 1-2.　*The Size of the Farm Buildings, and the Water Supply*

1. In . . . multi: supply *sunt:* " in the unnoticed measurement of the farm many have slipped". " Through failing to give attention to the measurement of the estate many (people) have made mistakes " or " have been at fault ".

quod: an unusual use is shown here in that *quod* introduces the reason for the statement already made and enlarges on it.

cum . . . sit: *utrumque* in the neuter agrees with the two statements *quod . . . fecerunt* and *alii maiorem (fecerunt): cum* with the subjunctive carries the meaning " since " but there is little more signification than a plain statement here.

pluris: genitive of price, " at too great a cost ".

sumptu maiore: " at too great an expenditure ".

fructus . . . disperire: " the produce perishes " because it cannot be satisfactorily stored and can be stolen if not under lock and key; this would apply especially to grain and wine. Sufficient space is needed for proper placing of sacks, wine barrels, etc.; and plenty of room for the circulation of air.

2. Dubium . . . ager est: some farms became specialised, so to speak, as a direct result of the nature of the soil and the topography.

sint: this is in the subjunctive because it is in a subordinate clause dependent on a main verb *sit* which is also in the subjunctive.

ut: " just as " comes in meaning before *ampliora*; enlarge, *ampliora ut horrea sint facienda.*

si non . . . proxime: this is a very condensed expression although the meaning is not in doubt. Enlarged it becomes " if the farm cannot be built so that it has water within the farmyard, then it should be built so that it has water as nearby as possible ".

primum ... perennis: the first kind which is within the farmyard is spring water underground which can be tapped by means of wells: it lies underground on harder strata of rock as is the case for instance in Campania, and runs out at some point or is tapped by means of boring; such water is usually cool and fresh. The existence of a water supply of such a kind would determine the location of the farm. This supply which is " on the spot " is the best. The second is a running stream, which is also perennial. This too could determine the siting of the villa: Varro's estate had a river flowing through (see below, III. v. 9). In Italy many streams and even rivers are completely dried up during the summer months so that well-water is a signal advantage. A good farm would be greatly favoured to have both kinds of water.

viva: " running water" sums up both spring and running stream water. The conservation of water was vital for the life of a large estate. The concentration in one place both of men and beasts together with the needs of irrigation of crops and orchards made heavy demands. This however was ensured by the Roman methods of conserving rain water in large cisterns as Scrofa recommends. Remains of cisterns are usually to be found on villa sites. They are sometimes large enough to seem like huge underground halls with the roof supported on pillars. Each farm then had its own water supply either natural or artificially conserved or probably both. Imprudent would that owner be who did not ensure an adequate supply against the drought of the scorching months (June-September) of the Italian summer. It is highly probable therefore that cisterns would be made even when the water provided by nature was in good supply.

xii. 1-4.   *The Location of the Farm*

1. Danda ... flabunt: the ideal setting for the farm itself is described: the wooded hillside would provide shelter, moisture, and probably running streams as well as a supply of timber, feed for swine, and covert for game, but at this point the emphasis is clearly on shelter. The pastures would be on the lower slopes, for sheep and goats. The Romans were skilful in their use of orientation with regard to winds. An east-west siting would have the benefit always of a cool breeze: it is noticeable for instance to anyone who visits the farm at Licenza, said to have been Horace's Sabine farm, that it is planned to catch the breezes which blow along the open valley with the result that every room enjoys a fresh current of air even in hot weather.

ponat: the subject is " he ", the person who is choosing the site for his villa: the use of the third person is perhaps to be compared with the general use of the second person singular with the present subjunctive signifying " you " in general.

ut contra ventos: *villam ponat* is repeated from the previous line, " facing the winds ".

ad exortos aequinoctiales: supply *ventos:* this is in an easterly direction.

cogare: = *cogaris:* the present subjunctive represents the general use of the second person for " you ", an imaginary person.

curandum ... ponas: Varro frequently uses the impersonal gerundive construction in the neuter in general statements expressing obligation: this is rare with the gerundive of transitive verbs and belongs to early Latin. *Advertendum* in the following sentence is in the same construction, and *vitandum* below, in 3. " Care must be taken not to place it facing (the river)." (See Woodcock, section 204.)

2. propter easdem causas: the orientation in connection with wind and sun is just as important where the ground is swampy.

crescunt ... minuta: the Romans attributed certain diseases to organisms so small as to be invisible: they reasoned that since diseases were caught from marshland that the ground bred unseen creatures which spread disease. There is no doubt that the reference is to malaria, which is now known to be carried by the *anopheles* mosquito. It does not seem that Varro attributes disease to an insect because he says through Scrofa that the disease-spreading creatures are not visible to the eyes (*quae non ... consequi*) but he comes near to guessing the cause. Malaria was especially rife at different times in history along the coast of Italy where there were long stretches of maritime forest and lagoons lying near the shores: the kind of terrain which can be a vast breeding ground of the mosquito. Since malaria is caught in marshy places and where there is standing water the statement is reasonable that something harmful is engendered in the swampy ground. The disease comes from outside and not from some disturbance from within the body.

difficiles: " difficult to cure "; malaria is a disease which continues to recur at intervals. Varro probably means this or something similar.

hereditati: predicative dative: " for an inheritance ": the use with this word is extremely rare.

quo minus ... noceat: " to prevent the disease doing harm ". *potero facere* is equivalent to a verb of preventing and so governs the construction *quo minus* and the subjunctive associated with such a verb but in the more familiar usage the main clause is negatived. Although that is not so in this passage a virtual negative is to be recognised in the question *quid potero?* to which the implied answer is in the negative, " There is nothing that you can do".

vendas: jussive subjunctive: the imperative might have been expected, but the subjunctive shows that Agrius speaks in general terms; so too *relinquas*.

assibus: ablative of price.

si nequeas: normally the future indicative would be expected in a supposition of this kind which makes an unqualified statement, but the clause is to be regarded as also potential and so has the subjunctive.

3. qui ... discutitur: " which, because the wind blows through it, if anything unhealthy is carried there, is more easily blown clear ", *qui* is the subject of *discutitur*.

inferatur: the subjunctive is to be explained as generic, " which could be unhealthy ", but below follows an almost identical statement *siquae ... inferuntur* with the indicative. Varro's use of different moods is very loose and at times inconsistent.

bestiolae: " insects ". It is not clear whether Varro refers to the invisible *animalia* already mentioned in (2) or to gnats and mosquitoes as well which are seen and known to be harmful, if only as far as Roman knowledge went, for their bites. They are harmful to beasts as well as to man.

prope nascuntur: they breed in the swampy land lower down, below the place where the farm buildings are situated, and cannot live in the drier atmosphere. Mosquitoes breeding in stagnant water can be eliminated by means of drainage.

4. Nimbi repentini: torrential storms are characteristic of Italian weather, taking place even in the summer months. When rain comes it is usually very heavy, causing rivers and streams to flood suddenly and dangerously. A small stream can become a swollen torrent in a very short time.

repentinae ... manus: these words belong to the *quod* clause: this placing is an example of *anastrophe*.

improvisos: this refers in general to the inhabitants of the farm: " those who are off guard " or " when they are . . .".

xiii. 1-5. *Directions for Farm Buildings*

1. ita ... caldiora: " in such a manner that the cowsheds are in the spot where they can be warmer in winter". The relation of the two clauses is difficult: an expression such as *in eo loco quo hieme possint caldiora* would have been expected; the use of the adverb *ibi* followed by the relative *quae* which relates back to *bubilia* does not allow a satisfactory grammatical explanation, yet, after Varro's manner the meaning is abundantly clear: *quae ... possint* is a final relative clause. *caldiora:* the comparative presents a certain difficulty, but perhaps it is to be understood to mean " warmer than if they were not well situated ", which comes to mean " in the warmest place " or it could carry the implication " warm enough ". The good farmer must have a care for his beasts, a point which comes out very clearly and frequently in Varro's work: their welfare must be considered in every way. The motive is more mercenary than humanitarian, perhaps, but Varro shows a certain sympathetic treatment of the animals on the farm and here is one instance.

loco plano: wine and oil can be stored on a level piece of floor, of beaten earth, in casks or jars but the floor must be *level* to keep the liquids in good condition and to allow the lees to settle evenly.

item ... faciendum: the reading is corrupt but the meaning is clear: appropriate jars must be provided for holding both wine and oil.

in tabulatis: " on a floored space ". Raised platforms of wood seem to be indicated so as to allow a passage of air under the dried beans or hay: it could serve too as a means of keeping vermin away. The plural may suggest that there should be successive floorings or storeys so that the produce could be put on shelves one above the other, an arrangement which would ease manipulation both in storing and in bringing out.

Familia: the slave personnel.

ubi versetur, ubi ... reciperare: the clauses are final, introduced by the locatival conjunction *ubi*. This place for rest and shelter would probably be a colonnade where there could be shelter from the hot summer sun, or where in winter the sunshine could be enjoyed. Varro's humanitarianism is seen at its best: he has a care for the physical well-being of his labourers even when they are only *fessi*, and not stricken with illness. His humane treatment is progressive and therefore worthy to be commented on. Shelter was needed as much from sun as from cold and it would be natural for the hands to take their siesta in this place. There can be no doubt that the custom was allowed to be practised especially during the hot months when no one can work in the heat of the day in Italy.

2. Vilici ... nemo: there appears to be one entrance only to the farm proper.

culina videnda ... admota: the gerundive is used agreeing with a noun here as a variant construction to those having an impersonal neuter gerundive followed by a clause, of which there have been several examples in xii: *Danda opera (ut)*, *curandum ne*, *advertendum*, *vitandum*, and in xiii. 1 *facienda stabula* and *providendum*. The example here represents an inversion not as simple as might seem: *culina* belongs to the *ut* clause which might read *ut culina sit admota*, but *culina* is placed outside it and before the gerundive which might have been, by analogy with the examples mentioned above, *videndum (ut culina . . .)*: the gerundive has been attracted to the gender of *culina*. This is another instance of an awkward turn of construction which is in effect perfectly clear in meaning. Varro's uses of gerunds and gerundives will not be commented on in future. The kitchen must be near the overseer's lodge; *admota*: " near by ". Adequate supervision of the kitchen premises is necessary and *admota* suggests that they should be so placed that they could be viewed from the lodge and were near enough for active inspection. They must be placed too with an eye to general convenience, easy to get to and in a good central situation for those who eat there. These seem to be the slaves' quarters not those of the owners. Precautions must be taken against the pilfering of food and materials.

quod ibi ... capitur: the preparation of food for the day which had to be begun before daylight is described in the *Moretum*, one of the poems in the *Appendix Vergiliana* ascribed to Vergil. The peasant's task begins with the rekindling of the fire which has smouldered during the night; he has no light until he finds the hearth and blows up the flame. He then brings out from a store the corn needed for the day, and grinds it on a hand mill. The coarse meal is passed through a sieve mixed with warm water and shaped into loaves which are then put in a brick oven to bake. Cakes are made too of cheese and herbs mixed with oil into dough. All this labour has to be done before the man goes out to plough. The passage illustrates Varro's statement that much has to be done before dawn. The day would always start early, perhaps about 5 a.m.

capitur: it is not likely that a great deal of food would be eaten before the day's work began, but a frugal breakfast would be needed. In Italy today a man does a day's work on one thick roll with perhaps a piece of ham and anchovy for flavour, together with coffee. The main meal is taken in the evening after the day's work is finished.

Faciundum ... inimicum: the thrifty farmer must take care of his equipment and keep it all under cover: it is not enough for it to be kept in the farmyard, but protection from rain is essential, nor is it enough to protect it from thieves only by a fence.

satis magna: the sheds provided must be adequate to house *all* the farm machinery and tools: this would include carts, waggons, ploughs, and tools both large and small, and all other small equipment.

Haec ... resistunt: " If these are kept inside a fenced enclosure inside the walls but in the open, they do not fear thieves, but they have no protection against harmful weather." The argument that if fenced round in the open they will be safe because they cannot be stolen, is not sound because they are still at the mercy of the weather.

non metuunt, non resistunt: " but " is required in English. The gist is that admittedly they will not be *stolen*, but they are likely to be damaged by weather, which could include, in addition to rain, excessive heat which could warp the wood of carts, ploughs, etc., and perhaps be injurious to wheels especially those which were shod with metal rims.

3. Cohortes ... semipiscina: this is a difficult passage and the reading *ut interdius* is probably corrupt. The general implication is, however, that there should be on a large estate two farmyards with a large pond, the first for watering the beasts and fowls, open and fed by a running stream (see above xi, 2. *quae influat perennis*).

una ut ... semipiscina: " one so that it has an outdoor pond, where there is running water, which when, (*i.e.* if, the farmer) wishes, can pass between columns to form a kind of fish-pond". The pond for the animals must always have fresh water and be open to the air, but part of the stream could be diverted, if the farmer wished, to make a kind of fish-pond surrounded by columns. A contrast is perhaps implied between *interdius* (in the open) and the columns which supported a roof; the cattle and farm animals would naturally keep to the open: *qui* refers to *lacum*. *cum velit:* a general term " if the farmer wishes ".

perfunduntur: the pond must be of large proportions if the cattle can bathe in the water. This is another glimpse of Varro's humanitarianism—he has a care for the cattle and other animals even to their comfort as well as general well-being.

qui: refers to *lacum*.

e pabulo cum redierunt: geese as well as pigs were taken out to graze near the farm.

In cohorte exteriore: the second farmyard where there is a pond for the soaking of fodder. This has no architectural adornment as has the first described above.

lupinum: the lupine, a leguminous plant, probably the *lupinus albus*, L., a tree lupine, was grown as a crop plant by the Romans and used both for manure and as a cattle food. Columella in the *De Agricultura* recommends that a yoke of oxen be given a *modius* of mast and chaff mixed with one of soaked lupines: and in December dried leaves or chaff with half a *modius* of lupines soaked in water. (XI. 2. 101.) (They had to be soaked in vats to get rid of their bitter taste.) He also states that to be effective manure lupines must be ploughed into the ground when in flower between mid-May and late June, but other passages recommend that they can be scattered on the ground about mid-September and ploughed in and he calls them a " ready aid " if other manure is not available. There were many advantages in growing lupines as a forage and manure plant: they could be sown in poor or even exhausted soil, required least possible labour, and they were beneficial to the land. They could be used as fertilisers even for vineyards. Storage in granaries was possible without detriment to their goodness. When boiled they were good winter fodder. They were best sown direct from the threshing floor and did not need to mature in the bin. The sowing time was in September and early October. It was best ploughed in where fodder had been cut and dried, but in case of need could be scattered on lean ground in mid-September and ploughed in. There is no doubt that the lupine was one of the most valuable crop plants known to the Romans: (see also below, I. xxiii. 1).

4. ex ea evehatur: cattle manure mixed with straw is the best of all for the ground.

secundum ... divisum: the Romans fully understood the uses of animal

manure. Any modern farmer might be speaking these words and the following.
There would be two manure pits which would be drawn from alternately.

Alteram ... agrum: the fresh manure would be put into a pit or part of it
if divided, the old (rotted) manure taken from the other on to the land.

quod ... quam ... melius: *quod enim melius (est) quod confracuit quam recens.*
The old manure is better for the land.

periti ... sucus: "experienced farmers, when they can, cause water to flow
into that place for this reason (in this way the goodness is best preserved)".
The manure pits must be kept moist.   This is especially true in a hot country;
during the summer months in Italy all such matter would be completely
parched if not kept covered.

5. secundum: " close beside " so that the grain could be tossed out on to
the threshing floor, and if it became necessary quickly tossed back again if
a rain storm threatened.

ab area: " on the side facing the threshing floor ".   This use of *ab* is to be
compared with the regular expressions, *a sinistra*, " on the left ", *a dextra*, " on
the right ".

nubilare: " to become cloudy ".   Varro indicates the derivation of *nubilarium*
from *nubes*; the shed is a protection against " cloudy weather " and hence against
rain and especially the rain-storms bringing torrential downpours which
are a feature of the Italian climate even in the summer months.   A torrent of
rain would soon ruin a harvested crop, so that on the appearance of threatening
clouds it would be necessary to get it quickly under cover.   A precaution such
as this is characteristic of the careful farmer.

Fenestras ... possit: windows would be necessary on two sides facing each
other so as to produce a through draught.   The grain would in this way be
kept cool and aired; *perflari:* an impersonal passive in the normal usage with
an intransitive verb.

possit: a final subjunctive.

xiii. 6-7.   *The Aim in Building a Farm should be Profit not Luxury*

6. anticorum: " of the men of former times ": a more familiar spelling is
antiquorum.

derigas: the subjunctive is due to the generalising statement in the second
person singular.

Illi ... hi: "the former (*anticorum*), *hi* (the latter) answers to *horum* which
means " of these men ", " the men of these days ".

libidines indomitas: " unrestrained love of affluence ".

Itaque ... contra: "the result was that their farm-buildings were more
prized than those in which they lived, but now they are prized usually in the
opposite way".   In former days more attention was made to the buildings of
good farmsheds and barns than to the dwelling-house.   The old-fashioned
farmer made no pretension to high living, but his resources were centred in
the agriculture of the farm and its needs.

maioris preti: genitive of value.

urbanae: the building in which the owner lived on the estate.

Illic: " in that time ", " in those days ".

Illic ... Italia: the farm buildings considered as satisfactory on an estate
in the old days are listed, showing the sound rustic economy of the ancients

without any of the added luxuries which had already become fashionable in Varro's time.

in lacum: the floor in a well laid wine cellar should be so constructed that if any jars burst the wine which ran out could be gathered in runnels. Here is yet another example of economy in the keeping of farm produce and care against loss. It does not seem that the floor itself would be on the slant but that the surface was so constructed that escaping wine could be collected.

cultura: the emphasis falls on this word. The various buildings in former days were constructed to meet the needs of the farm and its produce and not the desire for luxury and extravagance.

quae ... quaereret: the antecedent to *quae* is *cetera*. The subjunctive is one in a subordinate clause dependent on another (in this instance a " final " clause) which contains a subjunctive.

7. habeant: *ut* would normally be expected before *quam* but the latter does duty as a final conjunction.

pessimo publico: neuter of the adjective *pessimus* used as a noun: "to the great detriment to public (good) "; the ablative of price is used in an abstract meaning. The villas of wealthy subjects such as Metellus and Lucullus were harmful because they took up valuable land without being productive, and money was spent on luxury and not on true farming. The agriculturist's aim should be to produce food for mankind: for Lucullus see note on I. ii. 10.

quo: " to this end ". The relative is demonstrative in meaning.

hi: " the men of these days ".

ad frigus orientis: " facing the cool of the rising sun ".

ut antiqui: " as the men of old days "; supply *laborabant* from *laborant*.

haberet: subjunctive in an indirect question depending on *(laboraverunt)* understood from *laborant*.

in ea ... calidiorem: wine must be stored in the cool, oil in the warm.

Item ... villa: the house should be built on a hill for many reasons such as health, drainage, safety, and supervision.

xvi. 1-3. *The Neighbourhood of the Farm*

1. altera pars: It is difficult to decide to which part of the discussion this " second " part relates. It now concerns however what is outside the farm.

appendices: the conditions in the neighbourhood. *appendix* is used to mean " that which is added " and here indicates the conditions " added to " the farm, those outside, but affecting, the whole estate because of their nearness to its boundaries.

Eius: refers to *pars*.

infesta: if there are brigands in the neighbourhood, or if the locality is such that it could become a prey to them: (see 2 and note below).

quo: this is the adverb corresponding to *quis* after *si*, and comes to mean " anywhere (where) ".

si ... adportare: " whether there is anywhere where it is profitable neither to export our products, nor to import from there the things of which there is need". The consideration is whether conditions make marketing possible.

sunt: the subjunctive would be expected because the verb is in a clause subordinate to one already containing a subjunctive, i.e. *expediat* repeated from the previous clause.

tertium: a neuter noun. The third consideration to do with means of transport. Literally the word should be referred back to *species* and the feminine would be expected, but Varro has in accord with his use of variety or generalisations, reverted to the neuter and used the numeral adjective as a noun.

qua portetur: "by which way transport is possible", "which make transport possible". *qua:* the antecedents are *viae* and *fluvii* but instead of a relative agreeing grammatically with these nouns the relative adverb of direction is used.

portetur: impersonal passive is normally used with an intransitive verb such as *itur:* "they go". The use here is abnormal but the meaning clear: the act of transporting is thought of and not the people or the objects transported. The subjunctive could be explained as either generic or final.

ut ... prosit aut noceat: this is an instance of *ut* and the subjunctive expressing result. The local conditions might be such as to result in benefiting or injuring the land.

2. quis: the alternative form of the ablative plural of the relative *qui*.

refert ... necne: "it is important whether the neighbourhood is unsafe or not". *refert* is here used impersonally with an indirect question clause according to the normal grammatical usage, but a conjunction would have been expected, in this passage *utrum* because of the double question: (see 6 below, *refert*). If land is easily accessible to brigands through its geographical position it is a loss to attempt to work it. For instance land near a coastline with a flat shelving beach where boats could easily put in, such as the coast of Latinum and Etruria, must have been a prey to pirates: a mountainous region overlooking a plain could be equally insecure, if there were convenient hideouts and easy means of descent. Campania for instance through her history was always at the mercy of the mountain tribes, especially the Samnites. The land, however, was too fertile to be left untilled: some districts would be safer than others, open plains for instance at a distance from the mountains and the coast. Varro's own estate at Casinum was plundered at the command of Antony during the second triumvirate. (See Varro's Life.)

quae ... fructuosa: supply words in this way (*ea*) *quae vicinitatis invectos habent idoneos ubi vendant* (*ea*) *quae ibi nascuntur et illinc invectos opportunos* (*habent*) (*eorum*) *quae in fundo opus sunt, propter ea fructuosa* (*sunt*). Opportunities in the neighbourhood for buying and selling make the farm profitable. *quae* is shown to be in the neuter plural by the agreement of *fructuosa:* the masculine would have been expected agreeing with *fundus* but Varro often uses the neuter when writing in general terms.

Multi ... quibus: "Many have among their holdings some to which ..."
aliudve = *aliud*, + *-ve*.

desit: generic subjunctive, implying "may be lacking".

3. sub urbe: this statement is not limited to Rome but applies to any large town or city. Paestum, for instance, in Southern Italy was famous for its roses which flowered twice in the year (*Georgic* IV, 119), and Capua for a wild rose from which perfume was made: (see below I. xxxv, and note).

urps: *urbs*.

ubi ... venale: "where there is not a place (market) to which produce for sale can be conveyed ". *venale:* is in the neuter singular of the adjective *venalis,* connected with *venum,* " sale ", used as a noun.

divitum copiosi agri: " productive holdings belonging to rich owners ".

non care: " not dearly ", " reasonably ", " at an economic price ". *non* negatives the one word (*care*) and is not extended to the clause.

quae opus sunt: *opus* meaning " there is need of " is used as regularly with the nominative of the thing needed (here *quae*) as with the ablative. The subjunctive would have been expected in this clause since it is subordinate to another (*unde . . . possis*) already containing a subjunctive.

quibus ... possint: the antecedent to *quibus* is composed of the things listed in the *si* clause: " to whom the things which are left over can be sold ", " to whom your surplus products can be sold "; *venire* is the passive infinitive of *vendere:* (sell); it is compounded from *venum* and *ire* and is equal to " go for sale".

ut ... harundo: supply *venire possint* as for instance to certain (farmers) props or poles or rushes (could be sold). Varro is thinking of materials for vine-props: (see I. viii. 3).

Item ... possis: this long conditional sentence reads in a straightforward manner. The initial *si* clause (*protasis*) extends to *harundo,* and contains a relative clause, *unde . . . possis* upon which another *quae . . . fundum* is dependent. Then follows a second relative clause *quibus . . . possint* and a comparative clause introduced by *ut* and containing (*venire possint*) to be supplied from the previous clause. The main clause (*apodosis*) is itself followed by two comparative clauses introduced by *quam.* The plain fact can be abstracted: *si ea oppida aut vici in vicinia (sunt) . . . fructuosior fit fundus.*

colendo in tuo: " by producing (them) on your own estate (*fundus*) ".

xvi. 4-6. *Hired Labour: Profits Increased by Available Transport: The Neighbour's Boundary*

4. potius: this word is separated by *tmesis* (lit. " a cutting ") from *quam* two lines below.

Itaque ... habeant: " therefore, for this reason, farmers prefer to have 'yearly labourers near' whom they can employ than to keep men of their own on the estate". This means to say that farmers find it more profitable to make yearly contracts of employment with men who live in the neighbourhood rather than to keep their own slaves on the estate to do specialised work. The reason is given in the rest of the sentence: the death of one of these maintained by the owner is a financial loss to the estate. He has had to purchase him: he has had to feed, clothe, and house him. In a moment all this outlay is gone without any hope of repayment, and has to be replaced. It is more economical therefore to hire a worker on a yearly basis. Varro takes one aspect of the disadvantages of slave-labour and shows how the trend towards employing of freemen was already beginning.

quibus imperent: a relative clause expressing purpose; " on whom they can call ", " whom they can employ ".

medicos: these were often educated Greeks, although of the slave class.

fullones: " fullers ", " those who whitened or cleaned cloth ". Their work included the finishing of newly woven material as well as the cleaning

of worn clothing.   Fullers' shops have been uncovered at Ostia and Pompeii
containing troughs where the cloth was soaked and trodden.   The white
powder similar to that now called "fuller's earth" was used in the process.
To have such a shop on the farm would constitute an avoidable expense if
there were one kept by a professional *fullo* in the neighbourhood because
investment in both equipment and apparatus as well as in workers would be
saved.   The price of a doctor or skilled artisan also would be heavy, in excess
of that to be paid for a mere labourer whose work involved such unskilled tasks
as digging, hoeing, carting, etc.   There might not be enough work to keep
these professionals busy all through the year, and hence more waste would
follow.

fabros: "smiths".   Usually workers in metal, but also in wood; on the
farm they would make metal tools, shoe wheels, make carts, etc.

domesticae copiae: the *vilicus* and other responsible workers under him such
as the *promus*: [see below (5) and note].   Yearly contracts were entrusted to
him, and with him would rest the choice of making contracts or of buying the
necessary slave workers.

quos habeant in villa: "to keep on the estate": the subjunctive is final.

artifices: supply *parant* from the previous line.

familia: the farm hands who are slaves.

ne ... reddat: if skilled workmen are maintained as slaves on the estate, one
advantage follows: being in the position of slaves (*familia*) they can be kept
at work even on festal days.   Hired labourers on the other hand would take a
holiday.   At times of festivals the custom is not to give slaves days off from
work but to relax some of the heavier work.   This is another of Varro's suggested economies although he seems to think that in the main hired workers
are more profitable.

5. Sasernae liber: Varro has already referred to Saserna's writings—somewhat disparagingly, but now he quotes him as giving good advice: see note
on I. ii. 21 (text omitted).

nequis ... animadvertatur: the passage is not subordinated grammatically
to *praecipit*, but is a quotation from Saserna.   The wording is in the form of
direct commands and was probably meant to be part of a book of rules for the
vilicus which he was expected to have always at hand for reference.   The
present subjunctive (*exeat*) is one of command in the third person: so too *legat*,
abeat, animadvertatur.

promus: one who gives out, such as a "storekeeper", a "cellarer"; the word is
connected with *promere*, "to bring out", "to produce".   The *promus* on a
farm came next in importance to the *vilicus*.   His responsibilities might
require him to be able to leave the farm at times to replenish stores, to make
purchases, and to arrange marketing.

legat: generic subjunctive: "such as he may choose".

ne impune abeat: "he must not go unpunished".

ut ... animadvertatur: an unusual impersonal passive in that it is part of a
transitive verb.   The responsibilities resting on the *vilicus* and his control
over the lesser slaves is made clear.   It was only a person superior in education,
intelligence, and ability to use his own authority who could effectively use
such power.   The kind of person most suited for the position of overseer is
described in I. xvii. 4 in a quotation from Cassius' translation of Mago's manual
of agriculture.

ne quis ... fundo: this passage is in direct speech and is meant to be an improvement in instructions on that quoted directly from Saserna. In Varro's opinion the rule should be safeguarded. Saserna gives the *vilicus* too much freedom, both to go away and to stay away; he too should be under the master's control and he should not be able to absent himself for more than a day and not for any other reason than farm business. From these rules the conclusion may be drawn that the mass of farm labourers were confined always to the estate. Only those on whom responsibility rested could go outside its enclosure and even then not for any great distance. If the servant who was allowed to be absent had to return the same day then the distance cannot have been more than about fifteen to twenty miles and his errands must have been limited to business in some neighbouring village or town. The ordinary slave labourer was confined for life to the farm in which he worked and was never able to leave it or go elsewhere unless his master sold him to another.

nequis ... exierit: the perfect subjunctive may normally be used to express a direct command in the third person, although the present is more usual, and is translated in the present: " no one is to leave (the farm) ".

neque ... longius: supply *exierit*.

6. eundem fundum: " the farm, too ...". *idem* is used idiomatically to make an added statement.

qua: the relative adverb is used in loose reference to *viae* and in the next line to *flumina*.

navigari evehi, invehi: impersonal passives in the infinitive: the last two, in accordance with Varro's avoidance of strict grammatical usage, are from transitive verbs. This use will not be commented on again.

Refert ... habeat: the normal impersonal use of *refert* followed by an indirect question with the subjunctive is illustrated here. " It is of importance to the profits of the farm in what way ...".

possis: potential subjunctive in the generalising second person singular: " you could ...".

ut ... ferant: consecutive clause expressing the consequence of the presence of an oak grove: so also the following clauses (*ut*) ...*fugiant*, " *ut ... reclinent* " but the last *ut* with the indicative *solet* means " as ".

quercus iugulandes: these trees would impoverish the soil with their spreading roots.

xvii. 1-3. *Farm Equipment: Slaves*

1. De fundi ... quattuor: after discussing the natural endowment of an estate and its enclosures the dialogue now goes on to discuss the equipment needed for its working.

quae cum solo haerent: " which cling with the soil ", *i.e.* " which are concerned with the soil, or attached to the soil ". The use of *cum* with the verb *haerere* is unusual.

agri: comes in meaning inside the relative clause after *rebus*.

Quas res ... plaustra: in the first classification there are two categories, men and the aids to men: in the second men are ranked with the beasts and regarded as little more than beasts of burden who have the power of speech. In the first men seem to have a more dignified position. Perhaps it is that the two points of view represent the two different conditions of agriculture: in the first comes the smallholder working his own land himself with his family among the labourers (see 2 below). The second reflects the large landowner's attitude to those who work his land but with whom he himself does not work.

adminicula: " aids to men ". *adminiculum* derived from *ad* and *manus* means properly " that on which the hand may rest " and then in general " a prop ", " a support ". Its origin is to be found in the language of vine-dressers where it is used for a support to which the vine clings. Varro uses aptly a term from farming.

instrumenti genus: " the type of equipment ".

semivocale: describes the working beasts: they have a voice but cannot talk.

mutum ... plaustra: all the movable equipment of the farm is meant to be included. The phrase is not limited to vehicles, but covers machines and tools as well, in fact any object which can be moved or is portable.

2. ut ... progenie: it is not to be doubted that alongside the great estates such as Varro and his contemporaries possessed, the old Italian system of family smallholdings continued. *pauperculi* describes small farmers who made their living from their parcel of land on which all the family worked.

aut mercennariis ... administrant: " or with hired labourers when they conduct the heavier operations such as the vintage or the hay harvest with the hired work of freemen ". The smallholder would be obliged to find extra labour at these seasons just as modern farmers often employ travelling gypsies. The large landowner also found hired labour desirable and profitable. Farm operations such as the vintage and hay harvest cannot be protracted: both grapes and hay must be harvested at the right moment and with all speed, for otherwise they are spoilt.

obaerarios: the word *obaerarius* is derived from *ob* and *aes* and is used to describe a man who pays off a debt by working for a creditor: it is equivalent to *nexus* which has the same meaning, he who is " bound ", as it were, to the man to whom his debt is due. In early Roman law if a debtor could not repay his loan he could be seized, and compelled to work for his creditors until repayment had been made in this way. He did not in actual fact become a slave, but could be kept in prison and could even be hired out by his creditor: in this case he became a *mercennarius*, or hired man. This harsh system called *nexum* because it involved the " binding " of a man to the creditor, was abolished by the *lex Poetilia* in 326 B.C. (or according to some authorities in 313 B.C.). Thus as early as the fourth century the severity of the law of debt was mitigated. According to Varro, however, the system was still observed in other countries in his time.

3. gravia loca: " unhealthy places ". This seems to be the meaning of *gravia* as it contrasts with *salubribus* (" healthy ") in the next line. It probably refers to land which is malarial of which there was much in Italy and has been until very recent times: (see I. xii. 2 and note). A cogent reason for using hired labour would be the risk of illness which might disable and even cause death. The loss of slaves which were part of the assets of the farm would then be avoided (see xvi. 4 above where the point is made that if a slave dies, there is total loss to the farm). Furthermore, slaves when sick and incapable of work still had to be fed and housed, but the hired labourer could be sent away. Another factor in the trend away from slave labour to free is perhaps illustrated here. There is also the possibility that hired men were better fed, more resistant to disease, and ready to work because they received pay.

in salubribus ... messis: even on good land hired labour for the greater farm operations is more economical. The reason is that skilled labourers could be obtained for these seasonal operations for which extra hands were needed for a limited time and that these would have the incentive of completing the work so as to go on to another farm. The man who is free and paid for his work is a better worker than the slave who is without incentive and kept at a low level of existence. Poor workers might spoil the yield of grapes or hay for example, and in addition not gather in the full harvest. Treading of the grapes for the casking of the wine is implicit in *vendemiae*.

opera rustica maiora: supply *colere* from the line above.

Cassius scribit haec: this quotation from Cassius Dionysius appears to be taken word for word from his translation but is put into indirect speech. For this translation of Mago's work see the caption i, 8-11 above and notes (text omitted). Although Mago's treatise must have been written more than a hundred years earlier than Varro's it is noteworthy that his advice still held good.

qui ... possint: " such as can stand the work ". Only those who are really equal to the heavy labour of vintage and harvest should be employed. These are more likely to be hired men than slaves.

ad ... dociles: the hired workers must be adult, skilled, and teachable.

ne: probably *sint* is to be understood: " they are not to be ...".

Eam ... factitarint: " Such judgment is possible from orders (*i.e.* the way in which they carry out orders) in other matters, and by asking one of the new hands what they used to do for their previous master." This passage seems to suggest that gangs of hired labourers journeyed about seeking work at these seasons. Caution must be observed when the hired men are engaged both in regards to their physique and their character and experience. *factitarint:* is the perfect subjunctive shortened from *factitaverint* in an indirect question in primary sequence.

xvii. 4-7. *Farm Equipment: Slaves, their Treatment and Supervision*

Mancipia ... : the best kind of hired hands has just been described: Scrofa then returns to slaves, using a word which points the contrast: " those who have been bought " or " are owned ". Slaves were completely the chattels of their masters.

4. Qui ... dixi: " (those) who are in authority should be (men) who have been trained in letters (*i.e.* can read and write) and in education of some kind (who are) reliable (and) older than the labourers whom I have mentioned ". The slaves who are able to take responsibility and to be put in charge of others must be better persons than they are and must be educated. They would have to be able to read a number of things to do with the farm, such as rules, calendars, stock books, and medical books for the farm animals and doubtless many other written documents of this kind. Likewise the keeping of farm records and accounts of all kinds would be among their duties. Often they acted as business managers and so needed to be intelligent and well educated.

frugi: the dative form of *frux*, " fruits of the earth ", used predicatively as an indeclinable adjective, the root meaning of which is " fit for food, " hence " useful ", " fit ", " honest ".

Facilius ... audientes: " They obey them more readily than others who are younger." *audientes* agrees with *operarii* to be understood from the previous sentence.

dicto: from *dictum*. The case is dative governed by *audientes: iis* is also in the dative, ethic or possessive going closely with *dicto*. *dicto audire* is a regular phrase for " obey " and can carry a second personal dative (in this context, *iis*).

facere: the *vilicus* and others who hold positions of authority must take an active part in the farm operations and not stand by while their subordinates work. They must take a lead in showing what has to be done and how it is to be done.

ut facientem ... praestet: " so that a man can imitate him (the *vilicus* or one like him) as he works so that he can understand that he (the *vilicus*) has been put over him with (good) reason because he is superior in knowledge ". There is a sudden change to the singular in this sentence from the plural of the preceding (*iis* and *audientes*) where overseers and labourers are mentioned. Such variety is characteristic of Varro's style: the subject of *imitetur, animadvertat* is any supposed farm hand; *eum* is the *vilicus*.

5. Neque ... possis: it is made clear in this statement that the treatment of the slave-labourers was controlled by the owner of the estate. Varro prescribes a kindly attitude, an outlook which reflects an increasing humanitarianism (from whatever motives) of his time towards the slave population.

verberibus ... verbis: a play on words after Varro's manner. There are no two corresponding words in English, but an acceptable paraphrase might be " that they lash them not with whips but with the tongue ".

si ... possis: the sudden change to the generalising second person presents no real difficulty in meaning: " provided you can achieve the same result " (with words, that is, rather than whips).

eiusdem ... nationis: the majority of slaves who worked on the great landed estates were prisoners of war. As a result of the wars against Magna Graecia, Macedonia, Greece, and Carthage, thousands of captives swelled the slave markets to become a source of cheap labour. For example, after the fall of Tarentum in Southern Italy, in 209 B.C., 30,000 inhabitants of the city were sold into slavery as well as many others from Magna Graecia; during the wars with Philip of Macedon, the battle of Cynoscephalae in 197 B.C. added thousands of Macedonian captives and after Magnesia in 189 many Syrians were enslaved. In 167 in the final subjugation of Macedonia 150,000 persons went into servitude. These events took place more than a hundred years before Varro was writing his treatise but the acquirement of an even larger slave population had its origins in the conquests of Carthage and Greece. Towards the end of the first century and contemporary with Varro, Caesar's campaigns in Gaul sent thousands of strong northerners to the markets and Pompey's settlement of the near east caused large numbers of Asiatics to come into the possession of Roman masters. When Varro gave the warning that not too many of one nationality should be employed together he was perhaps thinking of the Servile wars which had taken place, before his time, and of other attempts at rebellion by disaffected slaves. More seriously in Sicily in 135 B.C., some 70,000 Syrians and others banded together to defy their masters and the government and again in 104 in the Second Servile War which lasted nearly four years, another threat came from hordes of Asiatic slaves. More recently in 73 B.C. in Varro's living memory the gladiator Spartacus at the head of an army of Gauls and Thracians made a stand on Mount Vesuvius and became a menace to the state. These and other smaller risings were in the end suppressed, but not without having demonstrated the dangerous potentialities of the servile races on Italian soil.   It is true that he speaks of domestic quarrels apparently on a small scale, but the wider danger lurked behind.   Those who spoke the same language could band together in cliques or larger companies and become a source of trouble to the master.

peculium: " some property of their own".   The root of *peculium* is *pecus*, " cattle ".   Slaves were allowed to possess beasts and to graze them on the owner's land (see I. ii. 17 and xvii. 7).   In this way they might accumulate some earnings and perhaps eventually buy their freedom.   The *peculium* was however in reality the property of the master since slaves had no legal right to any kind of ownership.   If a slave was sold and his *peculium* went with him, it could become the property of his new master.

coniunctas conservas: slaves had no marriage rights and therefore no legal marriage, but only cohabitation.   Varro seems to suggest however that this state might be a lasting one in some cases, although it was not protected legally. The children of such unions could only have the status of slaves, and eventually be added to the number of labourers on the farm.   Slaves could therefore have the semblance of a settled home and family relationships but without any legal guarantee of protection.   They would, however, as the next sentence shows, become more closely bound to the estate and be more settled if they were allowed to have wives under these conditions.

Epiroticae familiae: " the slave families of Epirus ".   Epirus was a district of northern Greece extending along the Ionian Sea (now the Adriatic) and bounded by Illyria, Macedonia, and Thessaly.   It is now the southern part of Albania.   The region is mountainous, cut by deep well-watered valleys, with plains on the coast.   It was devastated by the Romans in 168 B.C. and in 146 B.C. was incorporated in the Roman province of Macedonia.   There were two occasions during Varro's career when he could have observed conditions of agriculture under the Roman settlers who had received land there: (see Varro's Life).   It is a possibility that such a district divided by many mountain ranges with consequent difficulties of communication tended to make men more settled on the land.   Atticus, the correspondent and friend of Cicero when living in Athens from 88-65 B.C. bought estates in Epirus with his patrimony.   There he had large herds of cattle from which he acquired wealth and became one of the great landed proprietors.   Varro as a friend of Cicero was probably well acquainted with him and may be making a direct reference to his slave personnel.

6. honore ... habendo: " by showing (them) some consideration ". The men who are put in charge will respond to treatment which singles them out and makes them feel that they are thought to be of some importance.

minus ... domino: " they think less that they are looked down on and that they are held in some regard by the master ".   The slave-labourers who are superior will gain some self-respect from being picked out from the rest. *minus* goes closely in meaning with *putant:* " they are less likely to think ".

ita cum fit: " when this is done " or " whenever . . .".

7. cibariis aut vestitu largiore: Cato in the *De Agricultura* LVI-LIX states the amount of food and clothing to be given to the labourers: the field hands were to have four *modii* (pecks) or a bushel of wheat per month in winter and four and a half in summer.   The overseer, housekeeper, foreman, and shepherd three, presumably because they did less manual work.   Seven quadrantals of wine was allowed per person per year though more could be allowed: a pint of oil per person per month was also allowed and a peck of salt per year. For clothing the ordinary labourer was allowed a tunic three and a half feet long and a blanket every other year and a pair of wooden shoes. We can judge from this that the ordinary labourer's physical existence was kept at a low level, although better treatment might have been observed by Varro's time.

concessioneve: = *concessione* + *ve*.

peculiare aliquid: " something of their own ": slaves could have permission from their masters to graze cattle of their own: (see above, I. ii. 17; xvii, 5, where the *peculium* granted to foremen is mentioned).

ut ... in dominum: " so that for those to whom some rather heavy task has been imposed or some punishment inflicted in some way (*qui*) by kindly treatment in each case, it may restore their goodwill and loyalty towards the master ". *sit imperatum, animadversum* are impersonal passives. It is difficult to decide, what exactly is the subject of *restituat:* it could perhaps be explained as the rest of the sentence summed up as " it ". The construction does not hold together grammatically since the ablative of the gerunds *tractando* and *consolando* do not refer to a personal subject, who is the doer of the action as is usually expected. *qui* is the old ablative form from *qui, quae, quod* used with an indefinite meaning and going closely with *imperatum* and *animadversum*.

xix. 1-3. *Farm Equipment: the Animals*

1. De ... instrumenti: Varro (in Scrofa's words) has now reached the second division of the farm equipment as set out in xvii. 1 above, that which he styles the inarticulate. This contains all the farm animals, and is called *semivocale* because they make sounds but cannot talk.

boves trinos: " three yoke of oxen ", the distributive numeral shows that the oxen were grouped in three pairs; probably the same two always worked together, Varro advises below that when they are bought they should be equally matched so that a weaker is not worn out by a stronger (1. xx. 1).

Sed ... aliquem: Varro's considered opinion is that no fixed number of oxen can be prescribed for a given piece of land, because its nature must be taken into consideration.

Alia ... difficilior: some land is easier or more difficult (to work than another). *alia* in the ablative of comparison may be supplied to complete the meaning.

2. aliam terram: " some land...". This phrase corresponds to *Alia ... difficilior* above.

nisi magnis viribus: supply *sint*. The ablative is descriptive. Some land can only be broken up for the first time by especially powerful oxen.

saepe ... arvo: certain soil may be so heavy that at the first ploughing the oxen strain to the extent of breaking the plough beam. In a simple plough this beam (*bura* or *buris*) was a curved wooden pole some eight feet in length which connected the yoke *iugum* between the two oxen with the *dentale*, the sole of the plough, so called because it " bit " into the soil; it was protected from the constant friction and wear of the soil by an iron sheath which was the ploughshare (*vomer*). In a more developed form of plough the connecting pole was in two parts, the *bura* proper, the plough beam, which was fixed to the *dentale*, and the *temo*, the yoke beam which was jointed to it and fixed to the yoke. This pole whether in one or two parts was the most vulnerable part of the plough and being curved might snap off and become detached from the yoke and the *dentale*, and leave the share fixed in the ground.   Varro is probably describing here the " breaking " plough which had a flat symmetrical share and turned up the soil equally from both edges.   A broken plough would be an appreciable loss because the curved wooden beam had to be especially shaped. It is interesting in this connection to recall Vergil's description of an elm branch being trained while still part of the growing tree into the form of a curved plough (*Georgic I.* 169-70).

> *continuo in silvis magna vi flexa domatur*
> *in burim et curvi formam accipit, ulmus aratri.*

" Many a time in the woods bent with a mighty strain, an elm is tamed into the shape of a *buris* and accepts the form of a curved plough."

Quo ... quadam: " therefore we must follow on individual farms, while we are new to the work, by means of a threefold guide, the former owner's practice, that of the neighbouring owners, and some experimenting ".   This amounts to saying that a threefold rule must be the farmer's guide, but the turn of expression in the ablative is awkward.   It illustrates, however, a practical application of the principles on which successful farming is based, imitation and experiment.

3. Quod addit: the subject is Cato (*De Agri Culture*, X. 1): these words are taken up in *in hoc genere* . . ., the connection cannot be analysed grammatically, but the meaning is in no doubt: " as to his added statement . . .".

asinum molarium: the donkey-mill which originated in the second century B.C. was in general use by Varro's time and continued until the end of the Empire; one of the best was made of a specially hard grey rough porous stone, from the district of Pompeii, and consisted of two parts, the lower stone called the *meta*, so called because it resembled in shape the conical columns which stood at the turning posts of each end of the Roman circus (see above I. ii. 1 and note).   This in turn was fixed on a wider foundation of rough masonry. The upper stone, the *catillus*, was in the shape of an hour-glass and was fitted over the *meta* and rested at its narrowest part on the top of the cone.   On opposite sides of the waist of the hour-glass were two square pieces which had not been cut away when the stone was shaped.   Each contained a square socket and at right angles to these a round hole.   Wooden beams fitted into the sockets and secured by pins in the smaller holes were part of a framework which extended over the top of the *catillus* and was attached to the spindle, *spina*, an iron bar fixed into the top of the *meta*.   The hour-glass stone was separated by an iron bridge which rested on the top of the spindle at the narrowest opening between the two stones.   A slave or animal, usually a donkey but sometimes a mule or a horse blindfolded, was harnessed to the beams and trained to pace or run round and round thereby rotating the stone while grain was poured in at the top.   The flour was gathered up from the ledge below the *catillus* on to which it poured down as the mill was worked. Many examples of this type of mill have been found, in Pompeii, Ostia, and elsewhere.   At Ostia there remains a miller's shop containing ten such mills, some of which were worked by animals and some by slaves.   It is likely that the mills on Varro's estate at Casinum were made in Pompeii.

de pecore: " concerning other farm animals ".

agri colendi causa: this phrase seems to have its literal meaning of tilling the land. Scrofa talks of animals which are useful for the land and enrich it for the growing of arable crops.

pecularia pauca: the few animals which the slaves are allowed to graze for their own use: (see I. xvii. 5 and 7).

numero ... stercus: this sentence which is awkwardly expressed, amounts to saying that both the farmers who have meadows and those who keep animals for other reasons prefer sheep to swine because their manure is more beneficial. *pratum* is used for any kind of tilled land, whether for hay, grain crops, or vegetables.

parum: this word is used here as a strong negative.

xx. 1-5. *Farm Equipment: Working Beasts and their Training*

1. prima ... emuntur: it is unusual to find an indirect question dependent on a noun as here *probatio:* also remarkable is the indicative in *emuntur,* coming in a clause dependent on another which contains a subjunctive, *sint.*

qui idonei sint boves: oxen were the working beasts in ancient agriculture and were alone able to do the ploughing and to act as draught animals for heavy farm carts and farm operations: this is still true of most of Italy today, though machines are taking the place of oxen in some parts of the country. One of the glories of the Roman Campagna, of the Maremma, and all the country districts around Rome are the stately grey, or sometimes white oxen, dark-eyed, of immense strength, with heavy dewlaps and huge horns shaped like the framework of a lyre. For all their massive strength they are slow moving, entirely submissive to man, and of quiet disposition. They are to be seen ploughing the rich red earth of the Campagna and often met within the villages pulling carts loaded with farm produce. Varro's description of the ideal working ox with large horns, broad face, deep chest, and heavy quarters could apply to the ox as it is seen in the countryside today. When we see the fine beasts which are reared on the plains near Rome we cannot be certain that we are looking at the ox such as Varro knew: (see II. i. 14 where black horns are preferred). They belong to the herd called *Maremmana* by the Romans which derives directly from the large grey race of oxen which was diffused originally from Asia and spread into all southern Europe. A modern description of the ideal working ox corresponds in most respects to their build: it should have a big frame the body in the form of a truncated cone with the larger part in front, loins straight and muscular, a wide saddle, long muscular shoulders, wide thick joints, thick compact hoofs and shins. Thus the whole frame is stocky, muscular, and massively built. Its hide is grey, often white, with black points on muzzle, tips of horns, end of tail, and hoofs. In disposition it is steady, vigorous, and amenable.

ut ... sint: the sentence is " jussive ", " they should be ...". There are other instances in this chapter —

pares: the oxen must be equally matched: they were probably bought in pairs and always worked together.

ne ... conficiat: " to prevent the stronger tiring out the weaker in work ".

lata fronte, lato pectore, crassis coxendicibus: these are all characteristics which make for strength.

2. Hos veteranos ... vitandum: " those who have become adult on level ground should not be bought for heavy and mountainous country, and if the opposite happens to be the case, it should be avoided ". The ox must be put to work in the environment to which it has always been accustomed. *nec non . . . vitandum* is an awkward phrase, but the two negatives cancel each other and can be translated by a positive " in fact ", " moreover ".

veteranos: not " veterans " in the English sense, but " adult ".

Novellos . . . promi: instructions are still given to the farmer, but in a different grammatical form. So far the admonitions have been given by means of gerundives, and clauses containing *ut* with the subjunctive, now a change is made to a suppositional person *quis* with the future perfect in a *cum* and a *si* clause.

si . . . cibum: a method of accustoming the young oxen to the yoke is described. What is called *furca* (a fork) is to be put around the neck for several consecutive days. This will help the beasts who are to pull the plough to get used to the feel of the yoke on the shoulders and across the nape of the neck where the weight will rest. If they have been hand-fed (with specially tempting food) they will soon become docile. It is not clear what is meant by *furca* but the *furca* which was used for the punishment of slaves could not have been used for this purpose. This was a fork-shaped piece of wood in the shape of the letter " A ". The neck was inserted at the fork and the arms tied to the two ends. An instrument of this shape would not effectively fit the neck of an ox as it does not resemble the shape of the yoke. In all probability Varro describes a collar such as Vergil speaks of in his passage on the training of oxen for the plough (*Georgic* III. 166-9)

> *ac primum laxos tenui de vimine circlos*
> *cervici subnecte, dehinc, ubi libera colla*
> *servitio adsuerint, ipsis e torquibus aptos*
> *iunge pares, et coge gradum conferre iuuencos.*

" and first bind loose collars of pliant osier on their neck then when their shoulders free (before) have become accustomed to obedience, join two oxen suitably (matched) by these same twists (of osier) and train them to pace along together". Hand feeding is also advocated by Vergil: (*ibid.* 174-76). A method practised by the herdsmen on the Roman Campagna is part of the same process. The young oxen are driven into the farm enclosure, lassooed by their horns and tethered with their heads down and as close to the railings as possible. After being left for about half-an-hour in this position they are freed. This is repeated for several days in succession until they are used to being tied in this position. As a result they become quite docile and eventually accept the yoke.

subigendum: " the breaking is to be done . . .".

tironem . . . adiungant: a method is described which is observed on the Roman Campagna today.

eo levi: " with a light (plough) ".

3. Quos ad vecturas: " those for hauling ", " draught animals ". *Quos* is the object of *instituendum*.

Neque . . . requies: " the ox which you have put on the right should not stay all the time in that position, because if he changes to the left in turn, relief is given him when he works on alternate sides ". *feceris:* " will have put ". The future perfect indicative is used idiomatically, but here a general statement

is made in the present sense (*fit*) which could equally well refer to the future.
laboranti: agrees with *bovi* left to be understood.

4. in Campania: the soil of Campania is of volcanic origin, light and powdery and very easy to work.

eo facilius ... convehuntur: only a light plough is needed for the Campanian soil and this is suitable for cows or donkeys. There is no need to go to the expense of buying heavy oxen. The argument is that although you do not need the heavier oxen which also serve as draught animals, these lighter animals can be useful in other ways as well, the donkey can drive the mill and both donkeys and cows can haul.

vacca ... haec: the cow is more profitable because it supplies milk, cheese, and meat, although the ass is more economical to feed.

xxi. *Farm Equipment: Dogs*

Canes: Varro treats of dogs at length in II. ix. 1-16.
quos consuefacias: the subjunctive is jussive.
De ... stabulet: advice is given to the farmer who has no cattle and yet has meadows which need manuring. The practice of allowing another farmer to fold his flock on stubble is widely observed today. By this method the soil is enriched with animal manure.

xxiii. 1-6. *The Third Topic: The Crops*

1. illa ... quadripertita: "those first two of the fourfold division". Agrasius means to say that by now they have discussed the first two sections which are (1) a knowledge of the conformation and soil of the farm, and (2) the equipment needed for working the farm: he is waiting to go on to the third which will be the various operations which have to be carried out on the farm. These comprise the planting of the various farm crops and the kind of soil in which they have to be planted.

de tertia parte expecto: "I am waiting (to hear) about the third division."
Scrofa: Scrofa takes up the argument again, after speaking during the major part of the dialogue with only short interruptions from the others. These help to make a pause and break the monotony when one person speaks for so long. Agrasius' words suggesting a certain impatience to start on the next theme bring a touch of life into the scene and remind us of Scrofa's audience.

arbitror ... rem: "since I think that the produce of the farm is that which once sown is useful for some purpose ...". Scrofa's definition of farm produce and hence of farm profits refers to home production. Everything must be grown on the estate and receive careful planting.

quae ... serere: two indirect questions are condensed: expressed in full they would read: *quae maxime expediat serere, et quo quidque loco maxime expediat serere.*

ad pabulum: the plants included in the list which follows are all forage plants. They are legumes with the exception of the mixed crops *ocinum* and *farrago* which include cereals as well. Similar lists are given below (see xxxi. 4-5 and xxxii. 2). The natural order *leguminosae*, to which they belong, is the second largest family of seed-bearing plants and contains trees, shrubs, and herbs of various habits. Varro rightly derives the name *legumen* from *legere*, "gather", because the crops were "gathered" or "pulled" and not reaped as were cereals [see below (2) and xxxii. 2]. Those legumes which could be grown as crops provided nutritious fodder which was of the utmost economic importance throughout the ancient world. The Roman farmer knew from observation and experience that they also enriched the soil and so often made use of them for green manure, but he had not the scientific knowledge to enable him to understand the nitrogen-fixing properties of their root system (see below 3 and note).

ocinum: the translation often given for this as "clover" is incorrect: the word with which it is confused is *ocimum* which is a species of clover. The meaning is uncertain and we look in vain in Varro for a description of the crop, although he gives a suggested derivation of the name from the Greek (see below xxxi. 4 and note). Pliny however gives a detailed account showing that it was a mixed fodder crop. (*N.H.* XVIII, xlii, 143). He states that in old times there was a kind of fodder called by Cato *ocinum* which was used medicinally for cattle: it was obtained from a crop of fodder cut green before it seeded. He goes on to give the directions for planting: the old practice was to mix ten pecks of beans, two of vetch, and two of ervilia (bitter vetch) for each acre and to sow the mixture in autumn: Greek oats might also be mixed in. Cato (*De Agri Cultura* XXVII and Liv. 4) mentions it as a fodder crop. It is not certain however, whether Pliny and Cato write of the same plants. However, it is clear that *ocinum* was valuable fodder, and that it was a quick-growing crop cut green.

farrago: "mixed fodder". The word is connected with *far* "grain". The mixture was one of both cereals and legumes, and contained grain and barley with vetch and beans, sown very thick and fed green. A method such as this marks an advance in cattle-feeding in that it imitated the mixed feed of the natural pasture. It was economical of labour and also enriched the soil (see also below, I. xxxi. 5, and Varro's etymological comment).

vicia: vetch is a common wild plant in England. The variety grown as a crop plant by the Romans was probably the same as that seen in fields and hedgerows today, with delicate leaflets set on long ribs and pairs of rose or red-violet flowers. It is especially valuable for enrichment of the soil and was combined with cereals, to produce a mixed fodder: see *ocinum* and *farrago* above. See xxxi. 5 below where Varro suggests an etymological derivation of the name from *vincire*, "bind".

medica: the legume known by the botanical name, *medicago sativa L*, is now also called alfalfa or lucerne. It was perhaps the most valuable fodder crop in the ancient world, and continues to be of high economic importance in modern times. It is an erect perennial with a branched hollow stem one to two feet high and trifoliate leaves and a short dense inflorescence of blue or purple flowers: the pods are downy and coiled in a loose spiral. The tap-roots sometimes penetrate as far down as fifteen feet into the soil. The long roots make it resistant to drought and protect it from damage when the field is weeded. It is cut when the flowers first appear. One sowing can last for ten years and yield up to six harvests in a year. Lean cattle fatten on it and like other plants of the order *leguminosae* it enriches the soil. *medica* was in origin a native of the eastern Mediterranean and was grown as a crop by the Medes and Persians. It became the principal fodder of cavalry and chariot horses; there is no doubt that the excellence of the Persian cavalry was the outcome of its nutritious qualities. The introduction of the crop into Greece is attributed to Darius at the beginning of the fifth century during the Persian wars; for this reason it was named after the invaders, *medica*, "the Median plant" (from *Medus*, a Mede), and quickly became established. It is mentioned by Aristophanes in the *Knights* 606, πόια Μηδική, for instance as a usual feed for horses. The play was produced in 424 B.C. It was only during the second century B.C. that the plant reached Italy: there may have been two avenues along which it travelled, either from North Africa as a result of the Carthaginian wars, or from Greece proper after the Roman conquest of 146 B.C. Cato, writing about a hundred years earlier than Varro, does not appear to have known it. It is familiar to Vergil (*Georgic* I, 215-16). The use of medica as a crop plant became widespread along the old caravan routes eastwards to China and India, westwards to Greece, Rome, North Africa, eventually to Spain. During the Dark Ages it disappeared from Europe, but such is its unsurpassed usefulness that it is now widely cultivated in America, South America (the Argentine), and France.

cytisum: "shrub trefoil": Varro alone uses this form: it is otherwise always found as *cytisus*. The botanical name is *medicago arborea L*. The plant is a low shrub with trifoliate leaves and yellow flowers. It was especially valuable for cows because it increased the quantity and quality of milk: it was also an excellent bee-plant. Vergil writes of it in *Georgic* III, 394-5.

    2. non multo ... suco: "not much nourishment". The ablative is regularly used with *indigere* of the thing needed.

praeter cicer: "in addition to the chick-pea", the species cultivated by the Romans was in all probability similar to the chick-pea of the modern world known as *cicer arietinum*. It is an annual legume, growing to a height of thirty centimetres, with six to eight couples of oval leaves to the stem and small white or bluish-purple flowers. It probably derives from the Orient where several species are known. It has been a cultivated plant since remote antiquity. The nutritious seed, brownish in colour known as *cece* is a familiar ingredient of modern Italian food.

legumina dicta: although etymological studies are hardly part of a scientific treatise, Varro's love of words prevails. *legumen* is rightly, in his opinion, derived from *legere*, " gather", because such forage plants are not harvested or "mown" as is grain, but are "pulled" or "gathered".

In pingui: supply *terra seruntur ea*.

quae cibi sunt maioris: the genitive is descriptive, used here beyond the more familiar terms of reference: it normally serves to describe a person, such as, for example, *vir magni ingenii*, in relation especially to inherent qualities. (Woodcock p. 68, section 85, II.) The statement is the opposite of *quae non multo*.

    3. in annum prospicientem: "with a view to the coming year", "to next year's crop". Certain of these crops will be grown to provide green manure to enrich the soil for the next year's growth.

relicta: if they were merely left on the surface their nutritious qualities could not be fully effective: it is to be inferred that they would be ploughed in when the ground was being prepared for the next sowing.

cum minus ... cepit: "when it is just beginning to pod ": *minus* is sometimes used as a mild negative, "scarcely". The lupine must be cut at the stage when the pods are beginning to form, that is, just after the flowering. In the following sentence the same treatment is advocated for field beans.

ut ... expediat: this clause is consecutive, giving the result of the meaning of the *si* clause on which it depends.

pro stercore inarare solent: the roots of several of the leguminous forage plants have nodular swellings called tubercles, the cells of which contain bacteria which take in and store nitrogen from the atmosphere and then incorporate it in the soil in the form of nitrates. As a result when ploughed into the soil they are particularly valuable for enriching poor soil or for improving impoverished soil. As the roots decay the nitrogen is released into the soil thereby making up for deficiencies in this essential element, the most valuable and elusive element of plant food. The fact that the ploughing was not done until pods were forming shows that the ancients knew the critical date by long experience and observation. Modern laboratory experiments show that legumes of all kinds take in most nitrogen as they approach maturity so they had a sound basis for their practice. Legumes also play in this respect a useful part in the rotation of crops. The Roman farmer knew of this usefulness of leguminous crops although the underlying cause was not understood.

4. Nec minus ... diiuncta: the sentence is expressed in a roundabout way by the negatives *nec minus non, neque* which could have been dispensed with to leave a straightforward statement. Varro is speaking of two kinds of crops in addition to that which produces food for man: one, such as fruit and flowers, is for pleasure, the other for farm purposes.

propter voluptatem: "on account of the pleasure (they give)".

pomaria ac floralia: "orchards and flower gardens". Gardens were a part of every farm large or small and developed from the smallholding of early times when all necessary food was grown around the homestead. This was particularly true of vegetables. Later they became the nursery gardens which were cultivated on the outskirts of Rome and other large cities the produce of which was taken to the urban market. The Romans cultivated a number of fruits in orchards (*pomaria*) some native such as apples and pears and others imported from the east, apricots, peaches, plums, cherries, pomegranates, and citrus fruits. Flower gardens were in the first place planted to provide material for garlands as Cato states (*De Agricultura* VIII). These were constantly needed as well for private use to decorate shrines and sepulchres as for ceremonial occasions. With the increasing luxury of the end of the Republic flowers were demanded on a large scale and the pleasure garden became fashionable. These were often planted with evergreen trees such as cypress, laurels, and myrtles to give shade and coolness, and with flowers to give brightness and perfume. The Romans did not have a wide variety of flowers, nor did they cultivate special varieties: they had however, violets, roses in profusion and various kinds of hyacinths, lilies, marigolds, and sweet scented herbs. The small flower garden eventually developed into the great pleasure gardens such as those of Lucullus, of the historian Sallust, and the *horti Maecenatis* which were all in the immediate outskirts of Rome (see above I. ii. 10 and note).

neque ... diiuncta: "but are not separate from the productiveness of the land". These contribute to the economy of the farm although they neither provide food nor give pleasure.

4-5. Idoneus ... alia: the sentence falls into two parts, both expressed by a jussive construction. The first part begins with the gerundive *eligendus* and continues as far as *secuntur*. The second with *ut ... seras* which continues to the end.

5. salictum, harundinetum: the suffix *-etum* is used for a plantation, or a "bed", here "willow-bed" and "reed-bed". *salictum* is shortened from *salicetum:* compare *arbustum* from *arbos* and *-etum*.

sic alia: supply *eligenda:* "other plants are to be selected". The change from *locus* to *alia* which refers to things planted and not the place is awkward. After *contra, locus eligendus* is taken up.

segetes frumentarias: this accusative is governed by *seras* in the next line. Such crops as barley and *far* are meant.

item alia: (*seras*).

ut ... seras: the jussive use of *ut* with the present subjunctive.

corrudam: the asparagus plant belongs to the lily family (*liliaceae*), one species of which is a common vegetable grown for food. The plant prefers loose, light, sandy, and well-drained soil. It is a native of Europe and was valued as a culinary vegetable as much in the times of the Greeks and Romans as today. Since it grows wild in the warmer parts of the continent it may be that the Romans were able successfully to grow wild specimens and improve them by careful cultivation and irrigation for the table. A dark pencil-thin variety, which grows wild in the countryside, is sold in the markets in Rome today. Cato says in the *De Agricultura* VI, 3-4, "there sow *corruda* from which asparagus plants will grow".

unde viendo ... facias: Varro clearly indicates the derivation of *vimen* from the verb *viere*, "weave" "plait": "from which by weaving you may make such things as ...".

quid: is used after *ubi* in a general sense with the meaning "anything".

sirpeas: "wicker baskets", probably in this context, bodies which could be fixed on to wagons for carting manure, hay, etc.

vallus: "winnowing baskets", the usual meaning is "stake" but this can hardly be meant here. It is more probably to be understood as a diminutive of *vannus* a winnowing fan, or basket made of osiers.

crates: this is a general term for objects made out of wickerwork, but also of wood and straw; in this context it can be taken to mean hurdles which served several useful purposes. A hurdle, *cratis*, was made of a wooden frame intertwined with osiers. It could be used for carting manure, for making sheep-folds, and for harrowing; in this use it was known as a bush-harrow. When dragged over the surface of the soil it served to break up small clods and made the surface smooth, an important operation before a field was planted.

silvam caeduam: "spinney for leaf forage" in which the trees were grown to be cut. It could serve various purposes not only for fodder and kindling and for making charcoal, but also for the making of wooden tools needed on the farm. Elms and poplars provided fodder for sheep and working oxen. Green leaves of elms and poplars, sometimes oaks and figs, were used as long as possible in winter: even those of evergreen oak and ivy might be given to cattle if hay were scarce. Foliage could be dried for winter use. In Italy on the plains there is no green grass after about the middle of June until September without irrigation which meant that the feeding of the animals on the farm might at times be problematical. Sheep on the home farms might be fed on elm and ash foliage which had been stored. The foliage of the *silva caedua* was probably of greater importance than the timber.

6. aucupere: the shortened form of the generalising second person singular of the present subjunctive (deponent) used in a "final" sense with *ubi*: "where you may go bird-catching".

sic ubi . . . spartum: supply *seras*.

cannabim: this plant, "hemp", is an annual and is still one of Italy's most important crops. It needs deep, rich soil and can be grown on the same ground as the cereals, after they have been harvested, as it can come to fruition in a five months' cycle. Its properties of growth give it an important place in the rotation of crops. It is gathered between July and early August.

linum: "flax". This is an annual plant bearing small blue flowers, from which linen is made and thread (*lineas* in the following line). It rapidly exhausts the soil as Vergil states: *Georgic I.* 77.

urit enim lini campum seges.

" A harvest of flax burns the field."

iuncum: the rush needs marshy ground. Rope and cord can be made from the fibre.

spartum: "esparto grass" or Spanish broom is a species of broom, the Latin name of which is *spartum iunceum L*. It yields a fibre for making cord for weaving and for basketry. The long slender branches which drop their leaves very early are tough and cannot be easily broken, as a result they are useful in the vineyards as ties for securing vines to the props. The plant is cultivated on dry farm lands where willows will not grow. The flowers are yellow and attractive to the bees. This plant is of great usefulness to the farmer especially since it can be grown on dry ground.

bubus soleas: "shoes for cattle". Columella (VI. xii. 2.) gives us more information. In the passage describing the treatment of diseases of the hoof, he recommends that after the swelling has been treated by fomentation and if necessary, by lancing, the hoof must be covered with a *solea spartea*, "a shoe made of broom" so as to keep it dry. The extreme toughness of the broom would make it resistant to wear. It seems from this passage that the shoes were only used as a cure for lameness and were not applied in general use. The ancients do not appear to have thought of protecting the hoofs of working or draught animals: the metal ox-, or horse-shoe was unknown to them.

eadem: this word agreeing with *quaedam* is used in the idiomatic sense, when an added fact is given in addition to one already stated. "at the same time . . .".

recentibus pomariis: "newly planted orchards".

dissitis seminibus: "when the rooted cuttings have been planted out". *seminibus* (from *semen*) describes slips or cuttings which have been prepared and grown in a nursery before being set out in the place where they will grow into trees. This method of propagation is described below in I. xl-xli. Orchard trees were not grown from seed because the process would be too long to be economic.

in ordinemque . . . positis: "when the young shoots have been planted in order". Varro in all probability refers to the *quincunx* method of planting in alternate rows so as to give each tree as much sun and air as possible: (see above vii. 2-4 and note).

antequam . . . possint: the subjunctive following *antequam*, which in a temporal clause which expresses time only requires the indicative, here carries a suggestion of purpose. The farmer's intent must be to seize the right time for planting; he cannot prevent the roots growing too long but he can act in his own interests before they do so.

hortos: "vegetable crops", neither flowers or gardens are meant here as in 5 above. The planting of crops between rows of orchard trees is under discussion. The kinds most suitable would be low growing vegetables. This is still done in Campania and elsewhere.

ne violent radices: the planting of other crops between the vine rows was not approved of by the agricultural authorities. Without them the soil was clear for the frequent ploughings which were necessary when the vines were mature to ensure that the vine itself obtained all possible moisture and nourishment. Columella (XI. ii. 60) however allowed the planting of a green manure crop which grew quickly and could be ploughed in. A warning against the growing of other crops in the vineyard is even found in the Bible, *Deuteronomy* XXII. 9 "thou shalt not sow thy vineyard with divers seeds: lest . . . the fruit of thy vineyard be defiled". Varro's intent is to state what is allowable to be planted and at the same time to hint implicitly that it must not be done when the vines are fully grown.

xxvii-xxxvi. *The Farmer's Calendar*

A calendar is an invention of the human mind, a device for measuring time by means of which it becomes possible both to record the past and to plan for the future. In relation to agriculture it is to be regarded as a practical division of the year with a view to distributing the various operations on the farm and to ensure that these are carried out in an orderly sequence related to the seasons. The farmer needs to know by some calculation on which he can fully rely between what times the main activities of the farm should be carried out, when the seasons will come and go and for how long they will continue. So essential was this knowledge that there must have been a calendar in some form on every estate, for the guidance of the master, or in his absence, the *vilicus*. The question arises as to how such a calendar was calculated. A day by day count according to the solar year of three hundred and sixty-five days and a quarter was not workable, because the extra quarter of a day would cause such a calendar to become out of joint with the seasons, and the requirement of intercalary days which would not correspond to procedure on the farm. The farmer's need in any case was to know, not the separate *days* so much as the coming and going of the *seasons* so as to time with accuracy activities on the farm. For such practical purposes the stars were his proper guides. Since the constellations rise and set with absolute regularity and without deviation during the year, observation of their movement in relation to the sun could be used to calculate the changes of the seasons: for instance the beginning or end of a season could be recognised if a particular constellation were visible in the east just before sunrise or after sunset. Many estates probably had at least one skilled observer who kept watch on the constellations, an exacting task and one calling for accuracy. The relation of the stars to the sun and their rising and setting had to be observed from some particular point, perhaps a tree or a stake set in the ground. There may have been among the slaves on the large estates educated Greeks or Asiatics who had a knowledge of astronomy and could well have acted as star-gazers, or the task may have rested on the *vilicus* himself. Before beginning his account of the farmer's calendar proper, that based on observation of the stars, Varro gives, through Scrofa, a brief survey of the main operations in the round of work during the seasons, such as planting, harvest, ploughing.

In the next paragraph (xxviii) he divides the four seasons into eight divisions with reference to the position of the sun in the corresponding sign of the Zodiac. Within his account there is to be found an historical reference of great importance for dating the writing of the dialogue. *quae redacta ad dies civiles nostros* means when freely paraphrased "reduced to our official calendar as it is now in force". Soon after taking up power Julius Caesar reformed the official calendar which had been cumbersome and had needed frequent inter- calary days or even months to keep it properly balanced. This new calendar often called "the Julian" which is still in force today, took effect on January 1st, 45 B.C. We have proof here that the dialogue was written after this date and it follows that in all probability the usually accepted date for publication is roughly correct, that is eight years later in 37 B.C.

In paragraphs xxix-xxxvi Varro presents his own agricultural calendar with explanations for a reader unacquainted with the phraseology of the farm. It is evidently a well-known calendar and one which has been long worked and supplemented with copious comments for the technically ignorant readers. Varro gives full play to his deep interest in etymological derivations. The instructions for the *vilicus* are expressed in the infinitive, usually the passive, and are in few, direct words. The short straightforward instructions become evident when Varro's comments are removed, and the primitive method of calculation according only to two constellations and two positions of the sun easy to determine become clear, observations of a natural phenomenon are also included, the rising of the west wind. There is a probability that every owner of a large estate had his own calendar and the one Varro gives may be the one which was kept posted up for his *vilicus* to follow or for any to read who could do so. This calendar is calculated according to the rising and setting of two con- stellations the *Vergiliae* (the name simply means "spring stars") and the Dog Star (*Canicula*) according to their highest and lowest position on the horizon, and according to the position of the sun at the equinox and the *solstitium*. The equinox (derived from *aequus* and *nox*) occurs twice in the year, in spring and autumn: it is the time when days and nights are of equal length: the *solstitium* also occurs twice, in midsummer and midwinter. The times fall on the longest and the shortest day when the sun seems to stand still as the name implies from *sol* and *stare*. The word *bruma* used for the winter phenomenon is derived from *brevis* in the superlative form *brevima*, abbreviated from *brevissima*. In addition there is an observation connected with the west wind: it is expected to usher in the spring on February 7th: a calculation which cannot always have been exact but for the farmer it would be sufficient to know that the warm days of spring had come. Remarkably it is not related to the sun or the stars but is a natural phenomenon which also was easy to observe. Varro did not convert his calendar according to the new Julian but kept to the ancient well-tried form. Indeed, when we come to study the calendar we shall find that there is no connection between the calculations according to the four seasons and to the eight divisions in the Julian calendar.

xxvii. 1-3   *The Farmer's Calendar: The Four Seasons*

In this chapter there is a résumé of the old agricultural calendar calculated as all such must be, by the sun's course. This is divided into four seasons of about three months each, and then subdivided into eight periods of about one and a half months each. For each of the four seasons the main farm activities are listed: the agricultural scene is briefly outlined, ploughing, harvest, vintage, and pruning to show the annual round of the farmer's work. This summary account will be enlarged and commented on in Varro's own calendar which appears in full below (xxix-xxxvi)

1. tempora duorum generum: "times", or "measures of two kinds". The genitive is one of description: these are enlarged on by the words *unum* and *alterum* which follow. The sun has an annual course which is to be discussed first, the moon a monthly course, which will be discussed afterwards [see caption on xxxvii (text omitted)].

circumiens comprendit: "completes as it circles round".

Eius ... IIX: "Its (the sun's) annual course is first roughly divided into four parts of about three months each, and at the same time by a more detailed division into eight of a month and a half (lit. 'by months and a half')." The farmer's calendar described in xxix-xxxvi below is based on these intervals of approximately six weeks.

ternis: "in groups of three".

sesquimensibus: *sesqui* is a numeral adverb perhaps shortened from *semis-qui*, meaning one half more; it is frequently joined with other numbers or words to signify "one and a half".

2. Vere ... fiunt: "in regard to the spring plantings". This relative clause is not linked grammatically to the main clause *terram ... oportet,* but the relationship is clear. The function of the clause could be compared with that of an accusative of respect. The general intent is "to prepare for the spring plantings, the ground must be broken up".

terram rudem: "unbroken ground" may be *novalis* "fallow" described below I. xxix. 1 and note.

quae ... exradicata: "so that the (weeds) which have begun to grow out of it may be thoroughly rooted out, before any seed can fall from them". *oportet* is used in this paragraph and in the calendar which follows (xxix-xxxvi) for giving instructions and rules. In several instances below it is left to be understood with an infinitive.

quid seminis cadat: the partitive genitive is commonly used with *quid* (to mean "any") but the use of *quid* with *priusquam* is most unusual. The meaning here is equivalent to a negative and so is easily understood as carrying the notion of preventing.

glaebis ... percalefactis: "when the clods have been thoroughly warmed through". Although this is expressed by an ablative absolute, *glaebas* must be supplied with the adjectives *aptiores, faciliores,* and *relaxatas.* Normally an ablative absolute stands outside the construction of the sentence although the subject is usually the doer of the action referred to.

ad opus: *opus* is the work of breaking up and ploughing: the successive ploughings are described below (I. xxix. 2).

binis: "twice"; this form is to be explained as a kind of ablative of comparison of a neuter plural *bina.* The recommendation to plough the same land twice or three times should be compared with the instructions given below (xxix-xxxvi).

3. oportere: it is to be noted that the infinitive of *oportere* is here used to give a command, although *oportet* with the infinitive appears above: see the previous paragraph (2) and note.

siccis tempestatibus: "when there is dry weather". *tempestas* connected with *tempus* is frequently used to mean "weather", "climate" of any sort. Wet weather at the time of the vintage would be unlikely in Italy.

vindemias: supply *oportet* or *oportere fieri*.

silvas excoli: "plantations are to be cleared". Perhaps the woods mentioned are those grown for timber and fodder (*caeduae*) though *silva* can be extended to mean any kind of plantation: the instruction could indicate not only pruning but digging, clearing, and cutting fodder: (see above, xxiii. 5-6, and below, xxxiii and notes).

praecidi ... terram: *arbores* are probably the trees of the *arbusta* on which the vines were trained: if this is a correct surmise, the instruction seems to imply that the lower branches are to be cut off to leave the trunk clear so that the vine may climb unimpeded.

radices: superfluous surface roots are to be cut out from the trees on which the vines will grow so that their main roots may strike more deeply, and to prevent the growth of suckers. The second part of this sentence, *radices ... possit* can be regarded as an enlargement on *praecidi ... terram*.

nequid = *ne* and *quid*.

primoribus imbribus: "at the time of the first rains". The season is used for a point in time with the result that the ablative is used quite logically. It was essential to cut out superfluous roots in autumn because the ground remains warm and is conducive when combined with moisture to rapid growth.

Hieme ... glacie: the precaution to prune trees when there is no frost seems to have been written as an afterthought and as a rider to *praecidi arbores ...* It is well known to all agriculturists and horticulturists that pruning must not be done while frosty weather lasts because if frost infiltrates into the inner tissues of the bark it can damage the tree.

xxviii. 1-2. *The Eight Divisions of the Solar Year*

1. in aquario . . scorpione: the movement of the sun through four signs of the Zodiac is observed to ascertain the beginning of each of the four seasons. *in aquario:* "(when the sun is) in (the sign) *Aquarius*".

dies tertius et vicesimus: the beginning of each of the four seasons is the twenty-third day after the sun enters the sign; thus the sun can be said to be "*in Aquarius*", etc. The four dates are February 7th (spring), May 9th (summer), August 11th (autumn), November 10th (winter).

1-2. efficiat ut ver ... Nov.: according to Varro's calculation the seasons vary in length by a few days: there are ninety-one days in spring, ninety-four in summer, ninety-one in autumn, and eighty-nine in winter, amounting in all to three hundred and sixty-five days. The dates belong to the Julian calendar.

1. ad dies civiles nostros: this is a reference to the Julian calendar (see introductory note on the Farmer's Calendar above). The days are termed

civiles because the calendar was intended for official occasions such as the holding of elections, meetings of the Senate, religious rites, and ceremonies carried
out by the State.

2. suptilius . . .XLV: the eight divisions mentioned in the previous chapter
are listed in detail. It becomes clear that these are determined from fixed
points found by watching the sun and stars and natural phenomena. They are
of uneven numbers of days but add up to three hundred and sixty-five. There
is in reality no relationship between the Julian calendar calculated as it is by
the sun and moon (the lunisolar calendar) and the agricultural form based on
the sun and the stars combined with observations of nature.

xxix. 1-3.   *The Farmer's Calendar: The First Period*

Primo intervallo: the ablative is one of "time when" and is used for all
the eight divisions. The first period begins in February 7th and lasts until
March 24th; this gives an interval of some forty-five days. Varro considered
that *favonius*, the warm west wind of spring, began to blow about February
7th: the vernal equinox, which is the time in spring when day and night are
of equal length falls on March 21st. This gives rather more than six weeks
for a programme of heavy farm work, planting, pruning, manuring, trenching,
weeding. It could however be continued into the second period if the work
was not completed. In a farmer's calendar the dates are to be considered as
approximate. The implication is always that within such and such dates or
about such a time the various operations must be begun and may be continued
as needed.

inter favonium . . . vernum: for Varro the calendar begins with the spring:
such a calculation is arbitrary because in reality the ceaseless round of the
farmer's work can have no beginning and no end.

Seminaria . . . sariri: the wording of the calendar proper begins here, and
ends for this period with *sariri*. Apart from the *ut* clause the instructions are
expressed in the passive infinite which will be found to be the usual formula
throughout.

Seminaria: slips, cuttings, and seedlings of all kinds were grown in nursery
beds and then transplanted: both vines and olives were propagated in this
way as well as vegetables: for more details see I. xl. 2-4 below, where spring is
recommended as the time for planting in the nurseries.

putari arbusta: these were the trees which supported the vines. It was
important to see that their branches did not impede the upward growth of the
vine. The trees could however be allowed to grow high and the foliage to
sprout well above the height which the vine would reach, as can be seen for
instance in Campania today. *arbusta* is generally used for plantations of
living props for the vines, but it might also include in this context olives and
fruit trees. The trees on which the vine was trained could be elm, maple,
black poplar, olive, and several others.

stercorari in pratis: "the meadows are to be manured".

circum vites ablacuari: "(the roots of) the vines are to be dug around".
Such a process carried out in early spring was important to ensure sturdy
growth. The labourer dug around each vine stock and at the same time
cut away surface roots so that the deeper ones would develop more strongly
and be forced further down to the moisture which lay beneath the top soil.

The following phrase *radices ... praecidi* reads like an explanation inserted by
Varro to elucidate an unusual word but one which would be known to farmers.
prata purgari: "the meadows are to be weeded". Weeding involved a clear-
ance of the ground in preparation for sowing grass or vegetables. There were
not permanent grass fields as we know them in England because grass is parched
up in the summer months in Italy so that nothing remains but stubble. It is
possible however to rotate several crops on the same plot within a year including grass.
salicta seri: willows served many purposes on the farm: the pliant stems
were used for vine props and for tying vines, also for making farm equipment
such as winnowing baskets, hampers and others (see above xxiii. 6). The
willow was grown in damp ground (see above xxiii. 5).
segetes sariri: "cornfields are to be hoed". The crops which are to be hoed
in this first period in early spring will have been sown in the previous autumn
in the sixth period (see below xxxiv. 1) and will have sprouted or even be
well grown at this stage for the harvest will come in May or early June. Hoeing
is a necessity to ensure that weeds do not choke their growth and that the earth
is kept loose and aerated. That planted grain is meant and not bare soil which
awaits planting is made clear in the following sentence where *seges* is defined.
This kind of weeding was done with the *sarculum*, an iron tool with one large
prong or polished blade set at an acute angle to the handle.
quod aratum ... satum: "which (after being) ploughed, has been sown".
Varro seems to suggest that the word *seges* is connected with *serere*, "to sow".
novalis ... rursus: "fallow land is where there has been a crop (lit. where it
was sown) before it is renewed again by a second ploughing". *novalis* can be
used of two kinds of land: either unploughed grass land, or land that has been
tilled and is allowed to rest. The implication in the word *novalis* (connected
with *novare*, "renew", and *novus*, "new") is either the acquiring of a new field
not previously cultivated, or the renewing of a field which has been left fallow
for a season. The Roman agriculturist well knew the advantages to be gained
from leaving the fields fallow after cropping and so they were allowed to rest in
alternate years. This had the effect not only of giving a respite to the land,
but it also allowed the soil to accumulate moisture for the next crop. The field
which was resting had to be kept constantly under cultivation, so as to keep
the soil moist and clear of weeds. Several ploughings in the year were considered necessary: (see Vergil: *Georgic* I. 71-6).
proscindere: Varro recommends two, or better, three ploughings in order to
keep the soil entirely free from weeds, and to keep its consistency moist and
light. The fields which were to be fallow were ploughed first in September
(in Varro's calendar this ploughing is expressed as *arata offringi* in the fifth
interval, see below xxxiii), so that the ground was well soaked by the autumn
showers, then a second time in mid-winter so as to benefit from the rains and
frost which breaks up the soil and causes rain water to penetrate more deeply.
Then came a third time in spring at the time of the vernal equinox, from
March 21st onwards in Varro's second period, *boves terram proscindere* (xxx.
below). A fourth ploughing was carried out in mid-summer and was of the
highest importance because the breaking up of the ground at this season
prevented loss of moisture. Ploughing had to be thorough: the first furrows
were in straight lines and the subsequent ones were drawn across these ridges

so as to leave a level surface: clods were broken up by manual labour with mattock and hoe. Vergil well describes the need for fallow: (*Georgic* I. 71-2),

> *alternis idem tonsas cessare novales*
> *et segnem patiere situ durescere campum*

"you shall leave too the cropped field fallow in alternate seasons and let the corn land recover its strength in rest". *secuonda aratione* (1, above) means "when the land is ploughed anew" and so, brought back into cultivation.

2. Terram ... vocant: Varro, through Scrofa, speaks of two ploughings in connection with cornfields: the first by which the clods are broken up (this is the March operation at the time of the vernal equinox), and the second when the clods are broken down. The fact that the ground had to be twice ploughed is probably an illustration of the primitive type of plough then in use: the share was symmetrical and did not turn the soil as does the English mould-board plough. This developed type of plough did not appear in Europe until the end of the eighteenth century. The Romans knew none else than the flat type with symmetrical sides ending in a sharp point which was curved slightly downwards.

iacto semine: "where the seed has been sown". *iacto* from *iactare*, describes the action of throwing the seed broadcast. At the time when this was done the soil would be smooth and level after the second ploughing. A third was needed after the sowing.

boves lirare dicuntur: "the oxen are said to 'ridge'". A process such as this was necessary for drainage so that the field was crossed by alternating ridges of furrows and ditches. If rainwater were allowed to collect on an even surface the grain might rot. *lirare* is connected with *lira*, a "ridge" or "furrow", or, in its root meaning, a "track".

tabellis ... vomerem: "with boards attached to the plough share". The Roman symmetrical share could not form ridges as does the modern mould-board plough which turns the sods as the plough moves forward. Detachable wings in the form of straight wooden boards (*tabellae*) were used to achieve something of the same effect. When these were attached, the soil was thrown up on either side of the share with the result that a ridge was formed on each side with a ditch in between which became a furrow. By this method the seed was covered and a ditch made for the drainage which was essential for cereals grown in autumn. Otherwise the heavy rains of that season could leave the ground too wet. Although Varro gives this injunction in connection with the spring sowing he is making here a general definition suited to any season.

quo ... delabatur: *quo* introduces a final clause and refers in general to the process described.

ut ... praediis: the greater part of Apulia (see above I. vi. 3 note and Chapter II Transhumance) was given to sheep-rearing in Varro's time.

id genus: is an accusative of respect. The words are applied to the type of estate which was linked with ranching on a large scale. The method was lucrative enough but was at the expense of arable land as Varro seeks to suggest. The process of hoeing by hand shows in itself that the cornfields were small.

occare: the operation is explained below (I. xxxi. 1).

3. lacunam striam: "a hollow" or "a trench".

porca ... porricit: the connection in Varro's mind with the ritual use of *porricere* is forced, and is an instance, among several, of his failure at times to come to the most reasonable conclusion.

xxx. *The Farmer's Calendar: The Second Period*

Secundo intervallo: the second period lasts from March 24th to May 7th, an interval of forty-four days.

vergiliarum exortum: the spring rising and setting in early winter of the "spring stars" were important dates for the farmer in the ancient world because they fell in seasons when there was usually rain, especially in October, which caused them to be connected in thought with the rainy seasons. The common identification of the Vergiliae with the Pleiades is certainly wrong since these rise in early winter and set in April.

runcari: *runcare,* "root up". This may have been done with a *runca,* a tool with a scythe-like blade. Frequent weedings were necessary while the corn was growing to keep the rows clear and to prevent the soil from caking. At this stage winter-sown corn which had been previously planted in Varro's sixth period would be grown tall since the harvest was due from the end of May to the beginning of June. The clearing of the soil would need great care so as not to break down the growing corn. One weeding has already been done in the first period in early spring (xxix. 1).

id est ... expurgari: this remark is Varro's own, inserted into the formula of the calendar proper, to explain *runcari.*

boves ... proscindere: the first ploughing of fallow land is to be done (see above xxix. 2).

salicem caedi: the willow only lately planted, in the previous period in February or early March, would not be ready for cutting, but that of the previous year or even earlier could be gathered. The cuttings from young, tender plants would be used for tying vines, and stronger branches for vine props.

prata defendi: "meadows must be cleared (of cattle)". *pratum* can mean any kind of grassland or even vegetable plot, but in this context refers to grazing land. *prata* have already been cleared of weeds in the first period (see above, *prata purgari,* xxix. 1): the time has come for the grass to be left undisturbed by cattle so that a crop of hay may be grown. They must be driven off and kept in stalls or fenced enclosures, or possibly taken to other pastures: this process is called *defensio.* Varro explains in a later chapter, [xxxvii. (5 text omitted)] that any preparation to be made before a crop of hay can be grown consists in closing the meadows from grazing, and that the practice is usually observed from the time of the pear-blossom, that is, in early spring. Columella XI. ii. 27-28) agrees when he says that the days from March 1st to the 23rd are the time for clearing meadows and driving off the cattle (*a pecore defendere*). In some instances this might entail fencing the land if it was adjacent to another stretch of ground which was still being grazed. He gives the added details that *defensio* may be put into practice from January onwards in warm dry districts, but cold meadows go to grass only from the time of the Feast of Minerva (*Quinquatria*) which extended from March 9th-23rd. For Varro some meadows are to be irrigated. In the third interval in Varro's calendar at the time of the solstice, the hay is to be cut, and left to the last after other fodder crops such as *ocinum,* mixed fodder, and vetch. It would appear that the cattle were fed with this green fodder in spring during the hay season unless they were allowed to graze some pastures close by. After the first hay harvest the grass will grow again in irrigated meadows and a second cutting of hay can take place in the fifth period between the time of the Dog Star and the autumn equinox. Then it seems that fattened and nourished with the spring growth of legumes they can return to their familiar grazing grounds. It becomes clear that *defensio* implies not only the stabling and enclosing of cattle but also the feeding with fodder crops. The cattle who are cared for in this manner and for whom the hay is grown and stored are the working oxen, the donkeys, mules, and horses: they are not the sheep and goats who in the spring will be taken to the mountains for summer grazing not to return until the beginning of autumn to winter pastures.

Quae...fieri oportuerit: "what should have been done in the former period but was not finished should be done before (the plants) show buds and begin to flower". *oportuerit* is probably to be explained as a perfect subjunctive in a potential sense, in present sequence: *agant* and *incipiant* are in the subjunctive, although following the temporal conjunction *antequam*, because a sense of purpose is contained in the meaning. The injunction is to do with the planting of the nurseries mentioned in the first period*s eminaria omne genus ut serantur*: this is the time for vine, olive, and fig cuttings and slips to be planted in nursery-beds and these must be taken from the parent trees before the spring growth begins.

si quae ... inciperunt: "if any which drop their leaves have already begun to come into leaf". If deciduous plants have once begun to bud and show leaves they are not suitable for propagation. The taking of slips and other means of propagation must therefore be done at the very beginning of spring. The taking of shoots from a tree which was already budding would be injurious since the sap would be rising and the cut ends could "bleed". Varro's *caveat* may indicate a care for the parent tree rather than for the slips taken from it. Damage to the trees of a vineyard or oliveyard would be costly.

Oleam seri ... oportet: "The olive is to be planted and pruned into shape." This is the only mention of the olive tree in Varro's calendar, although olive cuttings are mentioned below (I. xl. 4), a remarkable fact in view of the great importance of olive culture in the Roman economy. It is perhaps however safe to assume that instructions for digging, trenching, pruning, and manuring the vine are to be applied equally to the olive: see for example above, xxix. 1. The olive is propagated by means of nodes, called by the Italians of today, *ovoli*, which are cut from branches where they are joined to the trunk, or at the joining of large roots on the surface of the earth. When planted out in the proper conditions they become rootlings. In modern practice the time for planting is the end of March and corresponds exactly with Varro's calendar. After about four years in the nursery-bed the young olives are planted out where they will remain permanently. The olives are pruned for shape, maintenance, and improvement; the first pruning aims at shaping the young trees into a vase, or lemon shape. The main stem is subsequently cut into the shape of a truncated cone about two metres high with three or four branches springing from it. In a few years, after several prunings, the desired shape is achieved. It is important that there is an open space in the middle so that the trees have the greatest possible amount of air and sunshine. Pruning should successively be done every year to keep the good shape of the trees, remove suckers, reduce height, and cut out non-bearing branches.   Although old trees need drastic pruning, *interputari* most probably describes the shaping of the young trees, especially since it follows so closely on *seri* and can be considered to imply "partly prune" or "prune here and there".

xxxi.   *The Farmer's Calendar: The Third Period*

1. Tertio intervallo: the third period falls between May 7th and June 21st, that is, from the beginning of summer until midsummer and contains forty-five days.

Vineas novellas: the soil around the cuttings or young shoots in the nursery beds and those newly planted in the vineyard must be kept loose and free from weeds.

fodere ... comminuere: the usual passive infinitives of the formula have been replaced by the active for no other apparent reason than variety.

occare: the meaning of this word is uncertain, but there is little doubt that it describes the operation of hoeing, as Varro's following remark seems to suggest. Any clods left after digging or ploughing are broken up and the surface levelled. The tool used was probably closely akin to the *rastrum*, a drag-hoe with prongs.

pampinari: an important stage in the training of the vine is described. The springing plants were to be trimmed: superfluous shoots, leaves, and tendrils were to be cut away to strengthen the vine for the production of the grape.   Young plants were trimmed too as is described in the next paragraph, so that only two or three main branches were left on the stock.   The process of *pampinatio* was applied to mature plants when old branches and leaves which had become superfluous were cut away.   If the vine were left to grow unchecked it could be weakened because the main stem could not provide enough sap for the fruit.

a sciente: "by a skilled (vine-dresser)".   Such a man could work a *iugerum* in a day.   He was probably a hired labourer (see above, xvii. 2).

neque in arbusto ... fieri: it is difficult to understand Varro's intent: the implication seems to be that the operation of *pampinatio* must be done in the nursery when the young plants are still small and tender, and not when they have been planted out and are climbing on the parent trees.   The Latin could also carry the meaning that *pampinatio* was not practised on vines wedded to live trees in a plantation (*arbustum*) but only to those growing in a vineyard on props but this does not make very good sense.   It is difficult to believe that the vines which grow around trees did not also need some kind of thinning out and trimming.   The question arises as to whether *neque* in this context can carry the meaning "not only".

2. Sarmento: the parent stock is called *sarmentum*.   Varro describes the trimming not only of young plants to form fruitful vines, but also the treatment of the mature ones when growth must be kept under control.

vitiario: the vine-nursery where cuttings and slips have been planted.

tota resicari solet: all the shoots of the young vine are cut back to strengthen the parent stem: as a result it will send out sturdier branches.   Such drastic pruning was usually done two years in succession after which growth was kept within bounds for the next four years: it was not good to allow the vine to grow too high at this stage.

3. Eiuncidum: *eiuncidus* connected with *iunx*, "a reed", describes a branch which is thin, slender, and reedlike. A parent stock which has become "leggy" is not likely to become a strong vine with a good yield of grapes.

quam ... palmam: "which, when it is at the smaller stage they call a *flagellum* (but) when it has grown larger, and the grapes are already beginning to form, a *palma*": *minorem* implies, "when compared with the later growth": *palma* is applied to the branch when it begins to bear. *flagellum* and *palma* describe the same branch at different stages of growth.

Prior ... flagellum: there can be no doubt that *flagellum* means a "whip" or a "switch" and from that came to be used for a young branch but Varro's love of derivations has run away with him. Instead of accepting what is a clearly logical origin for the word, he turns to a fanciful suggestion. Although his connection with *flatus* is unacceptable, it is repeated, strangely enough by Servius in his comment on Vergil, *Georgic II*, 299, *flagella dicuntur summae partes arborum ab eo quia ventorum crebros sustinent flatus:* "the highest parts of trees are called *flagella* because they withstand frequent gusts of wind".

Posterior ... palma: Varro has invented another derivation which is far from correct, and in doing so has also found for himself a word which does not appear to exist (*parilema*). *palma* in its original meaning is "the palm of the hand" (*i.e.* the whole hand when spread out). The *palma* of the vine is therefore that grown branch on which the leaves resemble hands, a meaning which has a logical reference to the appearance of the vine branch and its leaves in full growth.

4. ut in multis: supply *verbis:* there is no connection with *parire* as Varro suggests.

Hi ... dictus: "These are the parts by which the vine can hold that prop on which it creeps to grasp a hold, from which grasping the tendril is called 'the grasper'."

capreolus: Scholars do not accept Varro's suggested derivation from *capere* " grasp " but see in the word the same root as that of *caper*, "a goat", relating to its climbing propensities.

Omne ... secari: leguminous crops have already been mentioned (see above, xxiii. 1-2, and notes) where it is recommended that some should be sown with a view to the coming year. Those which are now to be cut at the rising of the *Vergiliae* from May 7th onwards have probably been sown in the previous autumn, in some cases after fallow. Sowing of legumes could also be done in early spring. Vergil himself says that the best time for sowing beans and *medica* is spring, *Georgic I*, 215-16. Varro on the contrary says that the best time for beans is autumn. Varro is alone in recommending the planting of these crops between the summer solstice and the rising of the Dog Star, *i.e.* in the fourth period in his calendar, between June 24th and July 21st (see below, xxxii. 2). This probably represents a green fodder crop quickly grown after the June harvest. Although autumn was the normal time for sowing legumes of all kinds, in Varro's calendar the bean is only mentioned in the sixth period. It is perhaps reasonable to conclude that Varro has omitted to list a customary farm operation and has left the other legumes to be included in the sowing of *faba*. Whatever is the answer to this problem it cannot be denied that the crops which were ready to be cut in May must have been in the ground already either since the autumn or the early spring.

novissime faenum: the hay is to be taken in last of all, probably to ensure as long a time as possible for ripening.

hoc ... purgentur: this explanation of *ocinum* is another of Varro's far-fetched ideas. If there is any connection with the Greek it is without doubt in reference to quick growth.

alvom = alvum.

Id ... siliquas: for *ocinum* see above, xxiii. 1-2, and notes. *ex fabali segete* is very difficult to explain, but could presumably mean "just as is the bean crop".

    5. Farrago ... coepta: "*Farrago* on the other hand harvested from a crop where barley, vetch, and legumes have been sown together to produce green fodder, either because it was cut with an iron knife (*ferrum*) was called *ferrago*, or for the reason because it first began to be sown in a grain or corn (*far*) field." The first derivation does not hold: authorities agree that the root of the word is *far*, "grain", and that the mixed feed represents a blending of legumes with the oldest cereals which formed the staple diet of both the Romans and their farm animals (see above, ii, 6 and note).

vicia: see note on xxiii. 1 above.

ut vitis: *habet* is repeated.

ad scapum ... vincire: "so that it may cling to the stalk of a lupine or some other plant, its habit is to bind it (*i.e.* to wind round it)". Vetch planted among other legumes climbs around them and becomes entangled in them: it is to be distinguished from the leguminous plants in *farrago*, which are not climbers, by the tendrils by which it becomes intertwined among the others.

si prata ... inrigare: this injunction to water the irrigated meadows implies that the operation begins at the time of the rising of the *Vergiliae* on May 7th but must continue through this and the following months until the cutting of the second hay harvest in July or possibly August. There could be no second cutting of hay (see I. xxxiii below) nor a productive summer sowing of fodder crops in late June (see I. xxxii below) without artificial irrigation during this dry hot season. In Varro's calendar the meadows are to be watered after the hay making in May to produce another crop in late summer and for a summer crop of legumes. It is not stated when these latter are cropped but there is no doubt that it was at the same time as the hay since they were cut just before the first harvest of hay (see above xxxi. 4). Although vegetables are not mentioned there is no doubt that irrigation of these went on continually in order to benefit from the several successive crops which were possible in the Italian climate. This was (and still is) especially true of Campania where the rich light fertile soil may yield as many as six different crops in a year. The watering of the meadows could be done in several ways. Where the soil was conservative of water, wells could be sunk in the fields where needed. This is characteristic especially of Campania where the soft porous soil allows water to collect on the harder strata below. Wells of a primitive type can be sunk here and there all over the cultivated fields. A pail suspended from a long pole is lowered and raised at will and the pole secured by a large stone when not in use. Another means of obtaining water was to make use of streamlets (*rivi*) drawn off from a larger stream or river, and allowed to run into fields as needed: these could be dammed when not in use with heaps of earth. Vergil describes this practice in *Eclogue III*, 111. Without irrigation the working oxen could not have been adequately fed, nor could there have been supplies of vegetables for human beings. Drainage canals with a network of small channels were also used.

simulac faenum sustuleris: the hay must be cut last of the green fodder crops (see 4 above) and then the land irrigated immediately to bring to growth

the second crop which will be cut during the fifth interval, in July and August.
Varro gives directions in Stolo's words [below, I. xlix. 1-2 (text omitted)],
for harvesting and drying the hay.   "I shall discuss next the harvesting of ripe
crops.   First the grass or the hay-meadows should be cut close with the sickle
when it has stopped growing or begins to dry from the heat and turned with a
fork while it is still drying.   When it is quite dry it should be made into bundles
and carted to the barn."

quae insita erunt: see below, xl. 5-6, xli. 1 for grafting. *poma* is probably
not limited to fruit trees but includes olives and vines: pomegranates are
particularly mentioned below, xli, 3 and 4: in addition the Romans had
apples, pears, almonds, figs, and cherries.

vesperi: the watering should be done in the cool of the evening to enable the
soil to conserve as much moisture as possible.

A quo . . . possunt: Varro's etymology is wide of the mark. *pomum* probably
contains the same root as is found in *pascere*, "feed".

xxxii. 1-2.   *The Farmer's Calendar: The Fourth Period*

   1. quarto intervallo: "In the fourth period between the (summer)
solstice and the (rising of the) Dog Star".   This extends from June 24th until
July 21st, and represents an interval of only twenty-seven days.

messem faciunt: the harvest was an especially burdensome operation in the
heat of the "Dog Days" and without any means of working other than with hand
sickles.   Although at the time that Varro was writing the *Res Rusticae* the
importation of foreign wheat into Rome had already begun, an owner of a large
estate worked by slaves must have found it profitable to grow his own grain
for home consumption.   That which is ready for harvest in this period has
been sown in the autumn before at some time between September 26th at the
autumn equinox and the winter solstice (see below, I. xxxiv. 1, where it is
stated that sowing begins and may be continued for ninety-one days).   After
that there is to be no sowing; neither must it be begun before the autumnal
equinox.   The labour in connection with the growing of cereals in Roman Italy
was heavy; it entailed at least three ploughings, two operations of weeding, and
sometimes irrigation: then came at last the harvesting with the most primitive
of tools; with the result that it could not be done in less than two stages—one
to cut off the ears, the other to clear the straw from the field: the latter comes
in the next period in the calendar in July to August, *stramenta desecari* (see below,
xxxiii, and note).

quod . . . maturum: the days are counted from the time when the sheath
first shows: the crop has reached this stage after the sowing of the previous
autumn and the winter growth.

cum sit maturum: "when it is ripe".   The subjunctive is one in indirect
speech dependent on *dicunt*.

Arationes absolvi: a warning is given not to prolong overmuch the first
ploughing which began in late March at the time of the vernal equinox.   The
aim should be to work the ground while it is still warm and to break up the
surface so as to preserve as much moisture as possible.

quae . . . aratur: "which become more productive in proportion to the
warmth of the soil".   The warmer the soil, the better will be the harvest.

si proscideris: "if you have broken up the ground".   The future perfect is used idiomatically because the injunction in *offringi oportet* which follows implies a future idea describing what will have to be done. *proscindere* is the first stage of ploughing when the large clods are turned up. If that has been done the next stage must be put into operation—the clods are to be broken down by a second ploughing: (see above, xxix. 2 and notes for an explanation of the different stages of ploughing).

2. Serendum: Varro is alone among agricultural writers in recommending a summer sowing of legumes. Since this follows immediately on the harvest, and seems to be a repetition of the crops which are to be taken in May, we may perhaps assume that Varro is recommending it as part of a policy of crop rotation. The second legume crop would be grown and ready to be harvested before the winter sowing of cereals and might even be utilised for enriching land which was to be fallow for the next year.

lentem: the lentil (*lens*) is another valuable legume (see above xxiii. 1-2, and xxxi. 4, 5). It is similar in appearance to vetch but has more slender leaves and tendrils and a white flower. Its cultivation goes back to prehistoric times. It is now grown in all Mediterranean countries and is prized for the highly nutritious seeds.

cicerculam: This may be a small species of *cicer* (the chick-pea), or merely a diminutive without any particular significance, a usage which would be possible in Varro's style. For *cicer* see above, xxiii. 2 and note.

ervilam: also *ervum*, is probably the bitter vetch, another of the many legumes useful for fodder. Vergil speaks of it as a nourishing fodder for oxen (*Eclogue* III. 100).

quae alii ... leguntur: Varro enlarges on the etymology already given above, (xxiii. 2) where he writes of the planting of different soils. The derivation from *lego* is probably sound.

Gallicani quidam: these are probably Roman settlers in Gaul rather than Gallic landowners.

occare: the soil around the vines must be kept loose and powdery to conserve heat and moisture as far below the surface as possible. Starting from the spring and the blowing of *favonius* this is the third operation to be carried out in vineyards. In February and March comes *ablaqueatio:* in May and June the vines are thinned and trained: now in the hot months of June the ground is hoed so that all clods are broken down. Before this, in the late autumn and winter the vines have been pruned. Again it is to be realised how many operations were needed to bring the vintage, as well as the harvest, to fruition. Care was needed all the year round for both grain and grape if the end was to be a success.

xxxiii.   *The Farmer's Calendar: The Fifth Period*

Quinto intervallo: this period extends for sixty-seven days, from July 21st, the time of the rising of the Dog Star, until September 26th. The corn has been harvested in the previous period, which began on June 21st with the summer solstice, but the straw which still remains in the field must now be cut and put in stacks to be kept for winter fodder. Later on (see below I. L. 1-3), three methods of harvesting are described. All involve two cuttings, one to reap the ears and the second to cut the straw. In ancient agriculture two operations could not be avoided since the reaping was done with hand sickles.

arata offringi: the second ploughing now takes place and must be finished as the word *offringi*, used of the breaking down of clods, suggests. After this the land will lie fallow for a year, or will be sown with other crops. Any injunction of this kind could also apply to land which has been fallow and is to be sown during the coming autumn. It was important that the ground should be throughly broken down while it was still warm and could absorb evenly and deeply the moisture which would come with the autumn rains, a needful preparation for the autumn sowing.

frondem caedi: in his directions for the planting of the farm Varro writes of the need for a *silva caedua*, "spinney", which will provide leaf fodder (see above, xxiii. 5). Now is the time to cut down the leaf foliage when it is at its greatest spread and can most easily be dried in the heat of these months.

prata ... secari: there has already been a hay harvest in May, at the time of the rising of the *Vergiliae* (above, xxxi. 4), which was followed immediately by irrigation of the fields (above, xxxi. 5) to ensure a second crop. Growth is rapid in summer provided there is a good supply of moisture, and irrigated fields would quickly produce a second growth of grass. The second hay harvest, is ready to be reaped in July and August and will be dried and kept for winter fodder for the animals on the farm, especially for the working oxen who are not taken out to pasture.

xxxiv. 1-2.   *The Farmer's Calendar: The Sixth Period*

1. Sexto intervallo: the sixth period extends from the autumn equinox, September 21st, until the setting of the *Vergiliae*, an interval of thirty-two days.
incipere scribunt ... serere: "they write what it is necessary to begin to sow". To this point the passive infinitive or in places the active has been used in the formula for the farm operations of the calendar. With *scribunt* there is a departure from the accustomed grammatical form, probably to be explained by Varro's love of variety for the sake of variety. Some scholars, however, consider that Varro quotes other writers at this point, but this is an unnecessary consideration: he had enough practical knowledge from the agriculture practised on his own estates to be able to draw on his own experience.
usque ad ... unum: the time to begin sowing grain was at the autumnal equinox, that is on September 21st. The operation could be done during any of the next 13 weeks but not later because that would bring the given time up to the winter solstice, after which as we are told in the next sentence it was not profitable to sow.
coegerit: the perfect subjunctive is probably to be recognised because the clause is in indirect speech depending on *scribunt* which is to be repeated from the previous sentence: so too *intersit* in the following line.
non serere: with this *scribunt oportere* is repeated.
quod ... intersit: *quod* refers to the warning already given not to sow after the winter solstice. *quod ... intersit:* "a matter which makes such a difference that ...".
ut ... existant: the clause expresses consequence "with the result that crops planted before the winter solstice sprout in seven days (but) hardly sprout in forty days if sown after the winter solstice".
minus idoneae: "unsuitable", "unfavourable", "harmful" (to the sown grain). *minus* is equal in meaning to a negative.

fabam ... seri: the wording seems to indicate a return to the familiar formula containing the accusative and the passive infinitive, but it could be governed by *putant* in the previous sentence. Beans and similar legumes were usually sown in the autumn in Italy, but they might be sown in spring. A summer sowing of mixed forage has already been recommended to take place after the harvest in June. An autumn sowing is recommended now but should be best done in the period which follows at the setting of the *Vergiliae* at the end of October. This mention of beans seems out of place: perhaps it is an illustration of Varro's hasty style, but he could mean that the sowing of beans should come at the very end of the sixth period. It is to be noted that Vergil advocated spring as the time for sowing beans: *Georgic I, 215.*

uvas ... occasum: the argument is that although the sowing of beans can be put off until the setting of the *Vergiliae*, the vintage must be gathered between then and the equinox and then it cannot be delayed because the grapes must be gathered immediately they are ripe and pressed without any delay. If they are allowed to hang on the vine after ripening they will lose their true flavour, and grapes once gathered do not remain fresh for any length of time. All the operations of the vintage must therefore be completed by the end of October.

vites ... propagare: immediately after the vintage comes another pruning of the vines and layering. The roots have been pruned and the ground broken up and trenched where necessary in the first period in early spring. In the third period the new vines have been dug, ploughed, and harrowed, the ones which will bear all this year's vintage have been thinned; in the fourth period between the summer solstice and the rising of the Dog Star the old vines have been harrowed, the new ones harrowed for the third time. In this review of operations Varro in the vineyard gives a clear description of the annual round from one vintage to another. Pruning and layering follow immediately on the vintage, and can continue into March: at the beginning of Varro's calendar and therefore of the farmer's year for him comes *ablaqueatio* in the first period; Varro then gives a second digging of the old vines between the summer solstice and the Dog Star, although along with a third digging of the new vines there must have been an earlier first digging which he does not mention. The young vines have been dug during the third interval, between the rising of the *Vergiliae* and the summer solstice. Work in the vineyard was continuous the year round, in digging, hoeing, pruning, thinning, and digging again.

propagare: for the methods of vine propagation see below, xl. 3-xli. 4.

serere poma: "you should plant fruit trees". The Romans knew several varieties of orchard fruits such as apples, pears, cherries, pomegranates, and several kinds of nuts such as hazel and almond: for methods of propagation see below, xl. 3-xli. 5.

Haec ... tempore: the operations which have just been mentioned, especially of propagation and planting should be postponed until the spring in districts liable to frost. This would be true of Northern Italy and Alpine regions, but not of Central or Southern Italy including Campania where the winters are warm enough to allow of planting in October. *aliquot* is an indeclinable adjective agreeing with *regionibus.*

materius, asperiora: Varro has a fondness for comparatives where the comparative meaning has little force.

xxxv. 1-2. *The Farmer's Calendar: The Seventh Period*

Septimo intervallo: the seventh period extended from the setting of the *Vergiliae* on October 28th until the winter solstice on December 24th, a length of fifty-seven days.

Serere lilium et crocum: flowers do not really have a place in an agricultural calendar which we should expect to be strictly limited to farm operations, but the growing of flowers was becoming more and more lucrative as the Roman life increased in luxury. Varro's attitude to the running of an estate is always with an eye to profitability and if flowers were profitable then they should be included. Even Cato (*De Agricultura* VIII) some hundred years earlier recommended that if the estate were near a city flowers should be grown for garlands of every kind, and shrubs from which boughs of evergreen could be cut, myrtles of various kinds, and laurels. Wreaths of every kind played an important part in Roman life both for public and private ceremonies and important occasions: benefactors were crowned with flowers, garlands were put on temple façades, on the shrines of the Lares, and on sepulchres: guests at banquets were crowned with roses and petals were scattered. Scented flowers such as roses and violets were used in the manufacture of perfumes and unguents (see above I. xxiii. 4 and note).

lilium: this was probably *lilium candidum*: a tall white lily: there was also a purple variety. The flowers were needed to stock *parterres* and borders. Vergil mentions this flower as growing in the aged Corycian's garden: *albaque circum lilia* (*Georgic IV*, 130-1) "white lilies around". The lily, being a fragrant plant, was useful too for bees.

crocum: this is the Latin *crocus sativus* or *colchicum* and is probably that which is called in England the saffron or autumn crocus; the leaves, long, fleshy, and pale green grow up in spring and die down before the flowers which appear in October and November. The colour of the beautiful cup-shaped flowers is of a delicate pale purple and in them are stamens of brilliant orange. The yellow saffron valued as a dye and as a flavouring agent is obtained from these. It is, like the lily, a good bee plant. It is propagated by seeds or offsets. It was grown extensively in the Middle Ages in England, especially at Saffron Walden in Essex to which it gave its name, for the dye which was used in connection with the woollen industry of the surrounding district.

rosa: the rose was a flower especially prized for the making of garlands and chaplets and for its perfume. Although the growing of special varieties which is so widely practised today by horticulturists was unknown to the Romans certain kinds were prized, especially those with a sweet scent and with many petals: rose oil and petals are mentioned by ancient medical writers; the rose is also mentioned as a bee plant. It grew prolifically in Campania, and those of Capua together with the roses of Praeneste were the most famous in Italy, especially for the number of their petals: the variety which had many petals, called *centifolia*, belonged to the district of Capua. The local wild rose may have been cultivated for the perfume industry for which Capua was also famous. Varro has already recommended the growing of roses on a farm which is situated near a city so as to take advantage of the demand for flowers and to supply the city markets: (see above, xvi, 3 and note).

Quae ... obruitur. a way of propagation of roses is described. "A rose which has already struck root, is cut from the root upwards into slips a palm-breadth long and set in the ground; when this has become a living root, it is afterwards transplanted." Small cuttings are taken from a growing tree, put first in a nursery-bed until they root and then are planted out where they are intended to grow.

violaria: violets were also greatly prized for their scent and were used largely in garlands. Varro has already recommended the growing of violets on a farm near a city where there is a demand for flowers: compare *rosa* and note in previous sentence (see above, xvi. 3).

Violaria ... macriorem: although Varro has recommended the cultivating of gardens including violet beds on a farm near a city so as to supply the local vegetable and flower markets, he goes on to give here an argument against the growing of violets, and at the same time gives a good example, but on a small scale, of impoverishment of the soil. They are plants which have to be grown on raised plots so as to facilitate irrigation by means of channels cut around them. The irrigation streams together with rain-water which will also flow along them wash much of the soil away. It is difficult to see why Varro so apparently contradicts himself, but he may mean that the loss of soil is worth it, where there is a good demand for the flowers. Violets are in these days one of the most prized flowers of Italy: they are raised for instance in rectangular beds in great quantity on the shores of Lake Nemi near Rome for the Roman market.

quod ... terra adruenda pulvinos fieri: "because raised beds must be made by heaping up the soil". *terra adruenda* is in the instrumental ablative: *fieri* is here the true passive of *facere*.

macriorem faciunt: "impoverish".

2. A favonio ... exortum: *Favonius*, the west wind of spring, according to Varro began to blow on February 7th: *Acturus* is the brightest star in the constellation *Bootes*, and one of three brightest in the Northern hemisphere: it is seen at its rising in line with the tail of the Great Bear (*ursa major*) and becomes visible at sunset on February 21st. This short period correctly belongs to the first period in spring which begins with the blowing of *favonius* (see xxix. i. above). The insertion of it here can only be explained as a kind of parenthesis following after the directions for planting the lily, crocus, rose, and violet. The underlying thought seems to be that these are all planted in the autumn but thyme is planted in early spring.

serpillum: Varro's etymological derivation from *serpit* is correct: the wild thyme roots easily and grows amongst rocks and on walls (Pliny describes a cultivated thyme which did not creep but grew to a palm in height: *N.H.* XX. xxii. 90). It was prized as a bee plant, used for flavouring sauces and fish, and as a medicinal herb. Vergil (*Georgic IV*, 30-31) describes it among the fragrant plants that should be grown near the hives as spreading: *olentia late serpylla* "wild thyme spreading fragrance afar". It is possible that the wild variety when cultivated grew more closely and profusely than in the wild state and that plots were planted to provide nectar for the bees. The plant belongs to the labiate family—a fragrant aromatic shrub with small leaves and whorls of small purple nectar-bearing flowers. The Romans knew also a cultivated variety, *thymus vulgaris*, which is probably the one described by Pliny. Varro may in fact include this cultivated thyme in his reference and may have introduced *serpyllum* for the sake of drawing attention to the etymology.

Fossas ... ne facias: the agricultural calendar proper now proceeds, in the customary formula (accusative and infinitive) with directions for farm operations which are quite apart from the flower growing with which the period is introduced.   There are three important farm operations for this period—*fossas novas fodere, veteres tergere, vineas arbustumque putare:* it is a time of digging and pruning.   Ditches could serve either for irrigation or drainage.   The season would be good for such work when the growth of vegetation, both crops and weeds, was at the minimum: the cool climate of the winter months would also ease the heaviness.

vineas arbustumque putare: the vines were to be pruned after the vintage in the sixth interval onwards from the autumn equinox.   It is now recommended for the seventh interval together with the pruning of the trees which supported the vines.   It will be seen that the same operation continues through the eighth period which begins at the end of December.   In February the trees are to be pruned again.   Clearly, pruning was continued for some five months during the autumn and winter and into early spring.

dum ... ne facias: several remarks have been inserted by Varro into the calendar by way of added instructions: this is another, with a *caveat* that there is to be an easing off of work for thirty days around the winter solstice. An interruption of this kind of the annual round may be due to some supersitition centred on that time of year.   There had already been the advice that there should be no sowing after the winter solstice: the reason given that seeds sown after would not sprout even in forty days does not seem to be the outcome of careful observation but rather of some supersititious belief.   *bruma* seems to be one of the few times in the busy agricultural year when some rest could be taken.   We are not to believe that nothing was done on the farm during these thirty days but that the main operations were not done.   It was a slack time; there was no heavy ploughing, no hurried vintage and pressing, no harvesting: activities were comparatively leisurely, and the weather was cool.

Nec non: the two negatives make an affirmative; the injunction is that winter is the right time for sowing certain things, such as elms.

ulmi: the elm as well as serving as a live prop for the vine was often grown on the boundary of the estate: in this case it could serve a double purpose.

xxxvi.   *The Farmer's Calendar: The Eighth Period*

Octavo intervallo: the eighth period extended from the winter solstice until the blowing of the west wind, *favonius*, on February 7th; it spread over forty-five days.   With this period we have come the full round of the agricultural year to arrive again in early spring, at the beginning of the first period.   The first fifteen days of the eighth period were a time of rest (see the end of the previous paragraph, xxxv and note).   The formula of the calendar continues, first in the passive infinitive and then in the active.

De segetibus: it was important that water should not stand in the grain fields. Cereal crops could be sown right up to the winter solstice (see above, xxxiv. 1) but if the soil was too wet the seeds might rot in the ground.

sarire: the sown field must be kept free of weeds.   This is a good time to remove them before they have made much growth.

putare: this is the fourth time that pruning of vines and in some cases the vine-bearing trees as well, has been mentioned. (See above, xxxiv. 2 and note.)

Cum ... hiberno: "when labour cannot be carried out on the land tasks which can be done indoors should be worked at". *antelucano* from *ante* and *lux* means literally "before dawn" (but here the inference must be that the work must start very early in the morning). During the days when the weather was unfit for outdoor occupations the hands could make farm tools of various kinds. Varro tells us in his list of equipment that nothing should be bought which can be made by the men on the farm from raw material grown on the farm, such as objects made of osiers and of wood, for example, hampers, baskets, threshing-sledges, winnowing-fans, and rakes: also others made of hemp, flax, rush, palm fibre, and bullrush such as ropes, cordage, and mats: see above xxii. 1 (text omitted). Cato (*De Agricultura* XXXIX) recommends that during bad weather manure should be carried to the heap, the animals' stables and the rest of the farmstead cleaned. Storage jars should be mended. There was only a little relaxation of work in bad weather and that the slave could not be called idle the indoor work which began at dawn would probably be continued until nightfall. Cato gives a warning that idleness must be prevented because expense goes on just the same.

ut vilicus norit: *norit* is in the perfect subjunctive (*noverit*). We have proof positive here that the *vilicus* had to be able to read. It is clear that he had to be an educated person as Varro has already stipulated (see above, xvii. 4). The calendar was only one of many documents which were in his charge. It was he who had to see that the farm operations were carried out at the right time and to watch the slaves and the seasons either with the help of skilful star-gazers or by his own ability. It was he who had to have enough knowledge and experience to interpret aright the instructions and to judge when the harvest or the grape was ripe. He it was who hired free labourers for vintage and harvest, who allocated work to the hands, who saw that the routine and round of the seasonal agriculture was carried out with the best returns and utmost economy.

Below is given the calendar extracted from Varro's account, with his remarks, etymologies, explanations pared away, in the form and bare outline in which it was probably posted up on the farm.

I. VER: FAVONIUS

seminaria ut serantur
putari arbusta
stercorari in pratis
circum vites ablacuari
prata purgari
salicta seri
segetes sariri

II. AEQUINOCTIUM VERNALE

segetes runcari
boves terram proscindere
salicem caedi
prata defendi
oleam seri interputarique

III.	AESTAS: VERGILIARUM EXORTUS	vineas novellas fodere, arare, occare vites pampinare omne pabulum secari prata inrigare in poma aquam addi
IV.	SOLSTITIUM	messem faciunt arationes absolvi offringi serendum legumina vineas veteres iterum novellas tertio occare
V.	AUTUMNUS: CANICULA	stramenta desecari acervos constitui arata offringi frondem caedi prata inrigua iterum secari
VI.	AEQUINOCTIUM AUTUMNALE	serere uvas legere vindemiam facere vites putare propagare serere poma
VII.	VERGILIARUM OCCASUS	fabam seri serere lilium crocum serpillum transferri fossas novas fodere veteres tergere vineas arbustumque putare
VIII.	HIEMS: BRUMA	de segetibus aqua deduci terram sarire vineas arbustaque putare

xl. 1-6. *Methods of Propagation: By Seeds, Slips, Shoots and Grafting*

1. Primum: this word is either an adverb meaning "in the first place" and referring back to the first method of propagation listed in the previous chapter, or what is more likely, an adjective agreeing with *semen* with the implication "the primary seed", that is the seed which originally caused growth to come into being.

duplex: Stolo speaks as befits his name (see above I. ii. 9 and note) and begins with the first of the four methods of propagation listed in the previous paragraph, that by means of seeds, but divides them into two sub-divisions, seeds which cannot be perceived and those which can.

unum ... opertum: "the one which escapes the notice of our senses, the other which is perceptible to our senses". *latere* usually means "lie hid", but here has the extended meaning with an object "lie hidden from ..." or "escape the notice of".

Latet... Theophrastus: Stolo takes Anaxagoras, the Greek philosopher, as his authority for the theory of spontaneous growth, drawing on a quotation from

208

his writings in Theophrastus (*Historia Plantarum*, III. 1. 4). According to
him the universe was made from atoms which were called seeds and seeds of
everything were in the air.   When certain seeds were carried by rain down to
the earth and became embedded in moist soil, they could sprout and grow into
plants.   Thinkers who held this view believed in the possibility of spontaneous
growth, that is that a plant could grow from a seed which had no parent.

Illud ... diligenter: Stolo does not dwell long on the primary seed, but goes
on to the second kind of seed, that which was perceptible.   The farmer should
know the appearance and properties of various seeds: some would be easier
to grow than others.   This is explained in the following sentence.

cupressi: the cypress tree is still today one of the glories of the Italian
landscape.   It is a tapering tree, sometimes reaching a height as great as
ninety feet; its branches are covered thickly with small shining evergreen leaves.
The cones which are sessile and usually in pairs are still called in Italy by the
name, *galbulo* (plural *galbuli*).   They are made up of hard angular scales
which form a globe-shaped berry: each contains a few seeds.

2. Primgenia semina: "the primordial seeds". Stolo speaks in philosophical terms, using an adjective which implies "that which is the first to
come into being".   These are the seeds which cannot be perceived by the
senses and from which comes spontaneous growth, according to the theories of
the Greek atomic philosophers.

Prima ... nata: the ring and repetition of words in this sentence is to be
noticed, *priusquam sata nata* repeated.   The first kind of seed existed of itself,
but in order to grow it had by some means to find its way to a suitable bed of
soil, the second, gathered from the first could not grow until it had been planted
by human hands.

Prima semina: *Prima* now goes back to the first method of propagation in
the list given in the previous chapter [xxxix. 3 (text omitted)] and not to the first
kind of seed mentioned above in this paragraph (xl).   The words refer to seeds
produced by nature from plants and sown by the farmer.

Prima ... adulterina: the injunction to the farmer to look carefully at the
seed which he will plant is now enlarged upon: [see above (1) *id videndum
diligenter*].   Good seed and fit for planting must not be old, and must not be
mixed with other kinds.   Importance is attached to this stipulation even today
by the Italian peasant.   In the autumn the visitor to any Italian village will see
old women and girls sitting on the steps of their houses sifting seed in large flat
wooden trenchers which are held on their knees.   There is little doubt that this
task could be one carried out by the slave-women on the estate.

ut naturam commutet: Stolo thinks that old seeds can change their nature
and produce other plants.   *ut* with the subjunctive here expresses consequence.

Nam ... brassicam: it has long been known by scientists that throughout
nature like can only produce like: the same principle was known to most of the
Greek philosophers, but Varro, in Stolo's words, accepts the theory that some
kind of radical change of species is possible.   Since *rapum* is a root vegetable
akin to the turnip, and *brassica*, cabbage, is a leaf vegetable, such a change would
be an absurdity.   The mistaken view perhaps arose from seeds of similar
appearance becoming mixed and then cultivated together, or when plants from
a former crop had grown again amongst a new crop.

3. Secuunda semina: "the second method of propagation".   It is to be
noted that *semen* is used for anything which when set in the ground grows into

a plant and includes as well as actual seeds, cuttings, shoots, and grafts. The second method is the growing of cuttings, or slips, taken from the branches of trees: it was the most successful way of propagating vines, olives, and figs and was universally practised in the Mediterranean region in the ancient world just as it still is today. Its usefulness was manifold: the growing time was far less than that needed for propagation from seed, and there was the certainty that the plant reproduced would be of exactly the same stock as the parent. Seeds do not always reproduce the parent plant exactly because some alteration in variety may have come about through cross-pollination. Cuttings were planted in nursery beds and carefully tended to enable an adequate root system to be developed before they were set out in the fields in the place for their permanent growth. Trenches were dug in the beds about three feet deep, covered at the bottom with a layer of stones for drainage and then filled with manure and earth. The soil and the exposure to sun and winds had to be closely similar to that of the field into which the rooted cuttings would eventually be transplanted. Regular irrigation might be necessary until a start had been made: the cuttings remained from two to five years before being transferred to a permanent home. In modern olive culture the slips are taken from pruned branches, which are smooth, tender, and about three centimetres thick: they are cut into pieces about forty centimetres long and paired off straight at each end. Then they are planted in rows in the nursery bed about fifty centimetres apart and put deep in the earth with about a finger's breadth above the surface of the soil: after three or four years they are transplanted before the trunk and foliage are too developed. There is no reason to suppose that there was any difference in the Roman method. When the young plant was set out care was taken to see that the sides which faced north and south still did so with exactly the same orientation as when it was growing in the nursery bed, so that, to quote Vergil *Georgic II*, 268, "the plants should not fail to recognise their mother (earth) suddenly changed".

mutatam ignorent subito ne semina matrem

This method of propagation as advocated by Varro is still today considered the best by modern olive growers. It is economical both because branches cut off in the course of the regular seasonal prunings can be used and because it does no harm to the tree. The olive tree by its nature provides another means of propagation which is practised in modern times and doubtless was known to the ancients. Nodes which sprout out new shoots grow on the tree near the surface of the soil at the foot of the trunk: these are called by the Italians *ovuli*: they can be cut out from the parent tree when they are in the form of a nut and planted by the same method as are the cuttings from pruned branches. New plants are easily obtained from them since they are the means by which the tree propagates itself.

unde tollas: this phrase seems to be redundant, but is an allusion to the trees from which cuttings (*secunda semina*) are taken. Exactly the right time must be chosen, and this will be different for different sorts and species. *tollas* describes the process of obtaining cuttings, not of transplanting.

Tempus ... idoneum: according to Theophrastus (*Causa Plantarum I. vi. 3.*), winter is excluded as a season suitable for planting. During the other three seasons there would be both warmth and humidity which were essential for growth. This is enlarged on in the following sentence according to the nature of the terrain. This is a good example of what the ancient farmer had discovered by experience and experiment. Olive cuttings were customarily, and still are, planted in March: this corresponds to Varro's *vere*. It seems however that others, possibly from olive trees, could be taken in autumn from July to the end of October: (the Dog Star rises on July 21st).

In sicco ... argilloso: the olive grows well on this kind of soil since it prefers a dry calcareous earth or a mixture of clay and sand with gravel below. Such soil does not become waterlogged and so can withstand the spring rains which, if too copious, would rot the young shoots. On the other hand Varro advocates a heavy warm soil for certain special varieties of olive [as in xxiv, 1 (text omitted)]. He adds that land to be suitable for the olive must face west and be exposed to the sun.

fere diebus XXX: an adequate explanation of this statement is difficult to find. The meaning could be that the best time for planting is either during the thirty days from the beginning of spring on February 7th when as has been seen in the farmer's calendar *favonius* rises (see above, xxix. 1) and from the rising of the Dog Star on July 21st, or the statement may have reference only to the latter period of autumn; the six weeks then counted would bring the time up to the middle of August. This is likely to be a time of drought and great heat necessitating constant irrigation. Another quite different meaning which at any rate seems logical might be that the cuttings could be expected to strike root in thirty days.

4. seminis: the third method of propagation is by means of a living shoot taken from a tree. The term *semen* is applied equally to this as to seeds and slips. The shoots were planted immediately in the place where they were destined to grow. The method was useful in that the labour of transplantation was dispensed with and sometimes quite large branches could be utilised thereby reducing the growing period.

quod defertur ... terram: "what is transferred from the tree to the soil in the shape of shoots", *i.e.* by means of taking shoots.

si ... demittitur: *demittitur* does not mean "buried in the earth" as some translators suggest, but "if (or since) it is planted directly in the soil" in the place that is where it is intended to grow. Shoots could not be buried, although cuttings might be, because they already had branches and leaves. See the use of *demittunt* below in this paragraph.

in quibusdam est ... oportet: the grammatical connection with the earlier part of the sentence does not hold, but *quibusdam* could be taken as referring loosely to *surclos* or to *arbore* understood now as plural.

ut ... oportet: "that the shoot be taken at the proper time". *deplantare* describes the act of removing a shoot from the parent tree: this must be done before the buds appear on the parent tree: otherwise if the sap were already rising "bleeding" could occur.

quid: "at all", this is an internal accusative with the intransitive verbs *gemmare*, *florere*.

quae ... defringas: the clause depends on *videndum* above: "you must see that what you take for planting from the tree, you tear off rather than break off". The full meaning of *deplantare* is to take a shoot from the stock in such a manner that some part comes away which will serve as a growing surface, a piece of bark perhaps with some inner tissues or, in the case of suckers, with roots. *planta* from which the verb is compounded is used for any kind of twig, graft, sucker, bough, etc., which will grow when planted. Another meaning is "the sole of the foot". The shoot must be taken in such a way that it has a "sole". A play upon words is perhaps to be recognised here, after Varro's manner. Vergil gives the same instruction:

> *hic plantas tenero abscindens de corpore matrum*
> *deposuit sulcis,*

(*Georgic II*, 23-4). "This man pulling slips from the mothers' tender stock, plants them in furrows." The drawback to this method was possible harm to the parent tree which would diminish yield in consequence.

in terram demittunt: the use of *demittunt* is beyond doubt: the shoots are set in the place in which they will remain and grow into trees: this corroborates the meaning of *demittitur* above which becomes a pointless statement unless the inference is that the shoot (*semen*) is planted once and for all where it is to remain.

in oleagineis seminibus: "in the case of olive cuttings". Stolo has reverted to the second method of propagation.

aequabiliter praecisum: the cuttings were probably not sharpened, as *praecidere* might be thought to imply, but cut straight across the top and base as is the modern practice. The words *in ... pedales* exactly describe the modern practice.

clavolas, taleas: both words mean "a small stick".

5. **Quartum genus seminis:** the fourth method of propagation is grafting. The great problem for the farmer was to get a good root system well established in a new plant. The climate of Italy where there is great heat during the summer months and the rainfall is not great, causes difficulty. Any tree, however, which was already firmly grounded with roots grown deep into the soil, even wild varieties, could receive grafts and could produce good fruit. Grafting not only made use of adult trees thereby reducing the growing time needed for the production of new plants, but offered other advantages. The farmer could cause barren trees to become productive provided good shoots were chosen to reproduce quality and quantity, and at the same time could preserve varieties and increase them. A plant grown from a graft would retain its characteristics unchanged, whereas one raised from seed might not be true to type. Other advantages to be gained from grafting were the repairing of a damaged tree, the replacement of damaged branches, the elimination of pests and blight, and general improvement of the tree. Grafting is still widely practised by the Italians in the propagation of the vine, the fig, and orchard trees. To quote from a modern manual "Grafting has the fundamental function of improving the production both in quality and quantity. It is necessary to pay great attention to matching graft with subject tree and to the choice of a shoot better than the variety to which it will be attached and to see that the join between the two parts is perfectly soldered."

Quartum ...... in aliam: *quod* is either *genus* or *semen*. There seems to be no particular reason for the subjunctive in *transit*: it might be explained as a clause depending on *videndum* which governs three indirect questions and so a clause inside virtual indirect speech or as a generic subjunctive with the implication "the kind such as to ...". The construction of the sentence as often in Varro is imperfectly connected. The clause *quartum ... aliam* stands apart and is almost a statement, but is incomplete without *est*. Some such translation as "with reference to the fourth method" or "in the case of the fourth method", is needed.

videndum ... obligetur: *semen* understood from the beginning of the sentence is the subject.   Both verbs are in the subjunctive in an indirect question. Varro gives three conditions for grafting: the first is the kind of tree to be grafted, the second the season, and the third the method of fastening the graft to the tree.   All these must be observed if success is to be reached.

neque enim si ... pirum: "not even if".   The apple will receive a graft from a pear-tree.   This statement seems to suggest that fruit trees will success-fully take grafts from other fruit trees.   Vergil speaks of the arbutus (the wild strawberry tree) grafted with nuts, planes with apples, chestnuts with beech, the elm white with pear-blossom, and acorns eaten by pigs beneath elm trees (*Georgic II*, 69-72).

> *inseritur vero et fetu nucis arbutus horrida,*
> *et steriles platani malos gessere valentes,*
> *castaneae fagos; ornusque incanuit albo*
> *flore piri, glandemque sues fregere sub ulmis.*

Hoc ... concepit: "Many people believe this who put their faith in sooth-sayers: these say that that which created the flash turns at one stroke into as many bolts as the varieties which have been grafted on separate trees".   The meaning is obscure, but seems to indicate a superstition that trees grafted with other kinds were more liable to be struck by lightning, that each graft would attract a flash.   The indicative in *concipit* is to be noticed; it can only be explained as a clause added by Varro to the saying to explain *tot fulmina fieri* or as a piece of careless grammatical writing.   The *caveat* amounts to saying "every one of your grafts will be separately struck by lightning".

Si ... non sit: the construction is that of the accusative and infinitive, but it is not clear on what it depends.   It could be explained as part of what the sooth-sayers advised, and as grammatically dependent on *a quibus proditum*, and then is to be regarded as an injunction of the soothsayers not to graft on a wild fruit tree, but this practical advice does not accord very satisfactorily with the warning about thunderbolts which amounts to pure superstitition.

inserueris is either in the perfect subjunctive as part of the statement in indirect speech, or in the future perfect in accordance with Varro's looseness of style.   *sit* follows the rule for the subjunctive in a subordinate clause in indirect speech.   Varro might be making this statement as a quotation from some writers, perhaps Theophrastas who is mentioned above (xl. iii).

6. dumtaxat ... malus: "for example if both are apple trees".   *dumtaxat* usually means "provided that" but here is used to make a supposition "given that".

ita ... arbor: "you should so graft bearing in mind the (resultant) fruit, (*i.e.* 'if you value results') that the shoot is from a better variety than is the tree to which it is grafted".   In accordance with one of the purposes of grafting, it is only worthwhile if the grafts are taken from the best varieties of trees.   *refer-entem* agrees with *te* to be supplied in a general sense.   *arbor* comes in meaning after *est*.   Columella (V. xi. 13) states that any kind of shoot (*surculus*) can be grafted on any tree if it is not different in respect of the bark.

altera species: the second method of grafting is when a branch not yet severed from the parent tree is grafted on to another tree close at hand, and only separated

when it has taken hold.   The usefulness would naturally be limited.   Columella
(V. xi. 13-15) describes it in order to show that any kind of slip can be grafted on
any kind of tree.

nuper animadversa: Varro's words suggest that the method of grafting one
live tree to another is a recent discovery: he may mean that it was not learnt
from the Greeks but was invented by the Romans and that fairly recently.

in arboribus propinquis: "when the trees (which are to be grafted) grow
close together."

vult: the subject is some person in general: "he wishes" is equivalent to
saying "you wish" or "anyone wishes".   Varro often uses this as a variant on
the generalising second person singular.   The same is true for *vult, traducit,
implicat.*

et . . . implicat: "and he inserts it (*ramulum*) in the branch of the (receiving)
tree which has been cut across at the end and split".

eum locum: The description jumps from the prepared tree to the graft.
There is a break in the sentence construction here but the meaning runs on,
"that which comes in contact with the place" (that is the prepared end of the
tree which will receive the grafted branch), is sharpened with a grafting knife
on both sides.   The end of the branch is pared off so that it has a long pointed
end which can easily be inserted into the receiving tree where a corresponding
slit has been made.

ita . . . habeat: "in such a way that on one side the part which will be exposed
to the weather will have bark accurately fitted to bark".   The bark of the graft
and the grafted tree must fit exactly to keep out the rain and as a safeguard
against heat.   They will be tightly bound together so that they become, as it
were, one tree.

Eius . . . curat: "He takes care to see that the point of that branch which he
will insert is set straight up."   If the graft is set in such a position, this will
enable it to take firm hold.   This seems to be the explanation of the difficult
words *ad caelum.*   The actual top of the branch is grafted so that it is necessarily
put in the receiving plant pointing downwards.   The injunction therefore seems
to imply that the graft must not be put in at an angle.

quem inseret: the generalising third person continues but is now changed to
the future.

Postero anno: the next year it will be possible to sever the grafted branch
from the parent tree.   It could be for instance that an olive branch now grows
on a fig tree as a result.

xli. 1-6.   *Grafts and Shoots*

1. Quo . . . transferas: "at what time (you graft), and what you graft".
Attention must be paid to the season in relation to what you are grafting.
Different times suit different plants as Varro goes on to say, because of the
nature of their fibres.

haec in primis videnda: "these plants must receive particular attention".
The neuter is used in a general sense.

nunc . . . solstitiali: supply *tempore inserebantur*, repeated from the previous
line.   Since the summer solstice is the season commended for the grafting of
figs, Varro writes as if this was a new discovery in agriculture, which may be so
since Cato names the spring as the only time for this operation (*De Agricultura,*

XL): he adds the requirements that the grafting be done after midday in periods of no moon when the south wind is not blowing. The fig ranked with the olive and vine as of the highest importance in the diet of both the Greeks and the Romans. It seems to have been indigenous throughout the Mediterranean region, being by its nature able to withstand the heat and drought of the summer months. Although the wood is soft, fibrous, and porous, the roots are long and spreading, able to penetrate down to the deeper moist layers of soil. The comparatively scanty foliage does not lose much water by evaporation. It grows best where the rainfall is light and the soil thin and dry: a well-drained hillside is the best locality. The domestication of the fig spread from the east, first to the Greeks who learnt it from Caria in south-west Asia Minor and from them to the Romans who considered the variety, *ficus Caria*, among the best. Several varieties were known to Varro (see above for the *ficus marisca*, vi. 4). These had increased by Pliny's time to twenty-nine and in modern times there are at least six hundred varieties known throughout the world. The figs ripened in June and September: the fruit of the later ripening could be left on the trees to be picked when needed. In order to increase the yield of figs the Greeks discovered the practice of caprification which is in fact a kind of cross-pollination. By attaching flowering boughs of the wild fig to female branches of the domestic tree, the minute insects (called by the Greeks *pselles* and now *blastophaga grossorum*) living in symbiosis with the flowers of the wild tree could pollinate the domestic tree. This was done in June and September. They also achieved the same result by planting wild figs in the fig orchards. The fig tree was of the greatest use to the ancients; it could grow in poor soil, it needed no special care, training, or pruning, its wood was tender, light and porous; it could be propagated by every known method. The fruit was easily preserved by drying. It played a large part in the normal diet together with bread, oil, wine, and grapes, and also figured largely in the food given to slaves, especially those working on the land: a large amount was provided for men to eat fresh at the fig-harvest. Although the fig loves warmth it can be grown successfully in England and produces ripe fruit in abundance if it is set in a sheltered sunny spot, preferably against an old high wall.

tenellum ... putre: the subject is *aqua* repeated from the previous sentence: *tenellum* agrees with *insitum* also repeated. "Because (water) quickly makes a very young shoot rotten." If moisture percolates in between the graft and the parent tree, before it has taken a firm hold, it can cause the graft to rot. *tenellus* is a rare word, a diminutive of *tener*. Varro uses it in one other place (II. vii. 11) for the hoofs of very young colts.

2. caniculae signo: the time for grafting the fig has just been given as the summer solstice, in June: the "Dog Days" when the Dog Star is in the ascendant, come in July and coincide with the hottest time of the year in Italy. The fig is safely and desirably grafted during the hot months because then excessive moisture is easily dried out from the soft wood.

ea: the fig tree.

Quae ... mollia: the harder wood is slow growing and so unless precautions are taken the shoot may dry before it coalesces and begins to sprout: "in the case of plants which are less soft-wooded". This clause is used in the same way as an accusative of respect—"as to those which ...". It is not grammatically connected.

vas aliquod ... alligant: the *vas* would be a terracotta jar probably similar to a plant pot, with small holes in the base through which the water could drip.

ne ... colescat: "to prevent the shoot from drying out before it becomes one with the tree".

Cuius surculi ... medullam: it is clear that the preparing of shoots for grafting called for skill. The bark had to be kept intact and the point for insertion so cut that the inner tissues were not exposed. This would need meticulous care and special tools.

ut non: "in such a manner that you do not ...". The clause expresses consequence.

Ne ... obligandum: as well as being closely tied the graft at its junction with the receiving tree is smeared with clay and bound round again with bark to insulate it against both moisture and heat.

3. inserant: there seems no reason for employing a subjunctive in this clause which is purely temporal.

ut qui ... inseratur: "so that what superfluous moisture is in it (the vine graft) may run out before it is inserted". *inseratur* is perhaps in the subjunctive because it depends on the purpose clause which contains the subjunctive *defluat* and is attracted to that mood.

aut in quam ... possit: an alternative method to drying out a shoot for grafting is described: "or into that (branch) into which they have inserted a graft, in that a little below the grafting point, they make an incision by which any superfluous moisture can drain away". This is done and left for some time, not specified, before the grafting is done.

continuo: "immediately": supply *inserunt*.

gemmam: this will become a growing point after the graft has taken.

4. tardiora: two plants are mentioned in this paragraph which grow slowly from the natural seed, and are therefore better propagated from cuttings: the first is the fig which has a very small seed, which would be slow in growing, and the olive which produces a nut the hard covering of which would be slow to germinate.

Fici ... edimus: the natural seed of the fig is inside that fig which we eat. *ficus* means both the tree and the fruit.

quae: is loosely connected with *semen* but attracted into the plural to agree with *grana*.

parvis: "because they are so small".

laxiora: "of looser tissue".

et fecundiora: supply *sunt*: "are at the same time of more rapid growth"; this contrasts with *ad crescendum tarda*.

ut femina ... item: "just as the female grows more rapidly than the male". This is, relatively speaking, true for plants also.

malus punica: "the pomegranate tree". It was so-called by the Romans because they received it from Carthage, but in all probability its original home was to be found in Persia and the neighbouring countries. The tree is small, with shining lance-shaped leaves and scarlet flowers. The fruit is full of a large number of seeds and an acid juice. This was prized in antiquity and was used both medicinally and for tanning.

5. in hoc umidiora quam aridiora: supply *sunt*: "in this respect the more moist are quicker than the drier". The plants which are moist and (therefore soft) grow more quickly than those which are dry. The use of comparatives without any special significance is to be noticed.

praeter ... nequeas: "except if you cannot do otherwise" (than plant seeds in order to propagate). The subjunctive seems superfluous but can be regarded as representing a generalising second person singular.

ut siquando quis ... vult: "as (for instance) if ever anyone wishes to ...".

tum ... pariant: after the second person singular in *nequeas* and the third person singular in *vult* in the previous sentence, we now have the third person plural; all three forms of the verb are used for giving instructions about how to plant.

ficos quas edimus: the actual fruit and not cuttings or suckers.

complicant: "they tie them together" or "pack".

inaruerunt: from *inaresco*.

ubi ... pariant: "so that they may grow into plants after being set in a nursery". In modern times seeds are considered as one means, but not the best, of propagating figs.

6. perlata: from *perferre*.

enascebatur: the imperfect tense carries the meaning "has been found to ...". "has been well known to" or "always ...".

l. 1-3. *Methods of Harvesting Grain*

1. Messis ... declinata: "the word is appropriately used in speaking of these things which we 'measure', especially of corn, and is said to be derived from that word". Varro is mistaken in his suggested derivation: *messis* is connected with a root *ma*, which means "to gather", "to reap". There is no connection with *metiri*, "measure".

esse ... declinata: the construction is the nominative and infinitive depending on *dicitur* above.

Umbria: this is a district in north-west central Italy, lying on the west of the Apennines between Tuscany and Lazio: it is crossed by the wide upper and middle valley of the Tiber and contains Lake Trasimene (*lacus Trasimenus*). Wide areas are given over to intense cultivation of which the best product is grain. It is one of the most important grain-growing districts in Italy. The chief town is Perugia.

falce: the tool used was the *falx messoria*, the harvest sickle, made up of two parts, a smooth curved blade, sometimes rather more than a semicircle and a short rounded wooden handle not larger than the grasp of a man's hand. The sickle was well-balanced and could be held comfortably for work.

manipulum ... terra: "they lay each sheaf as they have cut it on the ground". *manipulus*, compounded from *manus* and *plere*, means in the first place "a handful" then "a bundle" and in this context a sheaf, as much that is, as could be grasped in the reaper's hand. In the action of reaping in Varro's first method, the reaper grasped a bundle of stalks as near to the ground as possible in his left hand and drew the sickle through them and towards him with his right hand. By this method no straw was left in the ground for a second cutting but the ears had now to be separated from it. Cutting was easier when the straw was damp rather than dried out in the hot sun which made the surface slippery; it was often preferred therefore to reap in the early morning and in the evening to take advantage of the moisture from dew which might lie on the fields.

de singulis ... stramentum: they go over the sheaves where they have been laid on the ground and cut off the ears from each sheaf, picking up each as a whole, in the bundles in which they have been laid on the ground. This method of harvesting involves two operations, the cutting of the straw close to the ground with the ears attached and then separating the ears. The latter are taken to be threshed.

corbem: a basket of reeds, osiers, or willows made on the farm. It was amongst the many pieces of equipment which could be made in bad weather when the workers sheltered in the barns.

unde ... acervum: the straw which has been left in bundles on the ground, is later stacked. The threshing must be done first because the grain must not be left on the threshing floor to deteriorate or to become wet.

2. Altero modo: the second method requires another kind of tool for cutting off the ears as close as possible, a sickle with a serrated edge. This seems to have been a kind of saw with the blade set at an acute angle to the haft: with this the reaper worked upwards forcing the bunch of ears into the space between the handle and the curved saw, then when the stalks were held firmly the heads were easily cut off.

in Piceno: Picenum is the district in the east of Italy on the Adriatic in a central position. It extends on the west into the Apennines and is a highly fertile region. Varro's reference shows that it was an important grain-growing district.

fascem spicarum: the bundle which in this method is cut off is the amount which can be caught at one stroke in the curve of the sickle.

ut ... subsecentur: the method involves a second operation, the reaping of the straw from which the ears have been severed and which has been left standing on the field.

metitur: the form of the verb is in the impersonal construction from *metere*, "reap", "gather".

ut ... Roma: the district around Rome was part of Latium, the country to the south-west of the lower course of the Tiber which lay along the coast and extended inland to the Apennines and south-eastwards to Campania: later in Augustus' time Campania was included in Latium. The plain on which Rome stands and which is meant by *sub urbe* was, and still is, extremely fertile because of the rich volcanic soil with which it is overlaid.

locis plerisque: the wording seems to suggest that the third method of harvesting is the most generally employed.

ut ... subsicent: the clauses expresses "consequence"; the reaping is of such a kind that . . ..

medium: "in the middle".

summum: *stramentum* is to be repeated from the line before: "they grasp the top of . . .".

messem ... puto: Varro enlarges on his suggested derivation of the word, but it is erroneous: (see the note on l. 1 above).

Infra ... subsecatur: the stalks have been cut in the middle; what remains below the hand is afterwards cut off. The latter implies close to the ground.

2-3. contra ... defertur: "by contrast the straw which is attached to the ear (lit. 'clings with the ear') is carried in the baskets to the threshing-floor". Differently from the other two methods, a certain amount of straw is put on the floor.

3. Ibi ... palea: the suggested derivation of *palea* chaff from *palam* is untenable but that from *stratus* (itself derived from *sternere*) is correct.

cum ... dicatur: "in regard to this it is said that in this (reaping) one operation is approximately enough for one *iugerum* especially on easy land". *una opera* implies the work of one man in a day. Perhaps Varro refers to the second and third methods which involve one operation by which to reap the ears ready for threshing: in the first method both cutting the straw from the ground and cutting off the ears make heavy work for a day and are equal to double the task of the other two.

Messas ... debent: it is not clear whether it is meant that the day's work includes carrying the ears to the threshing floor or whether this is just a repetition of what has been said already. The carrying of the grain as well as the cutting would be very heavy work to be done in a day; the slaves probably worked from early morning until dusk.

li.   *The Threshing–Floor*

1. agro: this word seems to go closely with *loco*: "on raised ground".

quam ... ventus: a raised place for the threshing floor must be chosen which catches the wind. This is important as will be seen below, for the process of winnowing. Compare Varro's description with Vergil, *Georgic I.* 178-86.

modicum: the size of the threshing floor must be related to the crop.

extumidam: the floor should have a camber sloping from the centre to ensure good drainage.

si pluerit: the verb is in either (1) the future perfect indicative corresponding to the generally future tense in the main verb *oportet* which gives a command: lit. "if it will have rained" which is to be expressed in English as "if it rains", or (2) in the perfect subjunctive in a clause inside another (*ut ... non consistat*), which itself contains a subjunctive, in primary sequence.

omne ... extremum: "it is a fact that in a circle the shortest way is from the centre to the circumference".

Solida terra pavita: the surface must be compact, beaten hard, and smooth.

peminosa: another reading is *paeminosa*. Although there is some doubt about the meaning of this word not found elsewhere it seems to indicate that the surface of clay soil might become cracked in heat if it is not treated: a synonym therefore may be *rimosa*, "full of cracks", especially since *rimis* is used in describing in the same line the effect if the ground is not hard and dense enough. Clay being impervious to water, cracks in hot weather, and in wet weather becomes waterlogged. Precautions must be taken against conditions such as these.

ne ... aperiant: "to prevent ... their opening doors" or giving entrance to such pests as ants and mice. Even if the threshing could be done soon after the reaping and carrying, the work might not be completed for several days or perhaps not until the winter months. After that was finished the grain had to be winnowed, that is, separated from the chaff. A favouring wind blowing over the floor could be a great advantage and would reduce the labour considerably and for that reason some farmers might wait until such a wind blew. Whatever the conditions there is no doubt that the grain would have to be kept on the floor for some days and the necessity followed to guard against the vermin which might make inroads on it.

amurca: the dregs from the olive press as distinct from the oil. It was a thick grey fluid, served a great many purposes, and seems to have been invaluable on the farm (see I. lv. 7. below, where Varro speaks disparagingly of some people who let it run to waste). It was used as a fertiliser, a pestifuge, a weed-killer, as an impervious glaze for the interior of storage jars, for painting around vines and fruit trees to keep off insects, making plaster for walls, as a remedy for sheep scabs, and for many other purposes. It was easily stored and could be thickened by boiling down. Cato (*De Agricultura* CXXIX) gives instructions for the making of threshing floors which includes two dressings of *amurca*: "Dig the soil fine then, pour over it *amurca* and let it soak in thoroughly. Afterwards break up the clods. Next level the earth and beat it smooth with a roller or rammer. Afterwards pour *amurca* over it again and allow to dry.

2. muniunt ... pavimentum: if the earth is too soft to make in itself a hard floor, some farmers put a layer of stones below the surface to ensure good solidity, or they might even pave the floor. The second alternative was probably not to be recommended because the stone surface would be liable to bruise the grain.
in Bagiennis: these are probably the same as the *Vagienni*, a branch of the Celtic Caturiges who were inhabitants of the mountainous parts of Liguria, a locality subject to sudden showers in summer time.
id temporis anni: the genitive is partitive and *id* is to be explained as an accusative of respect.
Ubi ... meridiano: the instruction to build a shed nearby for the hands to shelter from the hot sun seems to be humanitarian but at the same time it is utilitarian. Slaves must have their siesta during the middle of the day when the heat is greatest and must be under cover: neither they nor the hired labourers who were employed to help at harvest time could be expected to continue working in the noonday heat.

lii. 1-2. *Methods of Threshing*

1. Quae ... spicas: some grammatical connection is needed between *quae* and *spicas*: this kind of anacoluthon (that is, a break in the construction) is common in Varro. A paraphrase is necessary in translation such as "ears of the largest and finest crop should be separated and taken to the threshing floor and kept apart". Literally the Latin reads "whatever crop will have been the largest and best the ears should be separated apart on to the threshing floor".
semen optimum: the next sowing must be provided for from the farmer's own resources. No doubt the seed was, even after being selected in this way, sorted again; this seed-corn of such importance to the farmer must be threshed separately. No doubt that there was a task here which the slave women could usefully do.
habeat: the subject is "some person" in a generalising sense: "the farmer".
e spicis ... grana: the grain must be separated from the ears on the threshing floor. After harvesting which involves at least three operations, the cutting of the ears and of the straw, the carrying in baskets to the threshing floor, there come two more before the grain is cleaned ready to be food for man. First the ears must be threshed on the floor so as to be separated from the straw and husks and secondly the chaff must be cleaned away from the liberated grain.
tribulo: the *tribulum* was made of a heavy board like a sledge studded on the underside with flints or iron points: a team of oxen dragged it over the ears

and straw on the threshing floor; to ensure good results, a heavy weight was put on it or the driver sat on it as he drove the beasts in circular movements around the floor. As the derivation from *terere* (*tritum*) shows, the rubbing action separated the grain from the chaff.

plostellum poenicum: "the Punic cart" crushed the straw by a revolving process. Varro must have seen this machine during his active service in Spain. It was a framework containing several rows of spiked rollers which crushed the straw and liberated the ears as they revolved. A chair was mounted on it for the driver. It was in reality an improved version of the *tribulum* using the same basic process. It is thought to judge from the name that it was learned from Carthaginian farmers in Spain. It does not appear to have been generally adopted after Varro's time although it must have been more effective than the threshing sledge and easier to work.

in eo ... iumenta: "some man sits on it and drives the team which pull it." The subjunctives are probably jussive, "a man is to ...." *quis* is rarely used in this sense unless it is preceded by certain conjunctions such as *si*, *ne*, etc.

2. grege ... perticis: "by means of a herd of oxen driven on (to the floor) and prodded with goads". They are made to go round and round the floor so that their hoofs tread out the corn. This is the most primitive kind of threshing employed before special machines had been thought of.

vallis aut ventilabris: when the threshing is finished the process called "winnowing" follows. There is now on the threshing floor a mixture of grain and chaff, but the grain must be further cleaned so that not even the finest chaff remains. The threshing floor has been purposely built on high ground where it may catch the wind, especially the gentle west wind which blows in the summer. The workers take winnowing fans (*valli*) and with them toss the mixture into the air across the wind so that the light chaff is blown away and the heavy grain falls back on to the floor. The wooden instrument which they use has been made from timber grown on the home farm (see I. xxii. 1 above, text omitted): it is a combination of a shovel and a fork and has the functions of both. Five prongs in the front enable it to be easily slid into the mixture and the scoop-shaped portion behind them holds it together so that it may be tossed into the breeze. The grain may also be separated from the chaff by means of a winnowing basket, *vannus* or *vallus* (feminine, declension 4). This, made from osiers (see above, xxiii. 5) is a large basket roughly in the shape of a wide shovel closed at one end with two handles on each side. By means of skilful shaking the grain is sifted from the chaff. *vallus* may be a diminutive of *vannus* (vannulus), the winnowing basket. This is the word used by Vergil (*Georgic I*, 166).

ut ... evannatur: the clause expresses consequence: compare *veniat* in the next line.

ad corbem: the grain lies heaped on the threshing floor cleaned of chaff and can be scooped up into baskets ready to be conveyed to the barn.

liii. *The Gleaning*

spicilegium: derived from *spica* and *legere*, lit. "the gathering of the spikes", "the gleaning". As the reapers have gathered up the cut corn, inevitably some stalks and ears have been left on the ground. A residue such as this can be sold, provided that the purchaser will supply his own labour.

compasci: if the gleaning is not enough to be sold, or to be gathered, it should be pastured. Labour costs are thereby avoided and the cattle benefited: (see below II. ii. 12 where the reasons for this practice are given).
Summa: "a consideration of the highest importance".

liv. 1-3. *The Vintage*

1. vindemiam ... ut videas: the limiting or concessive use of *ut* gives the meaning, "but yet you should see ..." or "although you should see ...". The complete idea is: the vintage may be undertaken, with the reservation that you must consider the kind of grape and the position of the vineyard. Some grapes, because of their situation, will ripen more quickly than others. In Varro's calendar the vintage could begin any time from August to mid-October, but gathering of different kinds of grapes could go on until the middle of November. The vintage is a time of rejoicing and revelry. The gathering has of necessity to be done in haste so that no grapes deteriorate on the vine; hired hands are needed in addition to the farm-slaves and the owner himself, if an absentee landlord, will probably come in person to superintend. Work is ceaseless for close on the gathering comes the treading of the grapes and the pressing so that the juice (*mustum*) may be extracted without any harmful delay. The colour, taste, transparency and state of the pips will show the expert when the grapes are ripe for the picking. At Rome the time for the vintage was signalled at the festival of the *vinalia rustica* by the *flamen dialis* himself for the festival and the cultivation of the vine was under the protection of his deity, Jupiter. On the first day of the festival, after the sacrifice of a goat, the *flamen* took a cluster of grapes and squeezing the juice into a cup offered it to Jupiter with a prayer for his favour on the crop. While these ceremonies were being performed, the crowd of worshippers chanted "*io messes et bona vina date*", "Give us good harvests and good vintage". The measures taken to prevent premature gathering of the grapes is shown by a law of the XII Tables of Roman law which forbids the vintage to begin before the leaves have fallen, in other words before the clusters are perfectly ripened. When the day came, the grape-gatherers, slaves and hired hands, assembled and their overseer led them to the vineyards: this had to be not too early in the morning because of the dew which might spoil the quality of the grapes and not at midday because then the grapes would be too hot and ferment too quickly. Some workers cut down the clusters with a small hook-shaped instrument, the *falx vineatica*. Others filled baskets (*corbulae*) woven of osier grown on the farm and made in winter or wet weather. The filled baskets were then transported on the back of a pack animal, an ass or a donkey, first to be trodden and then to be pressed. Some however went straight to the fruiterer to be sold for the table or hung up to be preserved, or put into jars to be kept until needed.
a quo loco: the grapes will ripen most quickly in the sunniest part of the vineyard and these will be picked first so as to allow the rest to remain longer on the vine. As the whole process of gathering and wine-making covers at least forty days there can be an extent of several days for the actual gathering to take place.
praecox: there were several table varieties of grapes of which one was *praecox*, which means literally "early ripener". It is not clear whether Varro names a variety, or uses an adjective to describe this quality.

miscella: a hybrid grape of poor quality but easily grown in any locality.

multo aute coquitur ... legenda: "ripens much earlier (than the others) and for that reason (lit. 'by which') must be gathered first".

2. In vindemia diligentis: the possessive genitive of an adjective is here used as a noun: "in (the course of) a careful (farmer's) vintage".

ad bibendum: "to be manufactured into wine".

eligitur ad edendum: "is selected for eating", *i.e.* for the table. *eligitur* contrasts with *legitur* to show that the best were picked out. The best variety of grape for the table was the *bumastus* or *bumamma*, the berries of which were exceptionally large and either white or black: they might be eaten immediately off the vine, or sold, or kept on the estate and preserved by being hung up in the fruit room or stored in jars.

lecta ... vinarium: the gathered grapes are taken to the press where the two processes, treading the fruit and pressing the pulp which are necessary for the manufacture of wine, will be carried out. The vat for the treading of the grapes, the floor of the press and the cellar for storage are side by side in the large barn-like shed which placing reduces the labour of transport to a minimum. The grapes are placed in a rectangular stone trough set on a rather high plinth with several channels grooved on either side through which the new wine (*mustum*) will flow into basins (*lacus*). Those who tread the grapes (*calcatores*) do so with bare feet often dancing to the music of the pipes to keep up a quick and steady rhythm. The grape-pulp becomes a slippery mass on which it is easy to fall: to keep their balance the labourers must either link arms or lean on crutches or hold on to ropes slung above their heads. Vergil describes a scene of vintage: *Georgic II*, 4-8.

unde ... veniat: the juice which flows from the vats is poured into large storage jars (*dolia*) which are fixed in rows in the floor of the adjacent wine-cellar, there to be left to ferment until the following year or even longer.

electa ... corbulam: the grapes selected for the table are put aside in separate baskets to be carried where they are needed.

unde in ollulas ... descendat: different methods of preserving grapes are described: some may be put in small jars then stored inside a large jar packed with wine refuse, or others may be stored in an amphora sealed with a coating of pitch (or resin) and kept under water in the farm pond.

aliae ... escendat: others which are to be eaten at once will be put in their proper place in the larder. No doubt this is in a cool place away from the sun.

Quae ... prelum: the treading of the grapes does not extract all the juice, but there remains in the mass an amount which can be obtained under greater pressure. The refuse from the trough is carried to the adjoining press heaped in a basket covered with a wooden lid. The type of press known to Varro consisted of two sets of heavy upright beams set in the floor opposite to each other. A long heavy beam (*prelum*) some twenty-five feet in length was attached to one set of uprights by means of a cross bar and slot; the beam was held at the other end between the second pair of uprights by means of a windlass and strong leather ropes: a stone weight might also be attached to the free end of the *prelum*. When the beam which moved up and down on the fixed end was lowered on to the mass, pressure was applied to it by means of the windlasses and what juice remained was thus squeezed out and ran along runnels into the same vats as those attached to the treading trough. The pressure was exerted aslant the basket which had to be turned about so that both sides were equally treated and all the juice possible forced out.   When the pressing was completed all the wine was poured into the great storage jars (*dolia*) which were sunk up to the rim in the floor of the cellar proper (*cellarium*) which was immediately adjacent.

3. circumcidunt: "some (wine producers) cut around the edges (of the refuse which remains) and press again".   There is yet some juice left on the outside edges of the mass and so these are piled up in the middle where the pressure is greatest, to be pressed again.

circumsicium: lit. "cut around".   The wine extracted at this stage is of inferior quality, and is kept separate because the taste has been spoiled by the iron blade of the spade.

Expressi ... hieme: here is another instance of the diet provided for the farm hands: drink of some kind was essential, but an inferior quality was given them, a kind of "after wine" made from the refuse of the grapes.   It was probably like the Italian Grappa, a strong inferior variety.

lv. 1-7.   *The Olive Harvest*

1. De oliveto: "In regard to the oliveyard": another interpretation is to take the phrase closely with the rest of the sentence especially with *legere* and *quatere* and translate as "from the oliveyard".

quam ... possis: "such as you can . . .", generic subjunctive.

quatere: the olive that is shaken down is bruised when it hits the ground and loses some of the oil.   It also deteriorates very quickly just as does fresh fruit when it has been bruised.

ea quae vapulavit: "that which has been hit": an olive which falls on the ground when the tree is shaken will suffer a blow.

digitabulis: authorities disagree about the meaning, translating either as "gloves" or as "pincers", both of which would fit the connection of the word with *digitus* "finger".   It is difficult to see how gloves even if made of hard leather could harm the tree.   "Pincers" made of metal would answer to *durities* and certainly could clip off pieces of bark to the harm of the tree: their use might also make for rapid and therefore careless work.   A cultivated olive is not usually higher than 20 feet and is frequently pruned and trained into a dome shape for easier gathering.

2. ad gelicidium: a piece of bark clipped off inadvertently, leaves a wound on the tree which will allow the cold of winter to penetrate the inner tissues. Since the olives are gathered at the end of the year from November to mid-December the damage cannot be repaired by nature until the following spring. There is very little frost in the Italian climate but a light hoar frost sometimes lies on the ground in the early morning and even this could have a deleterious effect on olive trees which only grow in a hot climate.

retectos: the suffix *re* is often a negative: "uncovered", "unprotected" and therefore "exposed".

quae ... poterunt: those that are out of reach.

ita ... ut feriantur: the limiting use of *ita ... ut* gives the general meaning that some olives have to be shaken down, but the branches may only be struck with a reed and not with a pole.   The pliant reed will not damage the twigs and branches as might the pole which is rigid.

gravior ... quaerit: "the heavier blow (from a pole) calls for a doctor". A humorous remark, meaning that first aid will be needed.

3. adversam: agrees with *oleam* to be supplied. The fruit itself should not be struck, because the shoot itself may then come away. The small white flowers are borne on the last year's wood in racemes springing from the axils of the leaves.

plantam: the small shoot which bears the ripe olives may come away with the berries and will not bear the following year.

Nec haec non minima: "neither is this the least (lit. not smallest . . .) reason": the negatives do not cancel each other as often. The care of the farmer for his trees is summed up: the two precautions to be observed, not to damage the bark and not to break off shoots if observed, will ensure a good crop the following year.

4. Olea ... cibum: the olive, just like the grape, returns by the same two ways to the home farm, one part for food (to be eaten) the other to be liquefied. *bivium* is a place where two ways branch off (as the word itself implies). Varro uses the expression metaphorically. The harvest is transported by the same way to the home farm where are store rooms and press rooms adjacent to each other as has been seen in the case of the vintage (see above liv. 2 and notes) and once there the olives are sorted out for either of the two purposes; they can, figuratively, go either of two ways.

unguat: olive oil was an essential fat in the diet of the ancient Romans performing the same function as butter in the modern world.

in balneas ... sequitur: the Romans seemed to have learned the use of baths in special buildings with cold, cool, and hot rooms from the Greeks. The oldest known in Italy are the baths at Stabiae on the gulf of Naples built about 80 B.C. The practice later grew of building baths on farming estates: there can be little doubt that there was such a one on Varro's own estates. The bath-house would contain three rooms, an *apodyterium* where the bather's clothes were left, then the *frigidarium* (cold room) and *caldarium* (hot room). After going through the successive stages of the bath, the bather would be rubbed down with oil to remove the effects of sweat and make the skin smooth. Large public baths usually called *thermae* which were built in Rome developed later than the time of Varro and are still among the greatest monuments of the city especially the Baths of Diocletian and of Caracalla. In his time, however, there seem to have been as yet no public baths; the first ones known to have been built were constructed by Agrippa after Varro's time in 25 B.C. *gymnasium*: oil was also much used by athletes as an embrocation to gain suppleness of limb.

5. Haec ... tabulata: "These from which oil is made, are accustomed to be heaped in piles from day to day on floors." *tabulata* are probably shelves or platforms made of wood. Stone might bruise the berries.

ut primus ... demittatur: *acervos* = *acervus*: the subjunctive in *demittatur* is governed by *ut* to be repeated from the previous line. Each day's picking is kept apart so that the first to be brought in is the first to be pulped. This ensures that no heap is left for too long or short a time and that the oil pressed out is of uniform quality. *primus quisque* implies "each heap".

per serias ... olearia: *seria* is usually translated as "jar" see below, III, ii, 8. *vasa olearia* are the jars in which the oil is stored, but at this moment they are being used for carrying the picked olives to the press room.

ad trapetas: the spelling is more usually *trapetum* and sometimes *trapetus*: there seems to be doubt about its true form. It was necessary first to crush the berries without breaking the stones and then to press them. The *trapeta*, which would be kept in the press room close beside the beam press, was an improved type of olive press. It consisted of a large stone basin, usually of grey lava of which the donkey mills were also made (see above xix. 3). In the middle of it was a solid column a little higher than the tip of the basin and of one piece with it. A square hole on the top of the column held a strong upright pin fastened with lead. Two millstones flat on the inner side and convex on their outer side were set upright in the basin and kept in position by a wooden beam which was threaded through them both and fitted over the pin. When the beam was pushed round, the stones performed a double rotation, going round the centre column and twisting on their own axis. The movement was sufficient to crush or pulp the olives but to leave the stones intact. This was important because crushed stones spoiled the flavour of the oil.

ex duro et aspero lapide: "made from . . .", describing the *trapeta* which is made wholly of stone except for the wooden beam. Olive presses of this type have been found at Pompeii and elsewhere.

6. oleum foetidum fit: the oil in the uncrushed olives if left too long becomes rancid. In every farming process care must be observed so as to obtain the best quality: the care of the olive berry is no exception.

in acervis iactando . . . oportet: "by tossing the heaps, you must give them air". The separate heaps of olives must be raked over or tossed to keep them fresh. Great care would have to be exercised to avoid bruising the berries; this would probably have to be done every day until they were crushed.

7. cuius utilitatem: Varro suggests that many farmers, through ignorance, do not make use of *amurca*. This is surprising since it serves a variety of purposes: (see above, li. 1 and note).

licet . . . fluere: "you can see it flowing out of the olive presses on to the fields". The dregs, so prized by Varro, are allowed by those who do not know its uses to run to waste out of the presses. *e torculis:* the olives have been pressed under press-beams just as were the grapes earlier in the season: the same apparatus may have been used although the wording suggests that these presses were out in the open. This is unlikely because shelter from rain would be necessary. Varro may mean that the dregs were taken from the press room and poured away outside.

cum . . . cum . . . tum . . . vehementur: the first *cum* governs *pertineat:* "since this liquid is of great importance not only for many purposes, but also . . .".

quod: as used here the word means almost "for example". The drenching of the tree roots with *amurca* as a fertiliser is not its only beneficial use.

ubicumque . . . herba nocet: "wherever harmful weeds grow". *amurca* was, amongst its many other qualities, a weed-killer. Varro gives a piece of observation here of what he has seen on other people's farms. In lxi below (text omitted) it is stated that good farmers store *amurca* in casks just as carefully as they do oil or wine. It should be boiled down to two-thirds as soon as it comes from the press and put in jars when cool.

END OF BOOK I

LIBER SECUNDUS

Praefatio. 1-6.

1. rusticos Romanos: the Romans admired the simple life of earlier times and looked back with nostalgia to those days when distinguished men farmed their own land and worked with "black hands". One such was Lucius Quinctius Cincinnatus who cultivated a smallholding in Rome across the Tiber. During the wars against the Aequi, he was summoned from the fields to the city and hailed as dictator. Yet, after his victory, he renounced all power and returned to his life on the farm. Livy's version of his story exemplifies the essential longing among the Romans for a simple rural life (III, xxvi, 7).

nonis modo diebus: "only on the ninth days". This is equal to saying in English "every eighth day". In the Roman method of reckoning the first and last day was included in any given period of time. A survival is probably to be found in the French fortnight, *quinze jours*. Varro is referring to the *nundinae*, the market day, the original meaning of which is "ninth day". From earliest times the year was divided into "weeks" of eight days, the last day of which (the ninth by Roman reckoning) was a market day. Farmers left their work on the land and came to the neighbouring city, Rome, or elsewhere, to trade their produce, and to transact public and private business including even lawsuits. The *nundinae* were marked on the official calendars by a special letter: in later times when the old eight-day week fell out, the word came to mean merely "market" and the connection with *novem* was forgotten. In the old ideal, the farmer lived in the country on his farm and only came into the city on business.

2. gymnasia: from earliest times there were *gymnasia* in the Greek cities where athletes could train for the contests at the great games at Olympia and other sanctuaries and where the ordinary citizen could take part in athletic exercises. The *gymnasium* was in reality a public sports ground provided for the use of all citizens. Running especially, riding and throwing the spear and the discus, ball games and other exercises could be practised. In addition there was a *palaestra*, a wrestling school, in the shape of an open quadrangle with sanded floor, and provision for undressing and washing. In the sixth century B.C. there were three gymnasia in Athens, the Lyceum, the Academia, and the Cynosarges. In Rome there were no gymnasia as such until the time of Nero: the nearest equivalent was the Campus Martius, a low, level extent of ground on the left bank of the Tiber, which was used as a sports ground. The aim throughout Greek athletics was physical fitness. The Greek institutions were carried on into Roman times but they did not play such an important part in the lives of the Romans as among the Greeks. In Varro's opinion bodily health could be obtained equally well from a country life and work on the farm.

Quae ... sunt: "Now hardly one apiece is enough". Because the degenerate farmers of Varro's day live so much in the town and do not work with their hands, they have to keep up their health by athletics and even have a private *gymnasium* at the villa in the country.

vocabulis ... Graecis ... oporothecen: after the conquest of Greece in 146 B.C. the impact of Greek culture became effective in all aspects of life.

Private tutors and teachers in schools taught both the language and literature with the result that all educated Romans knew Greek. Greek architecture and art also exercised powerful influences and even private homes were influenced by Greek models. It had become fashionable with them to use Greek words as a sign of refinement and elegance. *procoetion* is an ante-room leading into a larger one: *palaestra* a quadrangle for sports usually surrounded by colonnades for shelter. The *apodyterion* was next to the baths and is literally an undressing room where garments could be taken off and left while the owners took a bath. There were baths warmed by underfloor heating and hot pipes in many villas. The *peristylon* is an enclosed garden surrounded by colonnades as the Greek word indicates. *ornithon* is the aviary. Varro's own farm in Campania had a notable aviary described in Book III. v. 8-17. *peripteros* is a pergola, a succession of arches covered with climbing plant such as roses, to make a pleasant, scented, and shady walk. *oporotheca* is a room for storing fruit.

3. patres familiae correpserunt: Varro is attacking the absentee landlord who leaves his estate in the charge of a *vilicus*, "bailiff", who will even hire the paid workers.

manus movere in theatro et circo: we have here a glimpse of the many amusements provided for the populace of Rome at the end of the republic. In Rome games and chariot-races in the Circus Maximus and theatrical shows of all kinds were manifold, enough to keep a large indolent population well amused. Out of the old religious festivals had grown the system of holding *ludi* which might last for several days. These were organised by the state on a lavish scale. The *Ludi Magni* in Varro's time lasted from September 5th to the 19th. On the 13th there was a great triumphal procession to the Capital and thence to the Circus Maximus nearby where the games were held. In addition there were the *Ludi Plebeii* in November, the *Ludi Apollinares*, from July 6th to the 13th, the *Ludi Megalenses* in April, from the 4th to the 10th, the *Ludi Cereales* from April 12th to the 19th; lastly the *Ludi Floreales* from April 28th to May 3rd. On all these occasions there were shows in the Circus Maximus and theatrical shows in temporary theatres erected of wood all of which were free to all Roman citizens. These became more and more lavish as the demand grew for more magnificence to such a degree that it became the custom for the magistrate in charge of the *ludi* to supplement the state allowance out of his own pocket. It is no wonder that life in Rome became seductive compared with life in the country, which might be considered dull by comparison. *manus novere*, "to busy their hands": in the country this was done by working but in the city by applauding at the shows and in the theatre.

frumentum ... Chia: "we employ (a man) to bring us grain with which we can be filled". The first *qui* is a final relative with the antecedent (*eum*) left to be understood. The second *qui* is an ablative form of *qui* (*quae quod*) in a final sense. The importation of grain into Italy, in spite of the fertility of her own soil, began after the Second Punic War, towards the end of the third century B.C. when Rome gained command of the Mediterranean; the vast slave estates in Sicily, Sardinia, Carthaginian Africa, and Spain flooded the country with grain, cheap because transportation by sea was easier than on the mainland where roads were as yet difficult. For this reason the cultivation of grain at home declined. As more African territory was annexed, supplies of tribute and purchase grain increased still further. From 102 B.C., however, the seas were infested by pirates to an extent that imports were stopped. This state of affairs on the sea lasted until 67 B.C. (see Varro's Life), when Pompey suppressed the pirates around Sicily, Sardinia, and Africa and the Cilician coast. Then the seas were once more free and the grain trade revived. So it is that Varro can state that grain in his time was imported from Sardinia and Africa to the detriment and degeneration of the Italian farmer. The importation of foreign wines marks a desire for luxury and was not due to scarcity at home. There was never any lack of homegrown wines, but the best wines came from the Greek islands of the Aegean, especially Cos and Chios: varieties were developed and different flavours manufactured which pleased the Roman palate. *navibus vindemiam condimus*: the expression is sarcastic. The vintage is not carried in baskets as it would if it were gathered at home, but in ships: "we get in the vintage from ships". *condere* is "put in store", in the case of wine "put in the cellar". The sentence contains the condensed thought, "the vintage which we put in store in the cellars has been gathered in ships".

4. in qua terra: the phrase is shortened from *in ea terra in qua qui condiderunt urbem*: the antecedent to *qui* is *pastores*. In the legends concerning the beginnings of Rome the people concerned came of a pastoral community. Romulus and Remus it was said came of shepherd stock and they were adopted and brought up by the shepherd Faustulus who pastured his sheep on the Palatine Hill.

contra leges ex segetibus fecit prata: "in spite of the laws made pastures out of corn lands": see Chapter II, *Latifundia*.

opilio: "shepherd", the word is derived from *ovis* "a sheep".

si possunt ... armenta: "(even) if the herds can graze on the land". *si* is used to make an admission. The argument is that even if grazing takes place on the land, just as agriculture is work on the land, yet the two are not the same. The reason is given below (5). *armentum* is used as a collective word for all grazing animals (cattle, sheep, and goats).

armentarius ... bubulcus: the *armentarius* is the herdsman, the man who takes grazing animals to pasture: the *bubulcus* is the ploughman who has to do with the working of the soil and drives the working oxen: the word is derived from *bos*, "ox".

bos domitus: see above I. xx. 1-5.

causa fit ... novali: because of its manure the ox improves the soil and the produce grown on it, both grain and green fodder. For *novalis*, "fallow land", see above, I. xxix. 1 and note.

5. Alia ... pastoris: the skill and knowledge of the farmer is one thing, that of the shepherd another. They each have different functions, one is to produce from the land, the other to produce from the flock.

contra: is here used as an adverb.

societas: "association", "interchange". The farmer and the shepherd are not wholly apart from each other and their work is not unconnected; there is inevitably some overlapping. The farmer stays and works on the farm, the shepherd may range afar on distant grazing grounds.

quarum: the grammatical reference must be to *ratio ac scientia* but the more natural interpretation is to take this relative, used here demonstratively, as referring to *colonus* and *pastor*.

Quarum ... pastionis: the clauses in this long, but not involved sentence should be separated as follows:

1. *quarum ... magna:* supply *est.*

2. *propterea ... fundi: compascere* and *vendere* are subjects of *expedit* which governs the dative (*domino*).

3. *et stercoratio ... oppositum: quod* is to be repeated twice from the previous clause.

4. *qui ... praedium:* this relative clause looks on to the subject of *debet* which is the main verb: *is* should be supplied as the antecedent to *qui*.

pecoris: two systems (*disciplinae*) have been under discussion, that of the farmer and the other of the shepherd: a third is now added, that of rearing animals on the home farm, such as birds, fish, bees, etc.

6. E quis ... feci: "since out of these (three) parts I have written a book on agriculture for my wife Fundania because she owns a farm". At the beginning of Book I Varro announces his intention to write three handbooks for his wife's guidance in running the estate which she has just bought. This seems to imply that all three are dedicated to her, but there is no difficulty if Varro is understood to mean that the three books are considered to be for her use and that the first is dedicated to her. There is no inconsistency then, when Book II is dedicated to a friend, Turranius Niger, and Book III to another, Pinnius. Since Book II is written on the subject of cattle and sheep rearing, the dedication is apposite. Although nothing is known of Turranius, his name is probably connected with *taurus*, "a bull", which in Umbrian is *turu*: his name is also spelt Turannus. Varro seems to make one of his customary puns in the following remark: "*qui vehementer delectaris pecore*", "you who take a real delight in cattle".

E quis ... praeessem: a long trickling sentence such as this one appears in places in Varro: it is a characteristic of his uneven style. The clauses can be separated as follows:

1. *quoniam ... pecore:* causal clause.

2. *tibi, Niger Turrani ... de re pecuaria breviter ... percurram:* this, the main clause, contains several others. It is to be seen that *tibi* contrasts with *Fundanae uxori* in the previous line.

3. *qui ... pecore:* relative clause describing Turranius.

4. *propterea quod ... pedes:* a causal clause, explaining the previous statement.

5. *quo facilius ... ministres:* a final clause introduced by *quo* associated with the comparative.

quod ... equarias: causal clause. In translating, a stop might well be put after *percurram*, then some such wording supplied as "I shall take my discussion from ...".

in campos Macros: this was a district near the town of Mutina (modern Modena) and Parma, lying between the Apennines and the Alps in Cisalpine Gaul. The River Po was to the north and the Apennines to the south. Although the name means "the Lean Plains" the Gaulish sheep reared here were prized. It comprised a very fertile plain and mountains rising from it: a situation admirably suited for nomadic sheep rearing (see Chapter II, Transhumance). Crossed by the Via Aemilia, one of the highways which led out from Rome, it was easily accessible. The modern name for the district, Emilia, has preserved the ancient one. Varro's statement leaves no doubt that the road served as a trade route for cattle breeders.

in Apulia ... equarias: see below, II, i, 16-17 and Chapter II on Transhumance.

in Reatino: "in the district of Reate", now Rieti, in the Sabine country (see Varro's Life).

collatis: from *conferre*.

cum iis ... magnas: among the cattle breeders with whom Varro was able to converse was in all probability Titus Pomponius Atticus, Cicero's friend to whom many of Cicero's letters are addressed. He had a large estate in Epirus (see above, I. xvii. 6 and note).

cum ... praeessem: see Varro's Life.

Incipiam hinc: most scholars agree that part of the text has been lost at this point because the conversation (or dialogue) which follows begins without any indication of the place where the speakers meet together or any description of the speakers themselves.

The place and the occasion of the conversation in Book II is difficult to decide since there is undoubtedly a gap in the text after *incipiam hinc* (*praefatio* 6). In Books I and III the setting and time are fully described, but in Book II we are plunged immediately into a conversation which has already begun. So much is clear however, that the speakers meet on the day of a festival, because it is stated later that sacred cakes (*liba*) have been offered and sacrifice has been made: and at the finish of the conversation a freed man comes with a message from Vitulus for Scrofa and Varro asking them politely not to make the holiday (*diem festum*) shorter by being late [(xi. 12), text omitted]. Since the subject of Book II is cattle and sheep rearing some scholars hold that the occasion is April 21st, the day of the *parilia*, the shepherd festival and the birthday of Rome. The scene of the dialogue however is probably some town in Epirus (it is called *urbem* in the last paragraph of the book) and the date about 67 B.C. (see Chapter III). This is in accordance with Varro's stated intention (see Varro's Life), to reproduce conversations which he had had with cattle owners when he was on active service with Pompey in Greece. The place is assumed to be either Buthrotum or Cassiope, both on the coast of Epirus: Varro could have visited either or both when he was in charge of the Greek fleet during the war with the pirates. Three of the speakers, Atticus, Cossinius, and Lucienus, are known to have been Epirotes and Epirus was famed for its cattle. Those who take part are, in order of speaking:

Cossinius: he is perhaps the Cossinius mentioned in Cicero's letters as the friend to whom he had sent a book written in Greek about his consulship (*Ad. Att. I*, 19, 20; II. 1): his income was drawn principally from cattle farms in Epirus.

Murrius of Reate: coming from Reate he was a compatriot of Varro.

Marcus Terentius Varro.

Scrofa Tremelius: he is the "grand old man" of agriculture, who also takes part in the dialogue in Book I. His writings on agriculture had great importance but nothing now remains of them (see Chapter III.)

Atticus: his full name was Titus Pomponius Atticus: he was a friend of both Varro and Cicero and had large estates in Epirus (see also below II. ii. 1 and note)·

Vaccius: nothing is known of him and he does not appear in the passages selected for this book, but the choice of name for the sake of a pun on *vacca*, "cow", needs no comment.

Quintus Lucienus: a man of Epirus, a senator and race-horse breeder (see below, II. v. 18 and note).

There are six speakers but other characters mentioned are Menates who goes off just before the discussion opens [i. 1 (text omitted)], perhaps to make preparations for the dinner to which he has invited certain of the speakers (xi, 12), and Pomponius Vitulus whose freedman comes with a message to ask Scrofa and Varro not to be late for the festivities of the day (xi. 12).

i. 11-16.   *The Science of Stockbreeding.*

11. Relicum: the spelling is an alternative to *reliquum*.

de scientia pastorali: the emphasis is on *scientia*.

rerum ... palmam: "the palm in all agricultural matters": a palm-branch was often the prize in athletic contests and on the race-course.

quo melius potest: "in which he is able (to speak) better (than the rest of us)". The antecedent to "*quo*" is *dicendum*.

Igitur: it is not usual for this word to come first in the sentence. The sense is "Since you wish me to speak on this subject, I will do so".

parandi: the meaning includes not only the actual acquiring of cattle but the building up of a flock or herd.

ut ... capiantur: "so that the highest possible yield may be procured from it (cattle rearing)". It will be seen below that care must be exercised both in the choice and purchase of individual animals and in the assembling of flocks and herds.

nam ... fundamentum: a pun is made, after Varro's habit, on the derivation of *pecunia* from *pecus*, "cattle". This is a correct statement since before the invention of coinage wealth was measured by the amount of cattle owned. Some of the oldest Roman coins (*asses*) which were of copper were stamped with the device of an ox.

12. Ea partes ... ternas: Varro's method of presenting his subject by means of systematisation has already been exemplified in I. v. 1-4 (text omitted). There a scheme is set out in four divisions but it is not logically adhered to through the book. In the present passage in Book II Varro continues to follow this method further in speaking of cattle rearing. Here the subject is set out in three main parts of which each is subdivided into three divisions. Attached to these are nine sub-sections which apply to the three main parts of the latter: four have to do with the procuring of cattle, four with feeding, and one applies equally to all. In contrast with the arrangements in Book I, the scheme is closely adhered to throughout the book. The arrangement of the subject-matter may be tabulated as follows.

SCROFA'S DIVISIONS OF THE SCIENCE OF STOCKBREEDING

1. Smaller animals: i Sheep
                         ii Goats
                        iii Swine

2. Larger animals: i Oxen
                        ii Asses
                       iii Horses

3. Ancillary animals: i Mules
                        ii Dogs
                       iii Herdsmen

   9 Sub-divisions (1)   i Assembling   ⎫
                        ii Characteristics ⎬ Assembling
                       iii Breed   ⎪
                        iv Law of purchase ⎭

               (2)   i Pasturage: (a) locality, (b) time,   ⎫
                             (c) manner   ⎪
                        ii Breeding: (a) mating, (b) points in   ⎪
                             breeding, (c) control of suck   ⎬ Feeding
                       iii Feeding   ⎪
                        iv Health: (a) causes, (b) symptoms,   ⎪
                             (c) treatment   ⎭

A ninth point common to all is the number of beasts in a herd.

Tertia pars: the third part of the science of cattle-raising contains those animals which are not in themselves productive, but are needed for working purposes and could therefore be called ancillaries. They are the means by which the industry is carried on and are in fact indispensable to it.

muli: includes also asses. Their work on the farm covered haulage, sometimes ploughing in light soil such as that in Campania, work at the mill (see above, I. xix. 3 and note), and porterage in pack trains. They could pull carts and could also be ridden. They were useful little beasts of burden, very strong, very patient, and are still seen today as working animals on the Italian farms. Asses were pack animals for steep mountain paths and rough tracks which were everywhere to be found before the paved roads were made.

canes: dogs were indispensable for the farm (I. xix. 3). Attached to the shepherds and their flocks one or two served to guard the fold, assisted in driving the animals, and prevented them from straying: (see chapter on dogs below, II, ix, 2-16).

pastores: this word which means literally "shepherds" must include herdsmen in this context. The *pastores* are the slaves who have charge of any of the farm animals. They are included among the animals and come last in the list. It seems that Varro counted them as on an equal with animals, an attitude which has already been met above where slaves are enumerated among the instruments of the farm and described merely as articulate compared with animals which are semi-articulate (I. xvii. 1-2). Slaves were the chattels of their masters, being stateless persons, and could be bought and sold by the same procedure as could the animals. It is surprising that Varro thinks no more of them than this especially since he recommends a certain humanitarianism in their treatment (see I. xvii. 3-7).

novenas: "nine sub-divisions each". There are nine sub-divisions which belong to the three main parts alike.

octoginta et una: the divisions and sub-divisions under which the subject of

233

stockbreeding will be treated amount in all to eighty-one. This kind of tabulation and enumeration is characteristic of Varro's method of presentation. It is more successful in Book II than in Book I. He tries to put shape into a subject which does not in all its aspects allow of such rigidity although his treatment often makes for clarity and a certain crystallisation of his subject matter.

communis una: the ninth sub-division which is common to all is the proper number of animals to keep in a herd or flock. Considerations in relation to making this decision are the extent of pasturage available, the number of females to males and the number of young that should be reared [(II. i. 24), text omitted].

13. Primum ... expediat: instructions are given for building up a herd or flock: attention must be paid to the age of the animals. This point is of the greatest importance. The difference in meaning between *pecus, pecoris* (neuter), "a herd" or "flock", and *pecus, pecudis* (feminine), a "beast", with a collateral neuter plural form, *pecuda*, is well brought out in this sentence. In English there is no single word equal to *pecus pecoris*, with the result that it must often be translated as "herd" or "flock". *quamque* is from *quisque.*

minoris ... annicula: the genitive is one of price. *annicula* is an adjective agreeing with *bos* understood: "a yearling cow".

decem annorum: the genitive is one of description.

bima aut trima: these both agree with *aetate* to be repeated from the previous sentence.

fructum ferre incipit: "begins to bear". A cow can bear a calf when it is two or three years old.

14. E quattuor ... qualis sit: lit. "But of the four divisions the second one is a knowledge of the characteristic appearance of each several kind of animal, what it is." "The second of the four divisions is a knowledge of what is the characteristic appearance of each separate kind of animal." *forma* includes more than shape, but the various "points", the breed, general condition, the young, and so on: in fact what constitutes the ideal animal.

Magni ... fructum: "It is of great importance in what way each one contributes to the profit." *magni* is normally used in this meaning with the impersonal verb *interest*: the genitive is one of price.

bovem emunt: the present indicative gives a statement of what is usually done, and therefore amounts to giving instructions. For a description of a breed of present day Italian working ox which is thought to be similar to that of the Romans see I. xx. 1 and note.

Ita ... parvis: the characteristics to be looked for in a she-goat and a sow would indicate that they would produce more young than the others not so well formed. *ut sint:* "ut" is used in a limiting sense "provided that ...", "only in so far as ...".

asini Arcadici ... nobilitati: asses were the most prized of the products of Arcadia, the central district of the Peloponnesus in Southern Greece. To the west are wild, bleak mountains, but on the east there are fertile plains between mountains of lower elevation. The plateau of Arcadia in the central Peloponnesus contains several small lakes with natural underground drainage. They are subject to flooding in winter and after the spring thaw which brings melted snow from the surrounding mountains. When the water subsides the basins are surrounded with a green girdle which provides excellent meadows for pasture. All Arcadia was famous through antiquity not only for asses but also for sheep, goats, and swine. The term "Arcadian" has come in English to mean proverbially that which is connected with a peaceful out-door pastoral kind of life. *nobilitati* is an adjective.

Reate: Varro speaks from experience of his own native Sabine country (see Varro's Life.)

venierit sestertiis milibus sexaginta: "has been sold for sixty thousand sesterces". *venierit* is from *venire*, "be sold" in the perfect subjunctive after *ut* used in a consecutive sense. The ablative is one of price. Ll. Storr-Best, writing in 1912, calculated this sum at £480: it would of course be many times more now.

unae quadrigae: is used for a team of four and can be applied to the animals with or without the cart which they pull: *unae* must agree grammatically with the plural form, although it describes the group. It seems probable that asses thus sold had been matched and trained to work together, just as were the pairs of oxen. For a reference to an expensive ass see below III. ii. 7.

quadringentis milibus: the writer mentioned above calculated this sum at £3200.

15. de iure in parando: "concerning the law of purchase". The buying and selling of cattle, *emptio venditio* (this term includes all farm animals), was carried out according to a legal formula which protected both buyer and seller and could be applied to transactions of all kinds, but which contained special clauses to do with the market place. Originally sale was a simple exchange of goods and money between two parties done on the spot and then and there finished with: the agreed price was paid in money, the purchase handed over. Later came the development of the law which brought in the use of stipulation (*stipulatio*): by this the buyer stipulated, for instance, for delivery of the purchase, the seller for the payment of the price. At first no liability was attached to the seller for defects in what he sold; the risk lay with the buyer. It came to be recognised however that the seller was liable if he fraudulently concealed defects, and in the case of cattle it became customary for a stipulation to be required of the seller that the animals were in good condition. The buying and selling of farm animals must for the most part have taken place in the local market on market-days. Here such transactions were supervised by the curule aediles. Among their duties were those of the inspection of markets and in this capacity a set of regulations was drawn up and then added to at different times known as the *edictum aedilum curulium*. By this code various stipulations (*stipulationes*) became recognised by law (though according to Varro, not always compulsory) which could be incumbent on the seller: the ancient wording of that concerning cattle has been preserved in the Digest of Justinian probably in words which Varro knew: "*qui iumenta vendunt palam recte dicunto quid in quoque eorum morbi, vitiique est*", "they who sell cattle (this can be understood to refer to all beasts) must openly and honestly declare what disease or blemish if any is present in each one of them". The responsibility was on the seller to make defects known. Forms of stipulations such as those quoted by Varro in connection with purchase of she-goats and swine (see II. iii. 5 and iv. 5), (the text of the latter is omitted), are attributed to Manius Manilius, who was consul in 148 B.C. The stipulations recommended by Varro had been in use and had become legalised for nearly a hundred years. The monetary transactions including the public declarations of the soundness of the herd or flock must have been made in the presence of the aediles, whose duty it was to guard against fraudulent dealings in the market: then in the case of default on either side the case would be referred to them as *iudices*. Under Roman law payment of earnest (*arra*) was allowed, an action which showed a promise to pay, and for payment to be deferred. It is not clear whether the animals themselves had to be brought to market or whether the seller's word was his bond: this might differ according to circumstances. Another law concerned with slaves dates from republican times (see below, II. x. 4 and note).

Quod ... meum: "so that what belongs to another person may become mine". This implies how a change of ownership might be transacted.

necesse ... intercedere: "some intermediate process is required". A proper agreement must be drawn up as described above, before a *iudex*. It is possible that some agreements were made in writing.

stipulatio: see above for those needed in connection with animals and the following sentence.

solutio nummorum: the actual payment of money.

i. 16-20. *Pasturage, Breeding, and Feeding*

16. Alterae partes ... sunt: "the second group of four sub-divisions". The first group was concerned with acquiring a herd and flock: the second is to do with the care of animals.

eius: the word is redundant but illustrates Varro's roundabout way of writing.

qui: the old ablative form of *qui* (*quae, quod*) is used to mean "how", lit. "by what".

ut: this probably means "as" and should be associated with an indicative *pascis* to be understood from *pascas*.

Neque ... pascendum: "Neither are the same places suitable for all animals to pasture both in the summer and winter." This statement is explained in what follows.

Itaque ... committant: see Chapter II, *Latifundia* and Transhumance. "For this reason flocks of sheep are driven long distances from Apulia into Sanmium for summer grazing and they declare them (*i.e.* the shepherds must declare them) to the tax-collector, in case, if they have pastured a flock that has not been registered they commit (offences punishable) under the censor's law." *aestivatum* is a supine expressing purpose. *inscriptum* usually has a positive meaning, but here a negative is undoubted; *paverint* is in the future perfect tense, used idiomatically in relation to *committant* which has a future implication. This is one of several valuable statements in which Varro describes transhumanting sheep: see also II. ii. 9-10, II. x. 1-4, III. xvii. 9 (text omitted) The reference is to the Apulian shepherds who lead their flocks along the cattle-tracks (*calles*) into the Sabine hills.

17. Muli ... montes: see above II. i. 14 and note, and Varro's Life. The name Rosea is probably connected with *ros*, "dew". On this plain the asses and mules were reared, which migrated for summer pasture up into the mountains which overlooked it.

fit satura: "is filled with", with the implication that the animal has had enough to satisfy it when it has been fed with hay. The good herdsman will see that the feed is varied.

quod ... cytisum: the grammar of this sentence hardly holds together: the form *faba* is inexplicable when related to the gerundive *abiciendum* and the same is true of *medica*. A possible explanation is that the gerundives agree only with the neuters *hordeum, lupinum, cytisum*. Instructions for the planting and growing of leguminous forage crops have already been given (see above I. xxiii. 1 and I. xxxi. 4).

feminis ... dicuntur: "feed is reduced for cows because they are said to conceive more readily when they are thin".

18. Secunda pars: there follows the second division in the second of the two main divisions concerning the science of cattle-rearing.

fetura: "breeding" or "gestation". This comprises the knowledge of the times when mating should be arranged, and the length of time during which the farm animals carry their young. The times are different for different species and have therefore to be watched with care.

suillo pecori: the approximate dates for swine between which mating should take place are when the west wind begins to blow, a time which Varro gives as February 7th (see above I. xxix. 1) but this can only be an approximation, and March 24th. The litters born four months later in June and July would benefit from the summer warmth.

ovillo: The dates for the mating of sheep fall between May 13th, the setting of the constellation *Arcturus*, and July 23rd, the setting of *Aquila* ("the eagle"). Careful timing was essential in relation to transhumance (see Chapter II). Since the flocks migrated between May and mid-June to their summer pastures and returned to the plains in September or October, in accordance with the modern practice they would be mated after the shearing and just before going to the mountains. The ewes would become pregnant about the time of the summer migration. After five months (see below 19), the lambs would be born soon after the return to the plains. In this way they were spared the long trek for which they would not, in any case, be strong enough, but it was essential to ensure that the ewes reached the winter feeding ground before they lambed. Some unforeseen delay on the return journey might result in a disastrous loss of lambs.

habenda ratio quanto ... habeant: attention must be given (as to) how long they keep (*i.e.* should keep, the males apart from the females before breeding begins).

binis: "two months each". The distributive numeral implies "two months for every kind".

et armentarii et opiliones: the first are the men in charge of the herds of cattle, the second are the shepherds. Their advice is founded on knowledge arising from experience which is one of the essentials of good farming.

19. Altera pars: *Altera* relates back to *primum* in 18 above: the second division (*fetura*) of the second four sections (18) has been divided into two parts, the first concerning the time of mating, the second the length of gestation.

in fetura: "when pregnancy has begun".

alia ... solet: "some are accustomed to produce their young at one time, some at another". It is very difficult if not impossible to decide with what *alia* agrees, but the gender may indirectly be influenced by *pars*.

ventrem fert: lit. "carries the womb": "is pregnant".

quinos: the distributive numeral is used presumably because Varro speaks of two animals, *ovis* and *capra* together. Since the goat is a species of sheep it is natural that both should have the same period of gestation.

In fetura: the full implication in the expression which is terse after Varro's usual manner, is "while we are talking of pregnancy, let me tell you a surprising but true story".

in Lusitania ad oceanum: "in Lusitania (now Portugal) near the ocean (the Atlantic sea-board)".

Olisipo: the town mentioned is probably Lisbon.

e vento ... equae: "certain mares conceive from the wind at a particular time of the year". The belief that a female animal could bear young without being fertilised by a male was firmly held among the ancients. It was accepted by Aristotle (*Hist. Anim.* VI. i), Columella (VI. 27. 4-7), and Pliny, *N. H.* (VIII. lxxx. 166) without question. Vergil describes these mares racing before the wind, but does not give them a location (*Georgic III*, 272-7). The belief in spontaneous generation has already been illustrated in the chapter on seeds (see above I. xl. 1).

ut hic ... appellant: "just as in our country hens also are accustomed to conceive, whose eggs they call *hypenemia*". The last word, taken directly from the Greek, means literally "wind-eggs", "unfertilised", which produce no chicks. They are the eggs which hens lay as pullets before being fertilised by the cock.

Quae ... obterantur: the same care which was exercised in agricultural operations, such as the delicate handling of shoots, slips, grafts, the grain, the grapes, and the olives will be seen throughout the chapters on the rearing of animals. Their comfort needs and welfare are most closely watched and although the farmer's care will bring him profit it is difficult not to believe that something of humanitarianism and sympathy moved in Varro's mind towards the animals of the farm and of the pasture grounds. If we are not allowed to deduce so much at least the detailed attention given to each animal looks forward to the kindness of later ages. *ut pure et molliter stent:* "that they have a clean soft place in which to stand". It seems that they are kept in some sort of enclosure.

Dicuntur ... appellati: there seems to be a word or more missing between *intimis* and *vocant*, but the meaning is clear and is quite intelligible in Varro's style if a relative (perhaps *quas*, referring to *volvis*) is supplied. *cordi*, from the adjective *cordus*, sometimes spelt *chordus*, is an ancient agricultural word for an animal born after its time, "late-born". The etymology of the word is unknown although Varro seems to suggest that it is connected with *chorion*, the Greek word for any kind of membrane, but which can describe especially the inner membrane surrounding the foetus in the womb.

Tertia res est: the third part of the second group of four headings, that which has to do with feeding, especially the treatment of young animals, comes next.

20. et si ... mamma: "and if the mother is lacking in milk, he should put under the udder of another those who are called *subrumi*, that is 'under the udder'." Lambs or any other animals not yet weaned, were called *subrumi*. *rumis* (or *ruma*) from which the adjective is compounded is an ancient word for a breast that gives suck, a teat, or an udder. The name *Rumina*, the goddess of nursing mothers, was connected with it as Varro states later on [(II. xi. 5) text omitted] and informs us that at her shrine in the Roman Forum and near the Lupercal under the slopes of the Palatine Hill, sacrifices were offered with milk, instead of the wine customary in ritual, and that the victims were sucklings. Close beside the sanctuary once grew the *Ficus Ruminalis*, the sacred fig tree on the spot where it was said Romulus and Remus were washed ashore in the Tiber flood and suckled by the she-wolf. It is associated with some of the oldest strata of Roman beliefs and goes far to prove the antiquity of the word.

puri: "without blemish", as was required for sacrificial animals. These seem to have been sucking pigs.

quanti sunt porci sacres: "How much are pigs for sacrifice?" The passage comes from Plautus' *Menaechmi* ii, 2. 11 (288-90).

> *responde mihi adulescens; quibus hic pretiis porci veneunt sacres sinceri?*

"Answer me, young man: for what price are pigs sold here, sacred pigs, without blemish?" The form *sacres* is an alternative to *sacri*, but not found in classical writers.

ad sacrificia publica: many sacrifices were conducted and paid for by the state. Oxen together with pigs and sheep were the animals usually offered as victims at the annual festivals both in Rome and on the country estates. Those suitable for sacrifice had to be in good condition, without visible blemish and of the right age, sex, and colour, acceptable to the deity and of quiet disposition. At the annual *lustratio*, for instance, of the boundary of an estate which took place in May a procession of three animals, an ox, a sheep, and a pig was led around the estate to beat the bounds, which were then sacrificed. It was called the *suovetaurilia*. A white heifer was needed for the *feriae Latinae*, the Latin festival on the Alban Mount, an ox for the rites of Hercules at the *ara maxima*, a goat at the *Lupercalia*, and so on. All kinds of farm animals would be in constant demand for the many religious festivals in the City and it seems from Varro's remarks that they were especially reared and fattened for the purpose.

ii. 1-6. *Of Sheep.*

1. pensum: derived from *pendere*, means primarily, something hung, *i.e.* weighed. It is used especially for the amount of wool weighed out for a slave to spin in a day; from this comes the meaning "task", "duty".

limitata: "fixed", "settled": some authorities translate this word as "sketched in outline", taking a metaphor from drawing and painting.

nunc rursus ... re: "now again you tell us (the tale), you gentlemen of Epirus, about each separate detail". Three speakers in the dialogue are from Epirus or have estates there, Atticus who drew a great deal of his wealth from his cattle farms in Epirus, Cossinius and Lucienus.

pastores a Pergamide Maledove: the places are unknown and not mentioned elsewhere but are clearly in Epirus. *pastores* are the farmers or estate owners of Epirus who take part in the dialogue.

potis: this adjective meaning "able", "capable" is rarely declined in the positive form: its use is mostly ante-classical. The verb *possum* is compounded from *potis* and *sum*: the form given here, with *sunt* is archaic.

2. Atticus: the intimate friend of Cicero, who was born in Rome and lived from 102 to 32 B.C. and was the elder by three years. His proper name was Titus Pomponius, Atticus being added later owing to his long residence in Athens where he lived in retirement from 86-65 B.C. In 58 B.C. several years after his return to Rome he was adopted by his uncle, Quintus Caecilius and his name became Quintus Caecilius Pomponius Atticus. His daughter married Marcus Vipsanius Agrippa, Augustus' minister. Atticus corresponded for many years with Cicero and afterwards edited his letters for publication. He was a man of learning and a patron of letters and had great wealth mainly drawn from his estates in Epirus. A letter from Cicero (*Ad. Att. III.* 20) exists, written October 4th from Thessalonica to congratulate him on his adoption: it is headed: "*Cicero S.D.Q. Caecilio Q.F. Pomponiano Attico*". The congratulation was apposite because Atticus thus became his uncle's heir.

primigenia pecuaria: the most ancient kind of animal husbandry. Sheep are considered by Varro to have been the first animals to be domesticated; stockbreeding began with them.

si neque . . . aguae: "if they are neither old nor mere lambs". *vetulus* is a diminutive of *vetus*: *merus* has in origin the meaning, "pure", "unmixed" and is applied to wine which has not been mixed with water. The developed meanings are "nothing but", "only", "mere", and then "genuine" or "true".

Sed ea . . . mors: in other words it is more profitable to buy young lambs because although not profitable when first acquired they will eventually become so; there is no profit at all to be gained from old animals.

3. De forma: this word has a wider significance than merely "shape" or "appearance", which is the usual meaning: it includes also the various "points" which are to be looked for in a good specimen.

corpore amplo: the ablative is descriptive: several others follow in this paragraph and the next.

quae: refers back to *ovem*.

apicas: the word, a noun, is thought to be derived from the Greek, and to mean "without a fleece" but another interpretation, also from the Greek, is "unfit", "unnatural": the former is perhaps the more reasonable, especially since it could be said that Varro is making an etymological suggestion.

boni seminis: "of a good herd", "of a good stock". The genitive is one of description.

4. tortis . . . rostrum: "with twisted horns turned towards the muzzle".

Animadvertendum . . . agnos: this belief is also mentioned by Vergil: *Georgic III* 386-90, "and ever choose white flocks with soft fleeces. Although the ram be white himself, reject him if beneath his wet palate there lies hid a black tongue, lest he mark the fleeces of the unborn young with black spots. Look round for another on the full pasture." It was important that the natural wool should be white because the Romans wore white togas.

Ex progenie . . . formosos: "In regard to the young, you should notice (lit. it is noticed) if they produce fine lambs."

3-4. De forma . . . longa: the characteristics of both a sheep and a ram which Varro says should be found in good specimens, the robust body, the thick soft fleece covering the whole body, the short legs and long tail in the sheep and in the ram the forehead covered with wool, horns curving to the cheeks, the ears covered with wool, the sturdy chest, shoulders and hindquarters and long tail, all to be seen in the *Vissana* and *Sopravissana* sheep which are today bred on the Tuscan seaboard. They are a splendid strong breed fit to withstand the hardships attendant on transhumance. They are especially noted for the thick growth of fleece all over the head, and the rams for their twisted horns. These are probably not the descendants of the Roman migratory flocks but at least it may be said that a similar type of sheep might have travelled along the *calles* of ancient times.

5-6. De relinquo ... spondesne? this description of a transaction in the market-place whereby a flock is sold, is most vivid.   The presence of the aediles is not mentioned, but must not be forgotten.   The two parties in their presence first agree about the price, the buyer clearly asking to be assured of the exact amount and the seller (*ille*) pledging his word; the buyer then promises to pay the amount agreed, (*emptor* is the subject of *expromisit*) if he does not show the actual money: *expromittere* usually means "bring something out", but it is a rare word and we cannot therefore be sure of the exact meaning here.   It would be logical however for the buyer to show that he has the money even though he is not intending to pay for the flock until later.   Then, after the agreement has been made with the implication that the seller will deliver and the buyer will pay, there follows the *stipulatio* in the archaic words of the old law which Varro has already referred to above in II. i. 15 (see note).   In this passage it is quoted with direct reference to sheep and it becomes clear how the formula could be enlarged or altered to suit the kind of purchase involved.   The *stipulatio* puts the responsibility for defects on the seller, an advance on the old simple transaction of buying and selling on the spot.   The wording in this passage should be compared with the formula preserved in the Digest of Justinian and quoted in the note on II. i. 15.   The law requires that the seller should not knowingly or fraudulently sell any sheep which have defects of which he cannot fail to be aware.   In the *stipulatio* three exceptions are made because he might not know of three defects: whether any sheep were *luscae* (blind in one eye), *surdae* (deaf), or *minae* (without wool covering the belly).   Another meaning of *mina* is "deficient in milk in one teat": this would make better sense, but the explanation *id est ventre glabro* forbids this.   *habereque recte licere* is in the accusative and infinitive with *spondes*: "Do you guarantee ... that they may then be sold?" *i.e.* "that the title (to possession) may legally pass."   The beginning is seen here of the legal liability of the seller for title: this was expanded in the later law.   A *stipulatio* is made as a protection against *evictio* which represents a pledge on the part of the seller that he is the legal owner of the sheep and that he has the right to sell them: eviction would come into force if the sale were fraudulent, that is, if the seller were not the legal owner.   The real owner could then by law bring an action against the buyer in order to recover his possessions.

5. In emptionibus ... praescripsit: "in purchasing we make use of the formula by which the law has prescribed (how the transaction should be made)". *iure* seems to imply the *ius*, the generally accepted form of agreement, whether verbal or written, to which both parties pledge themselves under the law governing *emptio venditio*.   Varro seems to give directions for a sale which was not concluded then and there by payment of money.   A form of stipulation (*stipulatio*, see above II. i. 15 and note) in such cases was necessary.
excipiunt: the exceptions which the law allowed are explained in what follows.
pretio ... oves: when the price per head has been agreed.
quoi: an archaic form of the dative *cui*.
De relinquo: "For the rest," *i.e.* after certain *stipulationes* have been made.

6. adnumeratum: two explanations can be given: either the *flock* has been counted on delivery by the seller to the buyer, but it is clearly not necessary for the flock to be present in the market, or the *money* has been counted: if this is the correct interpretation then *expromisit* above means that the actual money has been put down, though it may not necessarily yet be paid over.   In that case "on delivery" is implicit in the statement.

nec non . . . pretium: "yet the buyer can sue the seller successfully under the law of sales if he does not deliver the flock, even though he has not paid the money, just as the seller can under the same law sue the buyer if he has not paid the money".   *pote* is the neuter of *potis* with *est* understood: the gender is difficult to explain and seems to be an example of Varro's arbitrary use of agreement.

In our reconstruction of the sale in the market-place first an agreement is made binding on both parties as to the price, which itself includes a promise to deliver and a promise to pay.   When the bargain has been struck, the buyer puts his money on the aediles' table as a proof that he has the means to pay.   Then follows the seller's guarantee that the flock is sound but actual payment is deferred until the flock has been delivered.   Both parties could sue each other if one of them fails to fulfil his obligations.

ii. 7-12.   *Of Sheep: Pasturage*

7. De alteris quattuor rebus: the second group of the sub-divisions under which Varro proposed to discuss each of the farm animals together with one other sub-division which applies to all in general, has been reached (see above II. i. 12 and note).

intus et foris: "within the fold and outside".   Scholars translate this as "indoors and out" but there are no indications that sheep were ever kept under cover.   It becomes clear from this remark and from what follows that a certain number of sheep might be pastured on the home farm.   Those kept at home would be needed for their milk which would also be made into cheese.   Cheese sold in the shops in Rome and in general in Italy made from sheep's milk, is called *ricotta* ("twice cooked") and is made in small rush baskets as it probably has been since earliest times.   A kind of sour milk probably similar to the modern yoghurt was also made.

Ubi stent: the subjunctive is difficult to explain but it may carry a "final" implication, "where they are to stand" which refers to the fold within which they will be enclosed.

ea uligo: the moisture of the ground if it is not kept well swept.   The floor of the fold must be dry, free from weeds because these when trampled on in an enclosed space mingled with the animals' manure would harbour dampness.

lanam: it would be harmful to the fleeces if the sheep were to lie down on damp bedding.

scabras fieri cogit: "makes them become scabby", in other words causes the common disease called sheep-scab.   Among sheep there are several forms of this disease which may give rise to scabs under the wool which causes it to fall out, or lesions in the mouth and around the lips and face, or again skin eruptions which become dry and scaly.

8. virgulta alia: "fresh bedding".   Although nothing was known about the causes of the infectious diseases from which sheep might suffer, the Romans knew that frequent change of bedding could be a protection.

incientes: "those that are pregnant".   The word *inciens* only occurs in the writers on agriculture, Varro, Columnella, and Pliny.   It is thought to be derived from a Greek word ἐγκυος and may be an agricultural term.

villaticos: an adjective for those which are kept on the farm, *i.e.* which are kept in the villa.

9. Contra illae: after instructions have been given for the pasturing of the flocks which are kept on the farm, there follows in contrast a description of the conditions under which the transhumanting sheep are to be grazed. They feed far away from home in different localities and so must be treated with care, and allowance must be made for differences in season (see Chapter II).

saltibus: the original meaning of *saltus* is "a woody uncultivated district used for pasture". From this it comes to mean "a woodland pasture". Varro uses the word for grazing grounds in general, but with good reason because the mountain pastures to which the transhumanting flocks were taken are likely to have been more thickly wooded in Roman times than at the present day. The meaning "a narrow pass", "mountain valley", or "ravine" is developed from the original. *saltus* was also a technical term for a portion of the *ager publicus* (publicly-owned land), four *centuriae* in extent, which amounted to some five hundred acres.

portant: grammatically the subject is *greges*, but in reality *pastores*.

cohortes: in origin *cohors* is an enclosed space as here: then it came to mean the group of persons enclosed in a space, and from that a company of soldiers, a cohort.

inquam: Varro breaks in eagerly at this point to remark that he used to own transhumanting flocks that migrated between Apulia and the mountains in the district of Reate.

cum ... pastiones: "seeing that between these two places, just as a yoke connects two baskets, in the same way the public cattle trails connect far distant pastures". The existence of the *calles* made possible the migration of flocks between Samnium and Apulia (see above, II. i. 16-17 and Chapter II): the trails which Varro mentions here still exist and are still in use.

sirpiculos: these are probably rush baskets: the word is connected with *sirpea* or *scirpea* which means "basket work made of rushes". The yoke was made of wood and curved so as to fit across a man's shoulders: a basket could then be suspended on each side.

10. Eaeque: Atticus continues after Varro's interruption. The passage in *Georgic III*. 307-338, in which Vergil describes the pasturing of flocks in the heat of summer, one of the most beautiful in all Vergil's Georgics, should be read alongside what follows here, and in 11 and 12.

ubi ... regione: though some do not migrate, yet different treatment at different seasons (*temporibus*) is needed: there follow rules for pasturing during the summer months.

aestate ... praestat: "because when they go out at daybreak to graze, especially because at that time the dewy grass excels in sweetness that of midday which is drier". There is often circumlocution in Varro's style. *pastum:* is a supine in purpose. *quod*: introducing the explanation of the statement *temporibus distinguntur* is almost equal to saying "for example". *cum* is redundant; *ruscida*: the form *ruscidus* is otherwise unknown but is familiar as *roscidus*, connected with *ros* and meaning "dewy".

Sole exorto: the sheep have been taken to graze at dawn, now when the sun is fully risen they are taken to drink at a nearby stream or to wooden water troughs.

potum: a supine expressing purpose: *cf. pastum* above.

redintegrantes: "being refreshed". Sheep in Italy frequently drink and are often to be seen wading into streams when the shepherd has brought them from pasture, and drinking eagerly. By contrast sheep in England in the much cooler climate are rarely seen to drink as there is enough moisture in the grass.

11. Circiter ... refrigeratur: in the middle of the day the flock must rest in the shade from grazing: the grass has become dry and lost its sweetness. Just as man takes a siesta during the hottest hours (see above I. ii. 5 where Fundanius says that he could not exist without his midday rest), so must the flocks, until the cool of the late afternoon. The subject of *defervescant* is *aestus*: *refrigeratur* is in the impersonal passive. The flock is taken to the shade of rocks (probably caverns) or spreading trees. In the latter case they crowd together with their heads inwards as close together as possible. In this way their head, their most vulnerable part, is protected from the sun and is in the deepest possible shade.

Aere vespertino: the second grazing comes in the evening when the air is cooler.

Ita ... molle est: "It is necessary to graze the flock in such a way that he (the shepherd) drives them with their backs to the sun (lit. with the sun turned away): for the head of a sheep is the weakest part". Sheep when feeding always face together in the same direction. They must not face into the sun. The singular in *agat* is to be contrasted with the plural in the previous sentence (*pascunt*), and the plurals in the following, *appellant, pascunt*, illustrating Varro's habit of frequent inconsistency.

Ab occasu ... contenebravit: there comes a second drinking and a third pasturing which cannot last very long because there is very little twilight in Italy; the sheep could not continue to graze after dark because of the difficulty of gathering the flock together and driving it into the fold. *ab occasu* implies "before sunset", *i.e.* "a little while before".

Haec ... observant: these methods of pasturing especially are to be observed during the hottest months of the Italian year. The shepherd's care is well illustrated: the flocks graze the grass at its freshest and sweetest in the early morning and again in the evening: the heat will cause thirst and to satisfy this they are taken twice to water: they have rest and shade during the hottest hours.

12. Quibus in locis ... factae: *in quibus locis*, "in places where the harvest has just been gathered"; the sheep may feed on stubble fields with advantage.

caduca ... stramentis: "they eat their fill of fallen ears and by trampling the straw ...".

stercoratione: dung was highly prized by the Romans as manure. It is the practice in England to graze sheep on an enclosed pasture to enrich the soil.

Reliquae ... mutant quod: "the rest of the grazing periods change this (routine) in summer and winter". A paraphrase is necessary in translation: "during the rest (of the year) the grazing times in the summer and winter are different from this, in that ...".

pruina iam exalata: whereas in high summer the flock goes to pasture as soon as it is light to benefit from the dew, in winter it must wait for the frost to be off the grass. There maybe a light sprinkling of hoar frost in the morning in winter in Italy but it rarely lasts for very long.

pascunt diem totum: darkness would fall in mid winter about four o'clock:

this together with the wait for the melting of the frost would limit the times of grazing to some five hours, perhaps with one hour for watering at midday. Throughout this passage arises the question of Varro's sensitivity: his words about the pasturing, leading to water, sheltering from heat have a touch of something more than the practical necessities of pasturage.

ii. 13-20. *Of Sheep: Breeding, Feeding, Health*

13. Quod . . . dicam: "As far as pasturage goes, I have said enough; as to breeding, I have to say what follows."

Arietes . . . sustinendum: the rams which are to be used for breeding should be especially fed for two months beforehand to strengthen them.

ab arcturi . . . occasum: this extends from May 13th to July 23rd, but a good chief shepherd would arrange for the lambs in a flock to be born approximately at the same time. This would be an important consideration in the practice of transhumance: lambs would not have to be born during the passage from one pasturage to another, nor too soon after the ewes reached the new ground.

vegrandes: "very small", "undersized". This is a very rare word compounded from *ve-* and *grandis*. The prefix *ve-* serves as a virtual negative and results in the meaning "not large": compare with this, *vecors* "senseless".

14. diebus CL.: the accusative of extent of time might well have been expected here. In Roman reckoning the first and last day is always counted, with the result that what was a hundred and fifty days to them means one hundred and forty-nine to us, but in any case the number of days is in this context within small limits, approximate.

exitu autumnali: in the case of transhumanting sheep, the lambs would in consequence be born on the winter pastures, a little while after the ewes had had time to settle and rest after the trek from the mountains in September or October: (see Chapter II). They would however have been obliged to make the journey while pregnant but the fact that it was all downhill, although rough, would have mitigated their hardships. Lambs are seen on the Roman Campagna in November.

primitus . . . herba: there is no grass during the hot summer months on the plains, only rough stubble interspersed with coarse weeds which are long-rooted enough to draw moisture from the lower layers of soil. Along the watercourses there will be green growth but nowhere else.

eadem aqua: it is not clear whether Varro requires both rams and ewes to drink the same water during the mating time or only the ewes.

arietes secernendi . . . obsunt: "the rams are to be kept apart, because they worry the sheep which are now pregnant and do them harm".

Neque pati . . . saliri: "You must not allow ewes less than two years old to breed." *saliri* from *salio* means literally "to be leaped on", "to be covered".

neque non . . . deteriores: the negatives cancel out: "and they themselves (the ewes) become weakened", "are spoilt".

Deterrent ab saliendo: "they prevent them from covering (lit. leaping) . . ." the rams can be prevented from covering the sheep by means of small rush baskets fitted beneath them. Cato (LIV) uses *fiscella* for the muzzle to be put on an ox to prevent it from biting at grass when ploughing.

ad naturam: "around the male organ". *natura* is used by the Italian peasants of today with this meaning.

15. eaque ... quaad convaluerat: "the ewes are driven into lambing sheds which are kept separate for that purpose, and there they put the new-born lambs near a fire until they have gained their strength". The ewes are kept under cover even in the equable climate of Italy. The Roman farmer knew the value of shelter and what is more, of warmth. It was stated in an article in *The Times* (October 16th, 1967) on mortality among lambs, that the weather also played a dominant role in the survival rate of lambs ... Much energy was lost after birth while the coat was drying and if the energy reserves were below normal at the onset the lamb might then find itself unable to maintain its body temperature. In cold weather it could die of straight hypothermia, but even in less demanding situations it would be unwilling to suck and a vicious circle was set up. ... There was not the slightest doubt "that if one houses sheep and lambs them indoors, prenatal mortality goes down to almost nil". The British farmer has only recently begun to realise what Varro knew two thousand years ago.

dum ... pabulo se saturent: "until they take their fill of milk", until they have learned to suck and obtain the proper amount of nourishment. *pabulum*, the word for the food of animals in general, usually means fodder: it is unusual to find it used for "milk". *dum* is used here in a "final" sense.

Deinde ... conculcentur: another instance of extreme care is shown in that the lambs are not allowed to pass the night in the shed with their mothers in case they are trampled on. The lambing sheds must either be partitioned, perhaps by hurdles, or there must be several of them. The writer once met with transhumanting shepherds from the Abruzzi who wintered some twenty miles out of Rome. Hollowed out of a tufa cliff was a great cave within which were other caverns which served as sheds. One was full of gleaming white lambs lying in straw, another adjacent one was for the ewes which were then out at pasture. They were to return in the early evening and would let the lambs drink their fill before they settled for the night.

Hoc ... lacte: the lambs drink twice in the day, first in the morning before the ewes go out to pasture for the day and again in the evening. *satulli* from *satullus*, is a diminutive of *satur*, "filled". The question arises whether Varro shows any feelings of fondness for the young of the flock in using this word; or is it the word merely of a profit-conscious farmer? There seems to be no strong reason why *exeant* should be in the subjunctive since *antequam* gives merely time, but Varro does not always observe the conventions usual with temporal conjunctions.

16. palos offigunt ... distantes: "they fix stakes (in the ground) and tether the lambs to them with bark or with some other kind of smooth material with intervals between". The tethering ropes must be smooth to avoid chafing the skin. The lambs are now put out to feed on grass near their shed.

ne ... membrorum: "so that frisking about together during the whole day, the frail creatures do not knock any skin off their limbs". Here is another aspect of the detailed care which Varro recommends for lambs. While the lambs feed near home, the ewes are separated from them for the whole day (*toto die*) when they are taken out to graze.

unguere buturo: butter was known to the Romans but not often eaten. The richest came from sheep's milk, and it was made by shaking the milk

rapidly in a tall vessel.   A little water was added to turn the milk sour: what
came to the surface was butter.   The stronger the taste, the more highly was
butter esteemed: it was astringent, emollient, flesh-forming and cleansing.
It was probably smeared on the lamb's lips to give it a concentrated taste of
milk.

olfacere labra lacte: if a lamb does not begin to suck, it should be made to
do so by giving its lips a taste of milk: this would be done by holding it to the
udder and pressing a few drops into its mouth.

obicere: the infinitive depends on *oportet* to be repeated from the previous
sentence.

viciam . . . teneram: vetch was one of the fodder plants: see note on I. xxiii.
1-2 *molitam* is from *molo, molere,* "grind": probably the vetch was pounded to
break it up and soften it.   If these are transhumanting lambs they would be
on their pasturing grounds on the plains and here the grass of late autumn
and early winter would be newly sprouting after the summer drought.   They
could crop the grass around their tethering posts but it was desirable for this
to be supplemented with cut fodder and grass.

    17. quadrimestres: the lambs suck for four months, in the meantime
learning to graze and receiving a supplementary diet of forage legumes.

quidam: "some (breeders)".

qui id . . . perpetuo: "they do still better who do not milk them at all the
whole time".

Cum depulsi sunt: "when they are taken from their mothers", "are weaned".
This is done when the lambs are four months old.

ne desiderio senescant: the time of weaning is a critical one and calls for
special precautions: care must be taken to see that they do not sicken through
missing the mother and the taste of milk.   In the wording here Varro shows a
sympathy and understanding of animals, their feelings, and their needs.

deliniendum: "they must be enticed".   A more usual form of the verb is
delenire but it appears in many manuscripts as *delinire*.

a frigore . . . curandum: another precaution is given which only the experienced farmer and one who understands the needs of animals could understand.
Just as at birth, now at the next stage and one involving changes, cold and
heat must be guarded against, "precautions must be taken to prevent their
coming to any harm from cold and heat".   The lamb's resistance is lessened
at the time of weaning, when they miss the solace of warm milk.

    18. oblivione iam lactis: "through forgetting the taste of milk".   They join
the flock only when they are completely weaned.   Transhumanting lambs would
be several months old and accustomed to move with the flock when
the time came in the following late spring and early summer for the
journey to the mountains.

calores aut frigora: at a critical time in the growth of the young rams,
precautions against heat and cold are again to be observed when they are
castrated.

Quos arietes . . . volunt: "those which they wish to rear as rams".   Some
are not castrated but kept for serving the flock.   The choice of lambs from
mothers which usually had twins shows that the Romans had begun in a simple
way to practise selective breeding.

ut sunt Tarentinae: Tarentum in the heel of Italy, on the north coast of
the gulf of Taranto, was in origin a Spartan colony founded by Phalanthus;

situated as it was on a fine harbour-site and dominating a fertile hinterland rich in olives and sheep-rearing, it so prospered as to become the first city of Magna Graecia in the fourth century B.C.   In 209 B.C. after the defeat of Hannibal whose cause it had supported, it was sacked and thirty thousand citizens were sold into slavery.   In 123 B.C., it was made a Roman colony and so was known to Varro.   The earlier riches of Tarentum were drawn in great part from the sheep which were pastured on the best sheeplands in Italy, in the rain-shadow of the Apennines.   Its location near the pastureland caused it to be a centre of trade in textiles made from the famous wool and for dyed woollen goods. The softness and pure colour of Tarentine wool were due probably to the importation of fine Greek herds by the first colonists in the eighth century B.C. The sheep were jacketed with skins to preserve the quality of their fleeces and to keep them free from damp and dirt.   It was said that Diogenes Laertius (the Cynic philosopher) on seeing the jacketed sheep, remarked that the children went naked, but the rams were clothed.   They are mentioned by Horace (*Odes* II. vi, 9-12).

> *unde si Parcae prohibent iniquae,*
> *dulce pellitis ovibus Galaesi*
> *flumen et regnata petam Laconi*
> *rura Phalantho.*

"if the grudging fates keep me away from there (Tibur) I shall seek the river of Galaesus sweet to the jacketed flocks and the rustic kingdoms ruled by Spartan Phalanthus".

quo minus ... putari: "in which case it cannot be either properly dyed, or washed and bleached".

19. Harum: *i.e.* "of the jacketed sheep".   The care given to these sheep is even closer than to the ordinary rough coated variety.   Their pens must be kept scrupulously clean.

lapide strata: the floor of their pen is paved.

His: the jacketed sheep: every care is to be shown to them, not only in the preservation of the quality and softness of the wool, but also in their feeding. They are not restricted to the legumes, but may have such delicacies as fig leaves, grape-skins, etc.   These were given to supplement their pasture, not instead of it.

Utrumque: "both kinds of feeding", *i.e.* too much or too little.

cytisum et medica: the fodder crops have already been discussed above: these also were given in addition to what they obtained from pasture: (see above I. xxiii. 1-2).

Nam ... lacte: "while it both fattens them easily and makes them abundant in milk".   Both *cytisum* and *medica* are the subject: the singular verb is to be explained as related grammatically to one of them or the two may be combined in one idea of "forage".   The ablative in *lacte* is difficult: *genere* (the archaic primary form of which is *gigno*) means "bring forth", "produce": the explanation of Varro's phrase is perhaps "makes them abundant in milk".

20. magister pecoris: the chief shephered is an educated man who can read and write.   He takes his book of instructions and medical advice with him wherever he goes: on transhumance it must not be forgotten, together with any remedies which might be needed.

Relinquitur de numero: it remains to discuss the number in a flock.   This is the last of Varro's divisions on the subject of cattle-raising, and the ninth, which applies to all alike (see above II. i. 12, *praeterea communis una*).

Illut fere: there remains the question of the number of shepherds to flock: the custom in Epirus is quoted, as it comes within Atticus' experience. Double the number of shepherds is needed for the jacketed sheep, in view of the especial attention needed for these animals who bear the finer wool.

iii. 1-10.   *Of Goats*

1. balasti: shortened form of the second person singular: the word is spoken in a joking way.

Faustule: the foster-father of Romulus and Remus.

Melanthio cordo: Melanthius was Ulysses' goatherd. While Ulysses was away and his return despaired of, Melanthius supplied the suitors of Penelope, his wife, with the best of the flock. For this action he was slain by Telemachus: (*Odyssey* XII, 217 and XXII, 474): for *cordus* see II. i. 19. It was a Roman family name: Varro employs it here as a pun.

quem . . . disce: "learn how a man should speak briefly". Perhaps a dig at Atticus who has spoken at length about sheep.

animadvertat oportet: the change of construction with *oportet* from that of the previous sentence is to be noted.

ut eam partet: " so that he may acquire animals of that age ". *eam* refers grammatically to *aetatem*.

utilior: it is to be noted that the advice in this sentence is a repetition of that given about sheep (see II. ii. 2).

2. nisi si glabrae: this probably means " short-haired ". *si* is redundant.

Melior . . . una: " The flock becomes better if it is not made up of animals obtained from different sources but from a single flock accustomed to run together." Such advice was not given in the case of sheep because by nature they herd together, but goats are accustomed to roam freely when grazing and owing to their nimbleness to climb to great heights. They are often, however, herded with sheep but not in greater numbers in the flock.

3. De seminio: for Atticus' remarks, see II. ii. 2, 3, 4. Cossinius says the same about goats that Atticus does about sheep.

hoc aliter . . . esse: " I say this differently ", " with this exception that the race of sheep is slower moving because they are of a quieter disposition: on the other hand that of goats is quicker in movement ".

in Originum libro: see I. ii. 7 and note on Cato.

Sauracti: the name is more usually spelt *Soracte*. This is a hill, some twenty-six miles to the north of Rome, which rises to a height of nearly one thousand, five hundred feet above the Tiber valley. It is in the Faliscan country near the modern Civita Castellana (*Falerii Veteres*) and is an isolated mass of hard limestone of the same geological formation as the Apennine ranges of which it is a geological part. It dominates the surrounding plains seeming like a grey petrified wave with its seven peaks. It can be seen clearly from Rome, especially from the Janiculum hill, now called the Gianicolo.

Fiscello: the *mons Fiscellus* in the central Apennines is probably to be identified with the modern Monti Sibillini. This range of wild and rugged mountains, one of the loftiest in all the Apennine system, runs between the Sabine mountains and Picenum, and stands to the north-east of the plain of Rieti.

caprae ferae sunt: although Varro quotes Cato, he may be speaking of wild goats which he has seen on these ranges. They are probably the chamois, the type now called *rupicapra*, which inhabit lofty mountains and are very agile and sure-footed. They have high backward curving horns and tremendous leaping powers, as much as four metres vertically, and seven metres on the flat. They feed in woods and brushwood among rocks near the snow-line. In Italy the race is in danger of extinction but a few are preserved in the National Park of the Abruzzi, a mountainous district between Opi, Civitella, and Alfidena. The Abruzzese chamois is of a particularly light colour and has exceptionally high horns. These are perhaps the survivors of the wild goats known to Varro in his own native mountains. In II. i. 5 (text omitted) he speaks of the wild goats in the district of Fiscellum.

4. propter . . . Caprasia . . . nominata: " an island off the coast of Italy is called *Caprasia* (Goat Island)", a more usual spelling is Capraria. The small island now called Capraria, lies between Corsica and the Etruscan coast of Italy, thirty miles from Populonia (now called Piombino). Varro enjoys the pun on *capra*, and as every rocky island which could not be cultivated could support goats, it is not unlikely that his derivation is correct.

Melia: the island cannot be identified for certain, but it is thought to be the same as Melos, an island in the Aegean sea, one of the Cyclades. There is no evidence that it was especially famous for goats but the rocky nature of the land was suitable and goats were reared on most Greek islands.

5. aliter dico atque fit: " I speak differently from what is usual ", " I give you a different formula ". Atticus in II. ii. 6 mentions the old law for the buying of sheep. This involves the stipulation that they are healthy and from a healthy flock. Cossinius explains that the stipulations made for sheep cannot be imposed on the sellers of goats because goats are never without fever: it is not possible to require a guarantee about their health: the *stipulatio* can only be that they are in good condition at the time of purchase and that the title is correct. The Roman belief in the unhealthiness of goats is without explanation. In modern times they are regarded as particularly healthy because they are known to be free from tuberculosis.

Manilius: Manius Manilius was praetor in Spain in 154 B.C. and as consul in 148 B.C. besieged Carthage. He was unsuccessful in the military field but became a famous lawyer. He was said to be one of the three who established the civil law (*ius civile*). He drew up a code of laws, or *formulae*, for purchase and sale (*venalium vendendorum leges*) which were generally used and known as the *leges Maniliae*. He appears as one of the speakers in Cicero's *De Republica*, (see below II. v. 11 where his laws are quoted for the purchase of oxen).

illut = *illud*.

Archelaus: he was king of Macedonia from 413 to 399 B.C. and is mentioned by Pliny as one of the kings who wrote on agriculture (*N.H.* XVIII, 3). Pliny also ascribes to him the two statements about goats, that they breathe through their ears and that they are never without fever.

curiosiores: " who observe more closely than others ".

6. ad hibernos exortos: " towards the sunrise in winter ".

alsiosum: goats are particularly susceptible to cold probably because in origin they come from hot eastern countries.

Id . . . lutulentum: " The stall, like most stalls, should be paved with stone or brick to prevent the goat-house from being wet and muddy." *Id* takes up a

noun understood from *stabulatur*.   *testa* is a brick or tile made of baked earth.
Varro's use of *quo minus* (more often written as one word *quominus*) is not
limited to verbs of preventing or hindering as is usual in classical Latin.   A
permanent stall such as is described here must have been on the home farm.
Foris . . . pernoctandum: " when they have to sleep out of doors ".   This
would be the condition under which they were penned when transhumanting,
or they might be penned out of doors on the home farm in the summer months.
in eundem partem . . . saepta: the summer pens must face the sunrise, just
as was necessary for the winter.   *saepta* comes in meaning before *in*: *spectent*
seems to imply a command " must " and is perhaps influenced by *oportet* in
the previous sentence.
ne oblinantur: care must be taken to keep both permanent and temporary
goat-houses clean and warm: this is yet another example of Varro's understanding of animals' needs and his attention to detail in matters of their welfare.
The same advice is given for the care of the jacketed sheep (see above II. ii. 19).
Non multo . . . quaedam: the care of this animal in pasturing is not far
different from that of the sheep, yet it has its own particular likes: the difference
between the two is that the sheep crops grass and low-growing plants and stays
with the flock, but the goat eats leaves, twigs, and undergrowth and the lower
branches of young trees and is inclined to wander away.
     7. in locis . . . carpunt: *virgulta* usually means " brushwood " or " under-
growth ", but is also used for planted shoots, as in *Georgic II*, 346

> *quaecumque premes virgulta per agros*

"whatever shoots you will plant in your fields".   These might be vines or olives,
figs or fruit trees of various kinds.   In this passage in Varro there is better
meaning if *virgulta* is taken as referring to young trees, rather than brushwood.
There is a warning in the words *in locis cultis* of the damage which goats can do on
cultivated land.   *virgulta* is apparently used in contrast with *fruticibus* which
are wild shrubs.
Itaque . . . nominatae: in fact from (their habit of) " nibbling " *carpendo*,
they are called *caprae*, " nibblers ".   Varro's suggested derivation is probably
not correct: there is thought to be some connection with the Greek word for a
wild boar, κάπρος.
Ab hoc . . . pascat: "in consequence of this (their habit of nibbling), in the
wording of the legal contract for the lease of a farm, it is usual for there to be an
exceptive clause forbidding the tenant to pasture the offspring of a she-goat on
the farm ".   In the early agrarian history of Rome there could have been few
examples of rented land because small holdings were owned by peasant
proprietors who worked their own farms with their families.   In the course of
time, however, as the old established form of cultivation was disturbed by the
need to have a large fighting force for Rome's wars and career of conquest,
ownership of larger estates became possible and gradually increased.   With
the passing of the *lex agraria* of 111 B.C. a great deal of public land became
private property.   Now it was possible for parts of holdings to be let on
definite leases to working tenants (*coloni*).   If goats were allowed to graze on
cultivated land, in vineyards for instance or among olives, they could do untold
damage to trees, even to the extent of devastation.   Goats must be kept away
from anything planted and allowed only to crop wild growth.   Plantations
could be ruined for future generations.

dentes inimici sationi: " their teeth are harmful to planted crops ". *satio*, from *serere*, " plant ", means anything sown and therefore cultivated and may include crops of all kinds, as well as cuttings and slips from trees. The damage that can be done and to what extent it was feared is illustrated by the taboo against goats practised in Athens and the sacrifice there of he-goats to Bacchus who discovered the vine: (see above I. ii. 17-20 and notes). The belief is also expressed that their spittle is poisonous: this is not so but they can cause denudation and destruction to cultivated bushes and trees to such an extent that this belief is easy to understand.

extra . . . excluserint: " they have excluded them from the circle of the twelve signs ". *excluserint* is a consecutive subjunctive. The argument is that goats are so destructive that although they are part of a constellation, they are not included among the signs of the Zodiac.

8. Quod . . . dictum: the he-goats are to be segregated for two months as was recommended for the rams.

Quae concepit . . . reddit: the she-goat which has conceived drops her young after the fourth month. The period of gestation is the same as for the sheep as is natural since both animals belong to the same general species (see above, II. ii. 13).

submittuntur: " they are introduced into the flock ". They are ready to run with the flock after three months: compare this instruction with those for lambs which seem to need greater care (II. ii. 18).

9. quae dispargant se: " they are inclined to ", " or such as to ". The subjunctive is generic, as are *congregent* and *condensent* in the same sentence.

in agro Gallico: see above I. ii. 7 and note.

plures . . . magnos: farmers in the *Ager Gallicus* keep more and smaller flocks rather than fewer and larger to guard against infection, and the consequent loss of large numbers of animals. This is illustrated in the example which Varro goes on to relate about a man who made a disastrous mistake.

existat: this word implies " to break out suddenly ". A sudden epidemic could spread quickly and devastate a large flock.

quae . . . perducat: " which could ruin the owner ", or " could ruin the flock ". *eum* could agree with a man in general, the owner understood from the singular of the subject of *faciunt*, or it could equally well go back to *magnos* with the same general reference in the singular after Varro's manner.

10. Quibus . . . Romano: " They think that that which came into the experience of Gaberius, a Roman knight, proves these conclusions. " " They " is used in a generalising statement. The translation might be better paraphrased as " It is generally believed that the experience of Gaberius goes far to prove these conclusions. " *Quibus* is probably in the dative rather than the ablative case: *adsentiri* is probably in the deponent rather than the passive form which was becoming rare: but if it is to be taken as a passive it must be regarded as an impersonal use.

equiti Romano: Gaberius belonged to the *equites*, the second rank of nobility in Rome, which enjoyed many political privileges.

in suburbano: " on his estate on the outskirts of Rome ". *suburbano*, only used in relation to Rome, means the estate itself and is comparable in usage with the names of estates taken from neighbouring towns such as that of *Tusculanum*, near Tusculum.

mille iugerum: the extent of this estate would be about seven hundred acres.

Land near Rome would undoubtedly be very valuable. Except for one instance II. i. 26 (text omitted), Varro uses *mille* as an indeclinable noun associated with the genitive case.

adduxit: the word seems to suggest that the goats were brought into Rome each day for milking.

sperans . . . mille: " hoping that he would collect a profit of a thousand denarii a day from his estate ". *denarium* is in the shortened form of the genitive plural, to be compared with *divum* (from *divus*) and others common in the poets. In 1912 a thousand denarii was estimated as equal to about £32, see Ll. Storr-Best.

in Sallentinis: " in the district inhabited by the Sallentini ". These were an ancient people inhabiting the modern district of Otranto in southern Calabria: the legends told how this region was first colonised by Idomeneus who had fought at Troy.

in Casinati: " in the district of Casinum ". This is Varro's own country: his villa was situated near the town (see below, III. v. 9 and note).

Menas: it is not known for certain who Menas was or why he should be especially mentioned. A freedman of that name appears in Suetonius' Life of Augustus (*Augusti Vita*, 74) as the only freedman whom Augustus ever invited to a formal dinner: he had received his freedom for his cleverness in betraying Sextus Pompeius' fleet. There is no evidence to show that this Menas is the one whom Varro mentions, but if this is so we may infer that Menas had been able to acquire land as a reward for his services.

Murrius: since Murrius came from Reate [(II. vi. 1 below), text omitted], Varro's own country, the mention of him approximates to saying that it is the practice in the Sabine country and must have been well known to Varro. Murrius is the chief speaker on asses for which Reate was famous (II. vi. 1-5).

v. 6-11. *Of Oxen*

6. in bubulo . . . quattuor: " there are said to be four stages of life in the race of oxen ". Varro points out that each of these has a different name and that each is further differentiated by a male and female form. Varro speaks of the working ox above in II. v. 3-4 (text omitted) in words almost amounting to affection and approaching the feeling for animals which pervades the *Georgics: Hic socius hominum in rustico opere and Cereris minister:* " this is man's comrade in his labours in the field and the servant of Ceres ".

prima: this agrees with *aetas* although an agreement with *gradus* would be more natural.

Ab eo . . . immolantur: " for this reason (*i.e.*, because a pregnant cow is called *horda*) a day in the calendar is called *hordicidia* because on that day pregnant cows are sacrificed ". *hordicidia* is compounded from *horda* and *caedere* and means literally the slaying of a *horda*, a pregnant cow. For an account of the festival which took place in Rome on April 15th and was dedicated to *Tellus*, see above I. i. 5 and note. The spelling with the initial letter *h* in both *horda* and *hordicidia* is archaic and dialectic. Varro gives the definition in the *Lingua Latina* (vi. 15) using the spelling *f*, *fordicidia a fordis bubus*. The use of *f* in place of *h* is probably Sabine. If so, Varro uses his own native version of the words.

in fastis: " in the official calendar ". The *fasti* were the official calendar of the Roman year which incorporated such dates as the Kalends, Nones, and Ides, the market days (*nundinae*), days on which state business might or might not be conducted and most important of all, the days of the various religious festivals. Eight years before the publication of the *Res Rusticae*, Julius Caesar introduced the reformed calendar, which is still followed today and on which the state calendar in Varro's day was based (see above, I. xxviii. 1 and note). The *fasti* were often inscribed on stone and set up in a public place for all to see: in Rome, for instance, a stone calendar was set up in 36 B.C. in the centre of the Forum near the Regia: or they might be painted on walls of buildings such as the one found at Antium (modern Anzio) of republican date arranged in vertical columns in black and red on a white ground. Ovid in *Fasti*, I-VI, gives a full account of the religious calendar up to the months of June with details of myth and ritual of the various cults.

7. Qui . . . expartae: the advice to buy young rather than old animals is given in every instance and may seem superfluous but a young animal is expensive because it has to be stalled and fed for two or three years before it becomes profitable: there might be the temptation to purchase older animals which might be cheaper but they might be past bearing and in any case the period of productivity would be proportionately less. The adjective *expartae* (*expartus*) does not occur elsewhere: the meaning " beyond bearing " " played out " seems undoubted, especially as it is the opposite of *integrae*, " sound ", " in good form ". The word may be an agricultural one familiar to farmers which leads to a surmise that it might be a local Sabine or Epirote term.

bene compositae: " well put together ", " of compact frame ".

compressis malis subsimae: with narrow jaws, rather snub-nosed: the implication is that the jaws should narrow towards the nose and that the nose should be flat. This would give a general shape to the ox's muzzle of a truncated triangle.

ne gibberae: " not hump-backed ". *ne* is the primitive Latin negative particle. In the time of Varro it had almost fallen out of use with a single word as here: it remained in such familiar usages as *ne . . . quidem*, *ne* with the subjunctive *etc.*

spina leviter remissa: the ox's spine should be slightly concave.

8. palea demissa: this describes an ox with a long heavy dewlap. In typical examples which are often represented in sculpture (see below) and painting the dewlaps fall in loose folds which surround the neck and extend as far as the knees.

inferiorem . . . subscrispam: the tail should have at the end a bunch of rather curly hair.

cruribus . . . rectis: " with straight legs, rather short ".

neque . . . displudantur: the description of the legs is continued: " and not such as to spread out as the animals walk ". *ingredientibus* is in the possessive dative. *displudantur* is a generic subjunctive.

nec cuius . . . divarent: " the hoofs must not be widely cloven ". Although the meaning is clear this verb does not appear in Lewis and Short's *Latin Dictionary;* it is undoubtedly closely related to *divaricare* under which this reference is given. The sudden change to the singular (*cuius*) after the plural (*ingredientibus*) in the previous sentence is very harsh.

molissimus . . . primus: "oxen of the latter colour (white) are the least robust, just as those of the first (black) are the hardiest".

9. Neque non . . . speciem: the two negatives cancel out each other: *debet* is to be supplied from 7 where it introduces the directions about the kind of oxen to be bought. *seminis boni* is a genitive of description. *et qui* introduces an admonitary statement: the young will be like their parents: the farmer, therefore, will be wise to choose good stock for purposes of maintaining quality. The Romans undoubtedly practised selective breeding and knew how to ensure the quality of the animals born and reared on the farm. The various points of Varro's well-bred ox go to make up a muscular animal of immense strength, well-suited by its sturdy robust frame to the plough, to breaking up hard sods and pulling heavy loads. It has thick shoulders where the hardest labour of bearing the yoke falls, its body is wide and the rump covered with strong muscles. The straight, short limbs are well-suited to working beside a fellow ox on the fields. This is the working animal of Roman husbandry and the friend of man. The best ox of Varro's day seems to have been black with black horns, different from the great cream and grey-coloured oxen with the tremendous horns which plough the Roman Campagna and in fact most Italian soil today, though Varro may have known another breed also, similar to this. The form of ox favoured by Varro, however, does not appear to have changed from Mago's time. Columella (VI. i. 3 and VI. 21) gives a description which he ascribes to the Carthaginian Mago and which corresponds in every detail with that of Varro. An inference might be entertained therefore that the Romans had this type of working ox, from North Africa after the conquest, but perhaps a more likely surmise is that it is a Mediterranean type. On many reliefs, sacrificial and others, from the time of Varro onwards is shown a sturdy ox of stocky breed with long tail, curled hair on the forehead, and heavy dewlap closely similar to what he describes.

Et praeterea . . . Italiae: the desire to improve stock was prompted by the limited pastures of Italy. For this reason good native herds were valued for their special characteristics such as those reared for working oxen, and even imported from Greece, as for instance cattle from Epirus, which were considered the best of all. The importance of Epirus for sheep-rearing has already been described, and for the excellence in general of its agriculture (see above I. xvii. 5; II. ii. 1; II. ii. 20). Emphasis was laid on the best strains for breeding purposes: Vergil's words bear this out: " after the birth every care is transmitted to the calves and straightaway they brand on them marks and names of the herd, both those such as they prefer to couple for increasing the flock, or to provide sacred victims for the altars or to cleave the earth and turn the rough ground with broken sods ". (*Georgic III*, 157-61.)

Gallici: fine working cattle were bred in the *ager Gallicus* in the Po valley, where there was an abundance of water-meadows (see above I. ii. 7 and note).

Ligusci: this name is more usually found as *Ligures*; they were the people who inhabited Liguria in Cisalpine Gaul, the ancient province roughly equal to modern Piedmont in north Italy.

10. victimas faciunt: some farmers bred cattle especially to be sacrificial animals, great importance being attached to size and colour.

qui sine dubio . . . praeponendi: " which without a doubt are to be preferred for ritual purposes because of the majestic appearance of their size and colour ". Horace speaks of the ox reared on Algidus or the Alban hills to be a victim, marked out as such from the first. " For the ox which feeds on sunny Algidus already dedicate amid the oaks and ilexes and flourishes on Alban grass, a victim for sacrifice shall tinge the priests' axes with its life-blood." (*Carmen* III, xxiii, 9-13). White oxen were highly prized: those bred in Umbria were used especially in the triumphs. Vergil describes the white bulls reared on the water-meadows by the river Clitumnus which rises in pools of crystal clear water shaded by fresh willow trees near Spoletum (now Spoleto), (*Georgic:* II, 146-8). A white heifer was sacrificed to Jupiter Latiaris at the feriae Latinae, the Latin festival held annually on the Alban Mount to which all the cities of the Latin league sent their deputies to receive distribution of the sacrificial flesh. An animal of great size, gleaming white, garlanded and decked with fillets as was the Roman custom would be a splendid sight in a ritual procession.

μέλανα κόλπον: " the Black Gulf ", now the Gulf of Samos.

stipulamur sic . . . praestari: " we ask for agreement in these terms: (do you agree) that these oxen are sound and that the buyer incurs no liability with them ? "

illosce = **illos** and -*ce* a deictic particle, implying "the ones in question" or if the oxen are present in the market " these which we are looking at ". According to Roman law if an animal had done harm, for instance by goring a human being, the responsibility for the mischief followed the animal and fell upon the person to whom for the time it belonged: the law term for this was *noxa caput sequitur*. The purchaser must be protected from such claims as far as possible and the vendor's word should provide a safeguard.

11. Manili actiones: there can be little doubt that the two legal formulae quoted for the purchase of oxen are from Manilius' code (see above II. iii. 5 and note and below II. vii. 6).

qui ad altaria . . . stipulari: the reason was not because no importance was attached to the state of the victims as the statement might suggest; the case was far otherwise. The priests themselves tested the animals. A refusal to eat or drink was regarded as a sign of ill-health.

Pascuntur . . . frondosos: Italy is deficient in pasture land, because of both the conformation and the climate. The Apennine ranges, though in themselves supplying fresh verdure on their lofty plateaux fed in the summer from the cloud caps and melting snows of the higher ranges, are small in extent. They occupy a large part of the country so as to leave in general only narrow coastal plains. These are parched to dry stubble in the hot months of June, July, and August. Since pasture is seasonally distributed, the system of transhumance results as well for cattle as for sheep (see Chapter II). Where the rich succulent grass which is the best feed for cattle was lacking the feed was supplemented by the Roman farmer with leaves of many kinds: Columella (VI, *III*. 6) recommends the foliage of elm, ash, and poplar and even in case of need, holm-oak, oak, bay, and fig. In the mountain pastures the summer change of diet to include the leafage of trees would be beneficial. Throughout Italy cattle were fed in the Ligurian Alps, the Sila mountain in Bruttium, in Samnium, and in Sicily where there were plains and upland pastures.

v. 12-18. *Of Oxen: Breeding, Health*

12. ne cibo . . . impleant: the cows' feed and drink should be reduced for a month before they are due to be mated.

pleniores: the bulls are given increased rations of feed for two months before mating.

fidem nostri: the constellation called " the Lyre " rises on May 13th.   It is mentioned by Plutarch in connection with the Julian calendar (*Caesaris Vita*, 59).

13. Tum denique: " then only . . . "; this refers to the end of the time of segregation and special feeding which begins with the rising of the Lyre.   A count of two months from that date, May 13th, would imply that the bulls are allowed to mate in July; this is supported by Columella (VI. 24. 1) who says that July is the proper time so that the calves may be born the following spring when the pasture is at its best.

mas . . . sinisteriorum: this statement is absurd but seems to have been generally accepted: it is even repeated by Columella (VI, 24. 3).   It is on a par with several other illogical beliefs which we have already met, such as spontaneous generation and the impregnation of mares by the wind.   The comparatives *dexteriorem* and *sinisteriorem* have no particular significance.

vos . . . legitis: this interlude gives a graphic touch to the dialogue: Vaccius makes a complement to Varro as a man of learning, one who studies the philosophers and can find out in his own reading in his own library.   The reference is to the *De Genere Animalium*, IV. i. 765a in which determination of sex is discussed, and the statement made that the male is on the right side of the womb and the female on the left.

a delphini exortu: the rising of the constellation of the Dolphin (which, mythologically speaking, received a place in the sky from Poseidon for having assisted in his marriage to Amphitrite), was on June 10th.   If the time be counted from the rising of the Lyre when the bulls were segregated from the herd for two months, a discrepancy becomes apparent.   According to Varro mating would then begin in June a month earlier than the time recommended by Columella.   This can only be an approximate date especially since he allows forty days from the rising which would extend at least to the end of July. We have already seen in the farmer's calendar how counting by the stars is often only a general indication of the length of time during which some particular operation takes place.

temperatissimo anni tempore: if the cows became pregnant in July and the period of gestation is ten months, the calves would be born in May of the following year.   The meadow grass would be fresh and newly sprung, and mingled together with the many wild flowers and herbs which grow in it would afford tempting feed.   This would be beneficial to nursing mothers: and when first the calves were ready to crop the grass it would be tender and succulent.

14. De quibus: this is another citation from Aristotle (*De Genere Animalium*, I. iv. 717b).

Eas pasci . . . cogit: throughout this passage the extreme care of cows in calf is remarkable.   Attention is given to their pasturing, the space for movement, precautions against flies in summer, their bedding, their special food, and protection from cold.   In such a passage as this which is to be compared with that about the care of lambs (see above II. ii. 15-19) there becomes manifest an understanding of animal behaviour and needs which perhaps exceeds that of a mere breeder who is anxious to see as large returns as possible for his care.   It is allowable to argue for some understanding and sympathy on Varro's attitude to his flocks and herds especially in the words *locis viridibus et aquosis*.

ne aut angustius stent: " that they do not stand too close together".   This might be when they seek shade at pasture or in the stables.   Steps must be taken to see that they do not injure each other, cf. *Georgic III*, 143-4

*saltibus in vacuis pascunt et plena secundum
flumina, muscus ubi viridissima gramine ripa,*

" they feed in clear glades and beside full-flowing streams where there is moss
and where the bank is greenest with grass ".

tabani . . . includere saeptis: cattle flies must have been a great scourge since
the Romans had no insecticides.   The farmers had to devise ways of protecting
the animals against these harmful insects.   Varro recommends that the
pregnant cows are kept in their stalls during the summer.   He has probably
left to be implied what Vergil tells us explicitly that they should be taken to
pasture in the early mornings and at evening because the flies are most active
during the midday heat.   They are not stalled for the whole time but only
during the hours of heat.   " This too for it is more active in the midday heat—
you shall keep away from the pregnant herd, and shall feed your cows early
when the sun rises or when the stars lead in the night." (*Georgic III,* 154-7.)

bestiolae quaedam: this probably describes eggs laid by the *tabani* under the
tails of the animals, where they incubate.

frondem aluidve quid: " leaves or some other kind (of bedding) ".

   15. fastidiosae: the provision of fresh fodder strewn close to the stalls to
tempt their appetites seems a quite extraordinary example of attention to
animals' needs.   Behind it all is the anxiety of the breeder for his profit but, at the
same time, the wording seems to reflect a sympathy for the animals.

   16. In . . . contuendum: " In (the matter of) rearing, care should be
taken of cattle as follows. " *contuendum* is either (1) deponent or (2) from an
active form of the verb.

Ad eas . . . pastu: the calves which are kept penned in are allowed to suck
twice a day in the morning and evening.

levandae matres: as the first step in the process of weaning, the calves are
introduced to green fodder gradually which is put down in the stalls alongside
their mothers.   They will taste and begin to eat it and so demand less of their
mothers.

Item his . . . : the cow stalls must be paved just as those for sheep (see above
II. ii. 19) and goats (II. iii. 6).   Varro is very insistent that stalls and pens
for the different farm animals should have hard floors, preferably paved, so as
to preserve their hoofs in good condition.

Ab aequinoctio autumnali: according to the time of mating given by Varro,
in June or July, the calves would be born in April or May; by this reckoning
they go out to pasture in October, when they are five months old approximately
and by now fully weaned.

   17. Castrare . . . fiunt: in spite of what Varro states here, the great
agriculturist Mago was in favour of castrating calves while they were still
young and tender, (Columella, VI, xxvi, 1.) It is remarkable that Varro should
disagree with his acknowledged master (see below, 18).

ut in reliquis . . . fructus: *reiculus* is a rare word and seems to be an agricultural term connected with *reicere,* " reject".   The numbers of the flocks
and herds must be kept fairly constant for this was especially important in
relation to the amount of grazing ground: the poorer animals must be weeded
out every year: not only weaklings born during the year, but animals which
have ceased to bear and those who are weak or ill or too old to be of further
service.   The expression *dilectus habendus* perhaps suggests that a list of the

cattle was kept: the head herdsman may have had a stock-book as do cattle-breeders of today.

Siquae . . . matres: " If any cow has lost her calf, you should put under her those whose own mothers do not provide enough. "

Semestribus vitulis: these are now fully weaned. Although they go to pasture now with their mothers, their feed must be supplemented with nutritious fodder.

18. armentarium . . . curo: " I take care to see that the head herdsman reads something of these frequently ". *armentarium* should in reality be expressed as subject of *legat* but it is made the object of *curo*: this use of the anticipatory accusative is common in Greek. Books of remedies for the different animals were kept for reference by the stockmen (see above II. ii. 20 for sheep, and below for horses II. vii. 16). If any of these originated with Mago the conclusion must follow that Punic husbandry was much in advance of Roman. No doubt many of the remedies given in Cato's *De Agricultura* and in Varro are among those learnt from Mago. A good stockman must keep them fresh in his mind.

Numerum gregum . . . alium: " one man makes the number of animals one (amount) (another man makes it another) ". The opinion of breeders differs about the size of the herds.

modicum: " a reasonable number ".

Lucienus: Quintus Lucienus, is one of the speakers in the conversation but joins the gathering late just as the discussion on pigs is ending (v. 1, text omitted). He is described as a senator, *homo quamvis humanus ac iocosus*, " thoroughly kindly and good humoured ", and a native of Epirus. He is reproached by one of the company for coming late; he cannot stay, however, because he has to pay his pence to the Lares: he asks Murrius to go with him as a witness that his due has been paid. Murrius is instructed by Atticus to give Lucienus, as they go along, a report on what has been said so far. Lucienus and Murrius return as the discussion on oxen ends (vi. 1, text omitted), which explains why Vaccius brings him in at the very end. We are probably meant to think of Vaccius turning to him or making some gesture. After him Lucienus as a breeder of horses is the chief speaker; he talks out of his own experience in Epirus. Murrius speaks next on the subject of asses because he is from Reate which is famous for the local breed and Murrius himself is a breeder (vi. 1, text omitted).

vii. 1-16. *Of Horses*

1. carceres: Lucienus takes a metaphor from the races in the Circus Maximus (see above I. ii. 11); the *carceres* were the stalls at one end of the Circus where the horses and chariots waited ready for the start of the race. At a given signal the barriers were opened, which were usually swing doors which opened outwards.

Q. Modius Equiculus: the identity of this man is not known. There may be a jocular reference, a pun or a joke, since the last name is an obvious derivation from *equus*, to some person known to them all. At least it is clear that Lucienus speaks of someone who values the mares which he breeds, and considers them as useful in war as the males.

patre militari: is a disputed reading, but can be understood to mean " as his father who was a cavalry man did before him ". Another reading, *in re militari*, makes better sense.   Mares were trained in war as early as the time of Homer (*Iliad II*, 763) where the mares of Pheretiades are described as the best horses which fought against Troy, as swift as birds.

in Peloponneso et in Apulia: the best places for rearing stud horses and selected breeds were districts where there were extensive marsh meadows, with a heavy rainfall and summer showers, or moist soils of lacustrine basins surrounded by or in the neighbourhood of high mountains.   In Arcadia, Elis, and Argos in Thessaly in Greece, horses were bred on the plains, and also in Italy, especially in Apulia in the coastal plains near the Gargano Peninsula. These two regions mentioned by Lucienus were among the most famed in Greece and Italy for their horses.   In the mention of the Peloponnesus the interest may be to bring in Arcadia especially because this has already been referred to in connection with the asses (see above II. i. 14 and note).   Later the plain of Rosea is coupled with these (see below 6).

2. Aetas: the horse's age can be determined from the incisors alone, the front teeth, six in each jaw which fall out and are replaced by permanent teeth. These new teeth are hollow when they first grow but gradually are filled in with ivory.

ungulas indivisas: " solid (*i.e.*, not cloven) hoofs " : this description includes asses and mules.

acutos medios: they lose the two front incisors in the upper and lower jaw.

et incipiunt . . . columellares: " and there begin to grow (those) which they call ' little columns ' " , these are the two small sharp-pointed teeth in each jaw between the incisors and molars which are usually called " canines ". *columellaris* is a rare word and occurs elsewhere only in Pliny (*N.H.*, XI, lxiv, 168).

3. amittere binos . . . completos: "it sheds two (in each jaw) though at the time when they are growing again it has them hollow (the teeth are hollow) but in the sixth year these are filled in; in the seventh year all its teeth have grown again and are complete in number (*i.e.*, forty teeth in all for an adult horse)".   The infinitives in this sentence are to be regarded as dependent on *solet*.   The sudden change to the singular from the plural in the preceding sentences is to be noticed.

Hoc . . . posse: *Hoc* is in the ablative of comparison and refers back to *septumo*. A horse's age can be determined by its teeth up to six years (the seventh year), but not after that until the horse shows signs of age which it is likely to do at sixteen.

brocchi: *brocchus* " with prominent teeth " is another rare word found in this passage especially applied to animals.

4. magnitudine modica: a horse should be of moderate size, not too big, not too small.

nulla . . . congruenti: " with no part of the body out of proportion ".   The shape should be harmonious and balanced throughout.

5. Qualis . . . appositus: the perfect horse is described in every detail. This passage is to be compared with *Georgic III*, 79-88.   (See below, 6 and note.) The general form and outline of the animal suggests strength with a graceful appearance.   It is the horse which will race in the chariots in the circus or serve a cavalry officer but it does not appear to have the finer qualities of the Arab horse or the highly-bred racehorse of today.   Its build is heavier and it is more stockily made.

iuba . . . implicata: the mane should fall over the right side because it was grasped by the rider who mounted on the right side of the horse.

spina . . . duplici: on an ideal horse the spine should lie between two ridges of muscles to give the maximum comfort to the rider.

toto corpore . . . appositus: bleeding as a cure for many illnesses was practised on animals as well as human beings. Blood was let from veins and it was a great advantage if they were somewhat prominent.

6. De stirpe . . . sint: " From what stock they come is of great importance." The Romans set great store on the special breeds which were developed in districts most suited to them after which they often took their name. In all that follows about the breeding of horses Varro has in mind the fine horses raised for riding and the racecourse: the selected breeds were regarded as a luxury and a status symbol.

a Thessalia: the Thessalian horses were bred on the plains watered by the river Peneus in the wide lacustrine basin where were some of the best pastures in Greece. The Athenian cavalry were mounted on Thessalian horses from the days of Peisistratus, and Alexander had horses for his cavalry from Thessaly for his conquest of Asia. His horse Bucephalus was bred at Pharsalus in the south. The Thessalian breed benefited from good feeding and free range over wide plains which also served as training grounds for races or cavalry manoeuvres.

in Italia ab Apulia: in northern Apulia between the Apennines and Monte Gargano is a broad, well-watered plain: it is not best suited to agriculture because of the underlying bed of limestone, but is watered from the Apennine streams and so is valuable pasture land (see above, II. vii. 1). A Greek legend associated Diomedes, the horse-tamer, with the region and the legend that the Apulian town of Arpi was said to have been founded by him points to the breeding there of horses by the Greek colonists before the Roman conquest. From here came the Apulian breed loved by the Romans and prized by Varro.

ab Rosea, Roseani: this is Varro's country in the Sabine country where he had his own stud farm. (See Varro's Life.)

contendit . . . sit: " it tries to rival (the others) in running or in some other way to show its superiority ". The good horse should be spirited, fearless, ready to take the lead and self-reliant. Vergil gives a description which can be compared with Varro's (*Georgic III*, 72-88): " your care must be exactly the same in choosing horses. Do indeed give particular care to those young horses that you intend to mate. Notice how the colt of a fine herd steps agile on the field and places his tender hoofs: he is first both to dare to take to the road and to plunge into frightening rivers and trusts himself to the dangers of an unknown bridge, and is not alarmed at an unaccustomed noise." Columella's account shows the same thought (VI. 29. 1.).

emptio . . . perscripta: " the terms for the purchase of horses is nearly the same as that for oxen and asses, because they change ownership by purchase on the same terms as those recorded in the formulae of Manilius " (see above II. v. 11).

7. ut partus . . . fiat: " so that the birth may take place at a favourable time ". If the mating takes place sometime between the vernal equinox (March 21st) and the solstice (June 21st) the foal will be born in spring or early summer when with grass at its best the supply of milk will in consequence be particularly rich.

Quae . . . nascuntur: see above II. i. 19 for lambs born late which are called *cordi*.

8-9. (Text omitted.)   The mating of horses is described.

10. ne aut laborent plusculum: the pregnant mare must do very little work.   The ox was up to the late XVIIth century the prime working animal, both for ploughing and for draught.   The horse on the other hand remained always the luxury animal for riding and racing although it might at times pull carts.   The horse collar which throws the pull of a loaded cart on to the shoulders and away from the neck and throat was not invented until Saxon times.   For this reason a horse could not be used as a beast of heavy burden: such work was suitable for yoked oxen where the pull came on the back of the neck.

ne frigidis locis: the same injunction is given for cows in calf (see above II. v. 15).

et umore . . . humum: *et* is to be coupled with *et* before *inter singulas* in the next line but one.   " You must both keep the stalls free from damp . . . "

inter singulas . . . longorios: " you must fix long poles to the mangers between the mares to keep them apart ".   The Latin is extremely difficult; *inter singulas* is not logical, but the meaning is quite clear: the mares must be penned separately to prevent injury to each other.

Praegnatem: the mare with foal must have a balanced diet.

11. Alternis . . . fiant: " Those who mate (their mares) every other year say that the mares continue to bear for a longer time and that the foals are better, and just as fields which are cropped every year are more quickly exhausted, so are mares which become pregnant every year. "   *restibilis* from *re* and *stare* means literally " that which is made to stand again ", and in agriculture a field which is tilled every year in contrast with *novalis*, " that which lies fallow " (see above I. xxix. 1).   *exuctiores* is the comparative of *exuctus*, more usually spelt *exsuctus*, the perfect participle of *exsugere* " to suck out ", hence it comes to mean " dried up ", " drained ", " exhausted ".

Quinquemestribus . . . factis: " when the foals are five months old ".   This is the age when they begin to eat, though they are not weaned until after another seventeen months.

farinam . . . furfuribus: the colts are gradually introduced to food by first being given barley meal mixed with bran.

12. eosque . . . exterreantur: " they should be handled at all times when they stand with their mothers (*i.e.*, while they still suck) to prevent their being frightened when they come to be separated from them ".   This is the very first stage of training: the foal must be fondled, patted, stroked, so that he becomes accustomed to man from his tenderest years and learns to trust him.

eademque . . . crepitus: the means by which the young horse is introduced to what concerns his future usefulness are gentle and gradual.   He is handled, he is allowed to see harness hung in his stall and to hear the jingle.   This is to be the concomitant of all his future life; whether he is eventually a cavalry or a race horse or performs some other service. *frenus* means literally the bit, but here includes all the harness connected with it.

13. Cum . . . consuerint: this would be the result of the handling (*tractandum*) recommended for the foal when he is not yet two years old. *consuerint* is either perfect subjunctive or future perfect indicative: the latter is more probable.

sedentem: the young horse is gradually accustomed to being mounted and to carry a rider.   It is noticeable how all his training is gentle and gradual.   It is a boy who first mounts him because he is lighter in weight than a man.   The boy does not immediately sit astride the horse but lies flat on the horse's back. The passage in *Georgic III*, 182-93 shows close similarities with Varro's account.

> " The horse's first lesson is to see the courage and arms of fighting men, to hear untroubled the trumpets, and endure a wheel creaking as it is pulled and to hear the jingle of harness in his stable.   Then increasingly to enjoy his master's caresses and to love the sound of his neck being patted.   And let him do these daring things at the very first when he is taken from his mother's udder and now and then let him put his mouth to a soft bit though he is still weak and trembling and knows not yet his coming strength. But after three years completed, when the fourth summer has come, let him begin to pace round the ring and make his hoofs sound in a regular tread."

sed melius post trimum: this coincides with Vergil's words quoted above. We learn from Columella (VI. 29. 4) that two-year-old horses may rightly be broken for domestic use: but for the races they must be over three: even then work begins only after the fourth.

farrago: for this green, mixed, nutritious fodder see above I. xxiii. 1.

14. quod . . . manendum: when the time (*i.e.*, the fourth summer) has come for breaking in, there is a careful programme of special feeding to be followed: for the first ten days *farrago* only, for four days barley, in increasing daily amounts; then for the next ten days the same amount.   This special treatment covers a period little under a month.

frigus: here is yet another warning against cold: see above in the case of pregnant mares (10 and note).   To have a fire in the stable in a warm climate such as that of Italy and in the summer points to extreme care.

15. alii ad admissuram: " some for breeding ".

quadrigarius: the driver of the four horse chariots (*quadrigae*) in the races.

neque . . . militarem: " nor does the man who wishes to train them for conveyance either for the saddle or the carriage, follow the same system as does the man who wants them for military service ".   Docile horses are needed for riding and driving, but spirited ones for cavalry.   *ephippium* is a word borrowed directly from the Greek, meaning " that which goes on a horse ", hence "a horse-cloth".   The saddle with stirrups was not known to the ancient world, therefore the way of riding a horse was different from that of today.   The rider sat close up to the horse's neck and held himself firmly seated by means of his knees.   The *raeda* (sometimes spelt *reda*) was a four-wheeled cart with an awning which would be essential in the Mediterranean heat, large enough to hold several people.   It was very much like a covered waggon and so roomy that it was possible for a traveller to have beside him his secretary and to dictate letters or take notes as the carriage went along.   Cicero (*Ad Att.* V. xvii) dictates a letter to Atticus in a *reda* when he is setting off on a two-days' journey: *hanc epistolam dictavi sedens in reda.*   " I dictated this letter sitting in my carriage."

Propter quod discrimen . . . equi: " To this difference (the difference that is between the quiet horse on the roads and the animated war-horse) the practice of castration is especially due."

16. quae ... oportet: " which the head groom should keep written down ".
Among the head groom's many duties was that of a veterinary surgeon. Varro
does not tell us much of this book but Columella gives a long list of diseases and
their remedies which is doubtless very little different from the records on
Varro's own estate (see Columella VI, xxx-xxxv).

medici pecorum ... appellati: " those who treat cattle in general, not only
horses, are called ' horse-doctors '."   The Greek word is the equivalent of the
Latin *veterinarius*, a cattle-doctor, one who looks after *animalia veterina*, beasts
of burden and draught animals.   The Latin word is probably connected with
vehere, "carry, convey".

ix. 1-16.   *Of Dogs*

1. Relinquitur: the treatise on dogs, the companion of man and of the
flock, shows throughout a deep and sympathetic understanding of the ways of
the working dog and his needs and care.

quod ad canes attinet: Columella (VII, xii. 1) says of the dog: " *quis famulus
amantior domini ?* ", " what servant loves its master more ? ", and describes it
as the guardian of the whole estate.   This latter shows how much the Roman
farmer relied on the dog for the safety of both farm and flock: the responsibility
that lay on it was certainly greater than today.

lanare: the fleece-bearing animals such as sheep and goats being small compared with the other beasts of the farm have not the strength to defend themselves.

eo comite indiget: the Roman sheepdog was primarily the guardian of the
flock against the marauding wolf: the sagacity of the modern sheepdog in
rounding up animals as is seen in many a sheepdog trial in England today is not
mentioned by Varro and does not seem to have been an important function.

lupus: the wolf (*canis lupus*) still exists in the high Apennines, in parts of
Tuscany, Emilia, Umbria, the Marche, the Abruzzi and Lazio, Campania
and Calabria.   Its habitat is in wooded rocky places in the higher altitudes of
mountainous districts: it is proverbial for greed and ferocity.   In winter,
wolves hunt in packs and even in these modern times during especially severe
weather may come down to villages near their lairs.   In Roman times wolves
were far more numerous and a real danger to flocks and sheepfolds.   Columella
speaks of wolves being driven off in the early morning or at dusk from the fold
(VII. xii. 3).   The sheepdog had to be strong enough to attack a wolf, catch
up with it, drive it off, make it drop its prey and then to bring the prey back.
Varro gives the dog full credit for its usefulness in the words *canes defensores*.

apris: the wild boar inhabited many forests and was an animal of the chase:
it is still preserved today in the Laurentian forest (*silva Laurentina*) which
clothes the shore near Ostia, some sixteen miles from Rome.

2. cum sciam: Varro mentions a particular instance known to him, of a
wolf being killed by a herd of mules.

circumfluxisse: the meaning is not quite clear but seems to make the best
sense if translated as " surrounded it at a rush ".   The mules must have
attacked in a circle otherwise a swift-moving animal such as a wolf could have
got away.

ungulis caedendo: " by kicking ".

tauros ... continuatos: this is not a particular instance as in the case of mules, but a generalisation as *solere* shows. Bulls will draw up in a circle so that they can look in different directions and stand close together so that their flanks touch each other.

genera duo: Varro gives two kinds of dogs, the hunting dog, and the sheepdog; Columella however distinguishes three, the hunting dog (*venaticus*), the watch-dog (*villaticus*) which guarded the home-farm and the sheepdog (VII. xii. 2). The *villaticus*, he says, should be black and the strongest but slowest of the three.

3. praesidio, praedae: the datives are both predicative.

3-4. Facie . . . leonina: the ideal sheepdog is described in most minute detail, a proof of the importance attached to having an animal with the right points. These will all be found to be adjuncts to its function as a guardian, a watchdog, and an attacker of wolves. In size it must match that of a wolf, its enemy; the shape of its mouth and the set of its teeth is made for biting and holding, with straight fangs and short lower jaws; the large head and thick shoulders are for strength, the straight legs and large feet for steadiness and quick running. The toes well separated will be a guard against the lodgement of stones and conduce to the easy removal of thorns. The loud bark is for protection and alarm. The sheepdog must essentially be white, both to be seen easily in the dark and to be different from the grey or reddish colour of the wolf. The form and appearance of Varro's dog must have been closely comparable with that of the Abruzzi shepherd dog of today. It is a large white dog with sturdy limbs and a thick shaggy coat, dark eyes, drooping ears, and long tail. This breed is seen not only in the mountains but also on the Roman Campagna today where it accompanies the flocks on their winter grazing grounds. It is an intelligent docile animal but can show fierceness to strangers. Varro's account of the sheepdog should be compared with that of Columella, (VII. xii. 3-9.)

3. neque resimis . . . subtus: " with the upper jaw neither too high or not drooping ". Varro seems to describe the kind of jaw that is best in shape for snapping and biting.

ex eo enatis . . . eminulis: " with the two teeth that spring from this on the right and on the left side projecting a little ". The canine teeth (four in all) should be rather prominent: these are the teeth made for biting and tearing.

superioribus . . . brocchis: " with the upper teeth (*i.e.*, those in the front of the upper jaw) straight rather than projecting ". Here is another characteristic which would help to ensure a sure bite and hold.

4. acutos . . . labro tectos: lit. " let the sharp teeth which they have be covered with the lips ". There is a variant in construction here as is often found in a long descriptive passage.

pedibus . . . displodantur: " with large and wide paws which spread out as he walks ". This shape would give the dog a better grip on the ground, especially advantageous on rough paths and tracks: it would also be easier to extract from the paw any material which might lodge in it. *ingredienti*, the present participle, is in the possessive dative agreeing with *ei*.

digitis . . . molli: these characteristics coupled with those described above show how every detail mattered. The big wide paw giving a good grip on the ground, which was helped by the well-separated toes and hard claws, were ideal for the roughness of the pasture and the transhumanting trek; the sole of the foot, on the contrary, had to be rather soft to help the dog to spring.

a feminibus . . . suppresso: " with the whole frame tapering near the top of the thigh ".

5. feminas . . . papillas: a bitch should be able to suckle each pup equally well.

itaque . . . Sallentini: just as the different breeds of horses are named after the districts from which they come so too are the dogs. The Laconian (or Spartan) hound was used for hunting but also seems to have been a sheepdog; it was white. Another breed was called *Molossus* after a people of that name in Epirus. They were large black dogs and were used as watchdogs on the home-farm. Their dark colour prevented their being seen at night with the result that they could more easily surprise and drive off intruders: Lucretius describes them as having open jaws, strong teeth, and a loud bark (*De Rerum Natura* V. 1063). A large black dog is still known and called by that name in Italy today. Dogs shown on mosaics in the entrance to houses in Pompeii which are black with very pronounced fangs are very like Molossians. Vergil mentions the Laconian and the Molossian dog, (*Georgic III*, 404-8) in connection with the guardianship of the sheep and of hunting: " the care of your dogs should not take last place; feed together the pups of the swift Spartan bitch and the keen-scented Molossian on rich whey (sheep's milk). While they are on guard you will never need to fear a thief in the night at the pens, or forays of wolves, or Spanish brigands creeping behind you ".

Sallentini: it is not certain where this breed came from: but it has been suggested that the *Salentini Campi* in Iapygia might have been their original home. The breed might therefore have been Illyrian.

videndum . . . sequentur: a dog will always do what it has been trained to do. The saying " you cannot teach an old dog new tricks " is well illustrated here.

Quare . . . fuerit: " It is better to buy from shepherds a bitch which is accustomed to follow sheep ", or " one which is (as yet) without any training ". The shepherd who needs a dog can then make it his very own and accustomed to his own ways.

Canis . . . pecudes: " Now the dog acquires any habit more easily (than other animals) and the attachment he forms for shepherds is closer than that to the sheep. " The dog is man's companion and he will therefore stay with his master rather than with the sheep, his charges. Varro goes on to relate a true story which illustrates this devotion. The incident gives a vivid glimpse of transhumanting flocks in the districts and the markets to which they were taken.

6. quibus gregibus . . . accessissent: " in (the purchase of) these flocks the dogs had been included, but not the shepherds ". By the agreed terms of purchase, the dogs were included in the price and were to remain with the sheep, but the shepherds had to accompany the sheep to Metapontum, stay there with them for the winter grazing along the shore and then to sell them with the dogs in the neighbouring market town of Heraclea. After this, according to the agreement, they returned home to Umbria leaving the dogs behind. This story gives a valuable statement of movement of transhumanting flocks over the peninsula and some notion of the distances covered.

Metapontinos saltus: " the grazing grounds of Metapontum ". This city, a Greek colony founded by Achaeans in the first quarter of the seventh century B.C., was in the " instep " of Italy, twenty-seven miles from Taras (modern Taranto) on the coast, near the mouth of the river Bradanus. In the hinterland was an extensive fertile plain watered by several rivers which was given over to corn and horse-rearing.   So wealthy was it that the early coinage was stamped with an ear of wheat.   These same plains were given over to pasturing sheep by the time of Varro.   The Greek mathematician Pythagoras died here: his tomb was still shown in Cicero's time.

Heracleae emporium: now is the modern Policoro.   It was a Greek colony founded from Taras three miles inland and thirteen from Metapontum.   The town was a market for the transhumanting flocks.   It was the scene of the victory of Pyrrhus, King of Epirus, over Rome in 280 B.C.

Neque eorum . . . praecepit: a dig at Saserna and a chuckle between Varro and his friends (see above I. ii. 25-7 and note).   Saserna's writings contained some strange prescriptions as well as sound agricultural science.

quod cognati . . . praesidio: " because those who are related are the greatest protection to each other ", or " especially protective of each other ".

7. fit alterius . . . est: " the dog becomes the property of another when it has been handed over by its former master to the next ".   The law demands therefore that an actual transfer must have been effected before the dog could be said to have changed ownership.   This is different from the buying and selling of oxen and sheep (see above II. i. 15 and II. ii. 5-7 and notes).   Varro invokes the laws or formulae of Manilius which applied to the sale of all kinds of farm animals: see the above citations for this sentence and the rest of the paragraph.

De sanitate et noxa: see above II. ii. 5-6 and note for the *leges Manilii* which applied to dogs equally with other farm animals.

stipulationes . . . pecore: there must be an undertaking on the part of the vendor that the dogs have done no damage and that the purchaser will not therefore be liable for that or any other action for which he might be held liable. By Roman law a claim for any damage done by an animal was transferred with it to a new owner.

hic: " here ", " in this case ", referring to dogs, because a special *stipulatio* has to be made with reference to them.

alii . . . ovis: the same kind of bargain was legal in the case of sheep and lambs.

8. Diligenter . . . providendum: sheepdogs must be well fed if they are to stay with the flock and give it all their attention.   If a dog were to be subject to hunger there were many temptations both at pasture and on the trek to lure it away to seek its own food.   If a dog is sure of its food from the shepherds it will not leave them.

9. nisi si . . . dentes: " even if, as some people think, they do not reach the point where they give the lie to the old proverb (dog doesn't eat dog), or even put the tale of Actaeon on the stage and turn their teeth on their master ". An old Latin proverb is quoted by Varro in the *Lingua Latina* VII. 31: " *canis caninam non est* ": " dog does not eat dog ".   *caro* is to be understood with *canina*: hence the meaning " dog's flesh ".   Actaeon was a mythical hunter who inadvertently came upon Diana, the goddess of hunting, at her bath. As a punishment he was changed into a stag and torn to pieces by his own hunting dogs.

10. Nec non . . . ut non: the first two negatives cancel out, the second *non* is incorrect and redundant: it seems to be needlessly repeated from the beginning of the sentence, possibly by scribal error.

ut . . . intritum: " though you must give it preferably crumbled in milk ". *intritum* is from *interere, -trivi, -tritum,* " rub into " or " break into pieces " or " crumble ". The milk is of course sheep's milk. It may be that the flavour especially attracted the dogs to the flock. Whey left over from the making of sheep's milk cheese was often fed to the dogs.

Morticinae: this adjective usually describes a natural death.

ne ducti . . . abstineant: " in case, lured by the taste, they do not refrain from attacking the flock ".

ius: " broth " or " soup ".

ea . . . contusa: the bones given to them must be broken up. Modern breeders say that dogs should not be given bones in any shape or kind because they can injure or block the intestines: Varro's description seems to mean very large bones because they strengthen their teeth and stretch the mouth.

Cibum . . . stabulantur: the sheepdog has two meals a day, whereas normally a dog has only one, in the evening. The morning meal for the sheepdog will have the effect of satisfying him so that he stays with the flock and close to the shepherd. *ubi pascuntur* is " when (the sheep) go out to pasture ". The dog may have to run long distances during the day so that for that reason he needs food before the day's work. *interdiu*, "in the day time ", but here the inference is " in the morning ", contrasted with *vesperi*. *ubi stabulantur*, " when the sheep are folded ". The dog will not be confined during the night, but has to be able to range freely and to be on guard against wolves and marauders.

12. **Quam paucissimos . . . lactis:** " The fewer you leave, the better they grow on account of the abundance of milk. " A statement of this kind expressed by means of superlatives rather than comparatives is very awkward, but seems to be one of Varro's peculiarities.

acus . . . aliut: " chaff or something else of the same sort ". Another example is here of extreme care for young animals, a care perhaps going further than mere utility.

neque . . . patiuntur: the young dogs when being trained to fight, must not be overdone because they then become overtired. Varro well understood the welfare of a puppy and a dog. He had probably trained many a one of his own.

13. **Consue . . . faciunt . . . vinclis:** " They are trained also so that they can be tied up at first with light leashes. " The dog was first tied with a strap made of leather which would later be substituted when the dog was grown stronger with an iron chain. The puppy would have to be taught not to gnaw its leash. Its teeth would not be harmed if it tried to bite at the leather and the weight would not be too great. *consue* and *faciunt* are more usually found as one word.

ut . . . perfrigescant: avoidance of chill seems to have been one of Varro's chief concerns: see above II. vii. 10 (mares in foal) II. vii. 14 (colts) II. ii. 15 (new-born lambs).

14. **nucibus . . . tritis:** a paste is made from ground almonds which is rubbed on their ears and between their toes. The Romans knew little of insecticides or disinfectants but did what they could to protect their animals from vermin.

15. **Ne . . . ferri:** " To prevent their being wounded by wild beasts, collars are put on them which are called *melium*; it is a band round the neck made of strong leather (studded) with nails that have heads: soft leather is sown under these heads so that the hardness of the iron does not injure the neck. " The nails are so set in the leather collar that the points stick outwards: the collar

is lined with a soft kind of leather to prevent chafing. Sheepdogs still wear these collars today: they are a protection against wolves, which may still be encountered in high mountain pastures, and against other dogs. An animal would not try to attack a second time if it had once been hurt by these spikes with the result that the other dogs belonging to the farmstead or flock are safe. It was not necessary for every dog to wear the *melium*.

quae ... tuto: " with the result that those (dogs) which do not have it are safe ".

16. quod, ... hiberna: more than one dog to one shepherd is needed for the transhumanting flocks who cover long distances and are exposed to the dangers of wild beasts: (see Chapter II).

Villatico ... gregi: some flocks remained on the home-farm and were not sent out to summer and winter pastures. It is probable that enough sheep and goats were kept at home to provide for the needs of the household in milk and cheese. Cow's milk seems to have been very little used. The size and number of flocks which were pastured around the homestead depended on the extent of pasturage available, the amount of green fodder which could be grown, and the needs of the household and slaves.

quod ... sit: " because the one becomes keener with the (companionship) of the other, and if one of the two is ill, this prevents the flock from being without a dog ". The understanding of a dog's nature comes out very clearly here.

x. 1-11. *Of Shepherds*

1. Cum circumspiceret ... pastores: " When (Atticus) was looking around to discover whether he had overlooked anything, ' This silence ' I remarked ' summons another (actor) to take up his cue. For there remains in this last act (to discuss) the number and type of shepherds to be kept '." Varro speaks in the language of the stage. Slaves appear last in the discussion; they are perhaps of less importance than the fine cattle and sheep which the farmer should breed. Their placing here shows, as nothing else can, their position in regard to the beasts: they are not better and of no greater value.

Ad maiores ... pueros: *pueros* in the accusative is perhaps to be regarded as the object of *habendum* to be supplied from *habendi* in the previous sentence.

qui in callibus versentur: " those who work on the cattle trails ". The transhumanting shepherds must be men of great strength able to withstand exposure to climate and to endure long, stiff climbs (see Chapter II).

(itaque ... armatam: the dangers which confronted the flocks on the trails from summer to winter pasture and back again were manifold: the flocks were at the mercy of robbers and wolves; in addition the shepherds might be caught in disputes concerning the right to pasture during their passage from one grazing ground to another (see Chapter II). The shepherds went armed on their long trek, even with javelins [see below (3)].

2. pascere communiter: this is recommended for safety.

contra pernoctare ... gregem: " each shepherd would make a fold for his own flock for the night and sleep with it ".

quod ei ... parent: " because the rest obey more willingly a man who is their superior in age and skill ".

3. Neque ... cordi: in a sentence such as this Varro almost touches poetry.

quod patiendum picks up in a neuter pronoun what has gone before, although grammatically *quae . . . patiendae* would be expected.

quibus . . . cordi: *sunt* is to be supplied. Cattle were fed on all kinds of leaves and foliage, because of the general lack of grazing land. It follows that they would sometimes feed in woods along the cattle trails: goats would range afar and climb in steep places to find their food.

Formae . . . iaculari: only an extremely tough type of slave could be equal to this work: *expeditis membris* implies that he must have supple limbs.

onera . . . possint: the baggage which had to be carried must have been considerable. The animals (*iumenta*) which accompanied the shepherd and flock must have been heavily laden: see below (5) where mares are mentioned and the pack-animals are called *dossuaria*, " which carry loads on their backs ". (See Chapter II.)

   4. quod neque Bastulus . . . iumenta: Bastulus and Turdulus are inhabitants of Granada and Andalusia in Spain. Varro speaks from personal experience gained by his service in Spain (see Varro's Life). Being of a northern stock the Gauls were of tall build, strong and powerful, and able to endure hardships.

In emptionibus . . . perficiunt: the law regarding property in relation to slaves is here usefully summarised. Six ways are enumerated in which a man could legally become a possessor of slaves: there were others less common.

hereditatem iustam: since a slave was a chattel of his master and was in law *res* (" a thing ") and not a person, he could be bequeathed to an heir along with the master's estate and other movable possessions; this method of acquiring a slave was called *testamento*. The will to be effective however had to be in writing and sealed by seven witnesses. On the other hand a slave was often given his freedom in his master's will so that his friends and relatives might appreciate the testator's generosity. Wills were under the protection of the praetors in Varro's time.

mancipio: *mancipium* (often called *mancipatio*) was a formal act of purchase, going back to primitive times. The seller and the purchaser met in the presence of five witnesses who had to be Roman citizens of full age and another citizen who held a pair of scales and a piece of copper (*aes*). The purchaser then put his hand on the thing to be acquired (in this instance a slave) and asserted his ownership. He next struck the scales with the piece of copper and handed it to the seller as a token of the price. *mancipium*, as the derivation from *manus* and *capere* shows, was in reality a form of conveyance, a transfer of ownership and the act was in Varro's time a mere symbol; in origin it appears to belong to the times before money was coined.

a quo . . . potuit: it was important that the seller should be the legal possessor and have the right to sell: otherwise *evictio* might follow; in that case the buyer might be deprived of his acquisition and the seller would have to pay him back double the price originally paid.

si in iure cessit: *in iure cessio* was a method of conveyance in the form of an imaginary suit. The two parties concerned went before a magistrate, a praetor at Rome or the governor in the provinces. The purchaser grasped the thing to be conveyed, in this case a slave, and declared that he was the owner. The magistrate, as presiding judge then asked the seller whether he also made the same claim. If he remained silent or answered " No " the magistrate then adjudged the slave to the buyer; the transfer of ownership was made by a pretence that the slave already belonged to the man who wished to acquire him. As in the case of *mancipium* the right of ownership had to be established to guard against *evictio*.

si usu cepit: this is the method called *usucapio* and by it was meant the acquisition of ownership by continued possession for a given time: in the case of a slave, this was probably one year, if the possession remained unchallenged.

si e praeda ... emit: in Varro's time the vast majority of slaves were prisoners of war, " by purchase at auction from war booty ". Great numbers were captured after Rome's various victories and sold to dealers. The sale of slaves thus became a recognised industry which was carried on by public auction in the market place. The slaves would be sold just like any other booty captured in war.

tumve . . . veniit: " and lastly when he is sold publicly together with the goods or at the sale of a (proscribed) person ". When a man was proscribed he was declared an outlaw, and his possessions were confiscated by the state, later to be sold by public auction (see Varro's Life). *sectio*, " distribution ", from *secare* is the technical term for such a sale.

5. solet accedere peculium: the *peculium* was a fund which a master allowed a slave to accumulate and to hold and to regard as his own. It was distinct from the master's property but in law belonged to the master because a slave being a person without rights could not in law possess anything. Varro's statement makes it clear that even when the slave was transferred to another he might not necessarily take his *peculium* with him. A stipulation could however be made so that the slave was not deprived of his *peculium*.

si mancipio ... simpla: " if the transfer is not by mancipation, double the purchase price is agreed to be paid (in case of eviction) or if (the two parties) have so agreed, only the amount paid ". A seller whose title to ownership might be doubtful would not use *mancipium* because he was bound by law to guarantee double the value of what he sold if *evictio* was proved against him. Without *mancipium* the seller need only guarantee to return the purchase money (not doubled) if both parties agreed. A safer method of sale was to transfer to the buyer and leave him to become the legal owner by *usucapio* after the lapse of a year. *simpla* suggests that if the seller were proved not to be the legal owner he would merely pay back his money to the purchaser. Such a precaution might be necessary in the case of slaves: among them there might be runaways, thieves, and low characters.

Magistrum ... pecudum: " The head shepherd must see that all equipment follows (*i.e.*, is transported) which is needed for the flock and the shepherds especially for the men's food and the medical treatment of the sheep."

6. nec hac . . . quaerit: " and the Venus of shepherds seeks no farther than this ". There could be no form of legal marriage for slaves since they were persons without rights. It was recognised, however, that a slave might find a mate with whom he cohabited and to whom he was as faithful as if to a lawful wife. On the other hand promiscuity was regarded as a normal condition. The children of slaves were themselves slaves and the property of the master.

Qui autem ... multi: the construction of the sentence in outline is *qui . . . iis mulieres adiungere . . . utile arbitrati multi.* Sometimes women accompanied the transhumanting flocks on the trails to cook the food and in general to be a solace to the shepherds.

casis repentinis: while on trek the only shelter which the shepherds had could be nothing more than a straw hut: hence Varro calls these *repentinae* because they were at best temporary structures and could be built quickly: in all probability there were many along the routes available for the shepherds of the passing herds, very like the *capanne*, made of reed or straw which are still to be seen in places on the Roman Campagna today. No doubt clusters of huts were there for the shepherds on the grazing grounds where they halted on route to let the flocks rest and feed.

7. firmas, nec turpes: " robust and not uncomely ". *turpes* seems to be used in contrast with *firmas;* a strong, well built woman is described.

non cedunt viris: " they do not give place to men ". They can equal the men in hard work.

quae ostenderunt . . . contemnendas: " who prove our women who have just had children who lie for days under mosquito nets to be weakly and despicable ". *eiuncidas* connected with *iunx* " a rush ", means " grown like a rush ", therefore " slender " and hence " weak ".

conopiis: *conopeus* derived from κώνωψ, the Greek word for a gnat, was a net of fine gauze draped around the bed as a protection against gnats and mosquitoes. Its use, which originated with the Egyptians, is mentioned in Herodotus (ii. 94) in a passage describing the Egyptian fisherman's use of it as a net for fishing during the day and as a tent at night. Even in Varro's time the use of such a net was regarded as effeminate, as is shown in Horace (*Epode* ix. 16) and Propertius (III. xi. 45) who both refer to its use as part of Cleopatra's luxurious living: the word has come down to us as " canopy ".

9. quem . . . putes: " to which you would think had not so much given birth, as had found it ".

quibus nos . . . habere: " whose custom does not forbid them to mate before marriage with those whom they please, to have freedom to wander about unaccompanied and to have children ". *denegavit* is a gnomic perfect, used to express a general statement which is true at any time, past, present, or future. The mention of *nuptias* shows that slave girls are not described here, but the daughters of the ruling classes. Marriage as such did not exist for slaves.

10. scripta habere oportet: just as a book of remedies had to be kept by the chief groom, herdsman, and head shepherd, the same must be done for the slaves. See above oxen (II. v. 17), horses (II. vii. 16), sheep (II. ii. 20).

rationes . . . pecuarias: the owner's cattle accounts. The chief herdsmen and shepherds were responsible for records and accounts (see above I. xvii. 4-7).

11. hirtas oves: " rough coated sheep " in contrast with the jacketed sheep, the flocks which needed more shepherds.

utique . . . abigere: " each of these men should have a mare for his own use to ride in those districts where it is usual to drive the mares to their stables (at night) ".

in Apulia, in Lucanis: both these districts were famous for horses bred on their level pastures (see above II. vii. 6 for Apulia).

END OF BOOK II

LIBER TERTIUS

The Place and Setting of the Conversation

The third book of the *Res Rusticae*, as is the first, is set in Rome: the occasion is the annual election of the aediles in 54 B.C., which was held in the Campus Martius. The place where the conversation takes place is the *villa publica;* here the seven speakers meet and discuss agricultural subjects while they wait to hear the result of the election. During the conversation there comes a dramatic interruption when a message is brought that a man has been caught in the act of tampering with the ballot box. Pavo goes off to investigate the complaint. The conversation ends when the result of the election is known. The candidate favoured by Varro and his friends comes into the *villa* wearing the purple stripe of a magistrate. This shows that he has been successful and the friends go their several ways. Those who take part in the conversation, in addition to Varro himself, are:

Quintus Axius, a senator: he comes from Reate in the Sabine country and so is a compatriot of Varro and of the same tribe for purposes of voting. They have probably travelled to Rome together for the election and both support the same candidate.

Appius Claudius, an augur: he is present at the election in his official capacity, and will have taken the auspices before the election; some matter of ritual or religious interpretation may arise for which his help may be needed by the presiding consul.

Cornelius Merula, a member of a consular family.

Fircellius Pavo, a compatriot of Varro, from Reate.

Minucius Pica.

Marcus Petronius Passer.

The names of the last four speakers are also those of birds: *merula* is a blackbird, *pavo* a peacock, *pica* a magpie, and *passer* a sparrow. In accordance with his love of making puns, Varro introduces them most aptly in relation to the conversation which is about birds. In English, " Peacock " and " Sparrow " are surnames common enough to need no comment. The choice of names to suit the situation has been found already in Book I where they pertain to agriculture in general (see above I. ii. 1) and in Book II. iv. 1 where Scrofa talks about swine and Vaccius in v. 6 about cattle: also in Book III in the account of bee-keeping, xvi. 1. Appius, the speaker, implies that his name is connected with *apis*, a bee. The time when the election takes place is during the hot months: no absolute certainty can be reached, but the remark about the heat, and the need for shade (2 below) points to a day in July or August.

ii. 1-6. *Of Various Kinds of Villas*

1. Comitiis aediliciis: " during the election of the aediles " (see above, I. ii. 2), where the aedile who has charge of the temple of Tellus is mentioned. The *comitia* were the voting assemblies of the Roman people which met in the Campus Martius in Varro's time for the purposes of holding elections. Up to the end of the republic they were assembled in an enclosed area called the *ovile* because of its likeness to a sheepfold. It was divided by wooden barriers into aisles along which the voters passed to a raised platform called a *pons* where they cast their votes. The *ovile* probably occupied the same site as the later *saepta Iulia*, a marble colonnaded enclosure which was planned by Julius Caesar, but not finished until 26 B.C. There were two classes of aediles: the plebeian aediles who were elected by the *concilium plebis* often called the *comitia tributa*, and the curule aediles by the *comitia tributa* proper. The voting groups were assembled according to the territorial tribes (*tribus*). The fact that Varro takes part in the voting in itself suggests that the election is that of the curule aediles, but determinative evidence in support of this conclusion is that a presiding consul is mentioned (see below, v.18 and note) and a consul customarily presided over the election of the curule and not of the plebeian aediles.

tribulis: "a member of my tribe". Varro and Axius both belong to the same territorial group; as natives of Reate they probably are members of the Sabine tribe *Quirina*.

candidato, cui studebamus: "the candidate whom we were supporting".

Dum diribentur . . . suffragia: "while the votes are being sorted". Each voter recorded his choice of candidate on a small wooden tablet covered with wax on which the named or initial of the one preferred could be written, and then tossed it into an urn (*cista*); these were collected and taken to the tellers (*diribitores*) in a room nearby (*diribitorium*) who counted the votes. The results were announced according to units, that is to the tribes in the *comitia tributa*, and not according to a majority of individuals. Tie votes were decided by lot.

vis potius . . . utamur: the subjunctive is jussive without a conjunction.

villae publicae: the *villa publica* was the only public building in the Campus Martius before the end of the republic. It was first built in 435 B.C., and afterwards restored and enlarged in 194 B.C. It was an antiquated building at the time of the dramatic date of the dialogue (54 B.C.). It was to be restored three years after the publication of the *Res Rusticae* in 34 B.C. The architectural form of it was a walled enclosure within which was a square two-storey building surrounded with open arches on the ground floor. No remains have been found, but the site was probably that of the present Piazza del Gesú. The *villa* which was the common property of the whole population, was decorated with paintings and statues. It served many public purposes as Varro himself states below. Officials engaged in levying troops, taking the census, mobilising the army, and presiding over public affairs were lodged here. Its nearness to the Campus Martius (see below, 3, 4, and 5 and notes) where large masses of men could gather increased its usefulness. Foreign ambassadors were also lodged here, and it seems clear from Varro that it could be used as a public park. Close by was the *ovile* where the voting for magistrates took place.

quam . . . nobis: the meaning of this passage cannot be entirely explained. The text appears to be corrupt, but the gist of it is clear, that Axius and Varro will be more comfortable sitting in the shade in the *villa* which is adjacent while they wait for their candidate than on some uncomfortable narrow bench where he is required to stay. The reference might be to the wooden enclosures in the *ovile*, sometimes called the *saepta*, made of planks. They are described by Servius *Ad. Eclog.* I, 34: *saepta proprie sunt loca in campo Martio inclusa tabulatis in quibus stans populus Romanus suffraga ifere consueverat.* "The *saepta* are spaces in the Campus Martius enclosed with wooden boards standing in which the Roman people used to cast their votes." The *ovile* seems to have been without shade which must have been very trying for the crowds in the hot months. A literal translation might be " than making one for ourselves out of our candidate's narrow plank ".

malum . . . pessimum: " bad advice is bad for the adviser ". The saying was proverbial, arising out of an incident, true or legendary, connected with Rome's early history. It was told how when a statue of Horatius Cocles had been struck by lightning the Etruscan *haruspices* (seers) were consulted. They purposely gave bad advice because of their hostility to Rome: later they confessed and were executed.

2. augurem: Appius belongs to the college of the *augures*, Roman diviners, which in Varro's time numbered sixteen. Their duties were to discover by observation of signs whether the gods favoured some particular action: this was often done by observing the flight of birds from an allotted space of ground or perhaps from the appearance of lightning. An election could not take place unless the auspices had first been taken and found to be favourable.

consuli . . . praesto: " ready for the consul ". Appius was attendant on the consul whose duty it was to conduct the elections.

siquid . . . poposcisset: *siquid* = *si quid*. The pluperfect tense of *poposcisset* presents a difficulty because it implies a future in the past, " any need should have arisen ". The verb *poscere* does not have a supine so that the future participle with *esset*, which might have been expected, cannot be expressed.

Sedebat . . . Passer: Axius and Varro find their five friends already sitting together on benches in the *villa publica*: two are on Appius' left hand and two on his right.

ornithona: *ornithon*, " bird-house " or " aviary ", is from the Greek and has the Greek form of the accusative. Axius speaks jokingly, because Appius is sitting among " the birds ", his four friends whose names are those of birds.

3. te: (i.e., Axius). This is the object of *recipio* to be understood from *recipis* above.

quoius . . . ructor: *quoius* = *cuius*. Appius answers that he is only too pleased to welcome Axius because he can still taste the fine birds which he served up to him at dinner during his recent visit to Reate when he stayed with Axius.

eunti: this present participle agrees with *mihi*.

de controversiis: Appius accompanied Cicero in 54 B.C. when he visited Reate to conduct a lawsuit against the people of the neighbouring town of Interamna Nahars (see Varro's Life). They both stayed with Axius: *vixi cum Axio* are Cicero's words in his letter to Atticus (*Ad Att.* IV. 15).

quam . . . nostri: there was no farm attached to the *villa publica* (see above, III. ii. 1 and note). It was a place for lodgement surrounded by an enclosed park: its character can be related to the later development of the simple villa into a luxurious country residence (though it was never a villa in the full sense of the word).

tua . . . Reatino: Axius is a wealthy landowner in the Sabine country, of which Reate was one of the chief towns. He owns two villas there, one mentioned in this passage as *perpolita*, " lavishly decorated ", and another on the bend of the river Veliuus which is no more than a farm (see below, 9-10). The two illustrate very well what a villa might be, either a luxurious residence to which a farm might or might not be attached, or a place wholly given over to farming.

4. Nuncubi . . . lithostrotum: the decorations of a rich man's villa were lavish and expensive.   Since the second century B.C. when the residential villa began to develop out of the simple farmstead as a country seat given over to pleasure, the buildings and the architectural decorations became more and more elaborate.   *emblema*, Hellenistic in origin, was a small design in brightly coloured mosaic, either a geometric pattern or a representation of flowers, animals, human heads, mythological scenes and many other subjects, and often surrounded with ornamental borders.   *Emblemata* could be set in a larger background of mosaic work, most often in a floor.   They were made in workshops, mounted on tiles or slabs of stone and could be exported, ready-made, to be inserted in a geometric floor (*lithostrotum*) made locally.   Although usually in the floor, *emblemata* were regarded as pictures to be looked at.   The word is derived from the Greek, meaning " something inserted ".   *lithostrotum,* literally " a laying down of stones ", is a word from the Greek for mosaic work. In Varro's time the design of mosaic used for flooring would probably be geometric, and would be confined to black and white or a few other colours as well.   The technique was the fixing of small cube-shaped pieces of stone or marble into a cement bed made usually of pounded marble or tile and lime. This made a hard smooth surface.

quae . . . contra: " in your (villa) everything is (exactly) the opposite ".   Appius says that in the *villa publica* there is none of the lavish ornaments which Axius has in his at Reate.

solius: the genitive agrees with *tui* to be understood from *tua:* " of you alone ", " that is yours alone ".

reliqui homines: Varro may have in mind foreign ambassadors and other visitors who were not allowed to enter the city: such as these were lodged in the adjacent *villa publica.*

consuli adductae: " brought before the consul ".

5. campo Martio: This was the level ground between the slopes of the Capitoline, the Quirinal and the Pincian Hills and the Tiber, covering in all about six hundred acres.   It is often called *campus* alone.   From the beginning of republican times it was owned by the state and was used for athletic and military exercises and for large gatherings of citizens for occasions such as Varro goes on to mention, as levying and arming troops and taking the census.

extremo: " on the edge of ".

quam omnes . . . Reatinae: " than all the villas possessed by everyone in the whole district of Reate ".   The question suggests that there were other rich landowners there besides Varro and Axius.

Tua . . . signis: *Tua* is the *villa publica*: it seems that it was adorned with paintings and statues, but was not lavishly decorated as were the villas of the well-to-do of the time.

Lysippi aut Antiphilu: the latter has the Greek genitive ending.   Lysippus, a native of Sicyon in Greece, was one of the great sculptors of the fourth century B.C.   He was a contemporary of Alexander the Great and pleased the king so much by his style that he would not allow any other artists to make a likeness of him.   He worked only in bronze, but Roman copies of his works in marble exist.   The so-called *Apoxyomenos* in the Vatican Museum in Rome is thought to be a copy of an original bronze and one of his best-known works.

Antiphilus, a painter, was a contemporary of Lysippus. His *genre* scenes were particularly famous, such as those of a boy blowing a fire, women dressing wool, and others.

sartoris et pastoris: Axius' villa, although magnificent and lavishly decorated, has a farm; *sartor* (a hoer) stands for tillage and *pastor* for sheep-rearing: the two together represent the whole life of the farm.

illa: *i.e.*, Axius' villa. It has a large farm attached to it and one which is beautifully kept.

tua ista: the *villa publica* about which Appius has been talking (see above 3). There was no farm attached to the *villa publica*, it was solely a place for public use (see above, ii. 1 and note) and for lodgings as needed.

6. tua: the *villa publica* where no agriculture was practised.

quam ... habebat: it seems that Appius had inherited a villa from his family: *avos = avus, proavos = proavus*.

ut illa: this refers to Appius' own estate.

faenisicia ... in tabulato: " a hay harvest on wooden floors ". A way of storing dried hay was to place it on several wooden floors one above the other. This arrangement allowed free passage of air around the hay and guarded against spontaneous combustion and vermin.

porta Flumentana: this was a gate in the republican wall of Rome, near the Tiber, as the name implies: it was near the Capitoline Hill and the Campus Martius. The district nearby called *extra portam Flumentanam* was occupied by houses of the wealthy.

Aemilianis: the district of this name outside the republican wall in the southern part of the Campus Martius was a fashionable quarter where there were houses of well-to-do citizens; nothing else is known about it.

ii. 7-18. *On the Profits from the Villas*

7. Quoniam ... villa: there are three aspects of the meaning of *villa:* in origin it was the simple farm of early times worked by the owner and his family and with no pretence beyond frugal living. From this developed the *villa* run for profit on a large scale, eventually the *latifundia* worked with slave labour (Chapter II), the property of such agricultural tycoons as Varro himself. Then elaborate country houses became fashionable which were the pride of the wealthy, but had little or no dependence on agriculture. The conversation which follows emphasises these various meanings.

a M. Seio ... villam: Appius is thinking of buying a seaside villa. The coast-line south-eastwards from Ostia at the mouth of the Tiber, was becoming lined with such pleasurable places which were set on the sea's edge and served as an escape from the heat and clamour of Rome. Cicero, for instance, had a villa at Astura, midway between Ostia and Antium (modern Anzio). Marcus Seius was a landowner and successful farmer, and a friend of Atticus and Cicero. He was well known for his home-feeding (see below, 13, 14), the self-sufficiency of his farm, and his hospitality. Appius praises his profits from his farmstead; he excelled especially in the rearing of birds, peacocks (III. vi. 3, text omitted), geese (III. x. 1-2, text omitted), and was an authority on bee-keeping (III. xvi. 10). He died in 45 B.C. At the dramatic date of the conversation in Book III, 54 B.C., his reputation in the agricultural world was firmly established.

asinum . . . emptum: " the ass bought for forty thousand sesterces ". For the rearing of fine asses in the district of the Rosea see above II. i. 17; vii. 6. Appius was shown this valuable ass when he stayed with Axius in his villa at Reate (see above III. ii. 3).

Quod si . . . non habent: the argument is that unless a building is part of an agricultural complex it cannot be termed a villa. That which Appius is thinking of buying from Seius on the seashore is not therefore a villa because it has no farm, is not part of a farm, and can only be called a house, which is an absurdity, for *aedes* means an urban dwelling. It could only be dubbed " the kind of house which Seius has ".

 8. cum . . . fuisset: " when he told me that he had never been entertained in any villa which he had enjoyed more, after he had stayed with him (Seius) for several days ". Merula had enjoyed a visit to his seaside villa.

nec tamen . . . trapetas: Seius' villa was neither an elaborately decorated dwelling like the fashionable ones of the time, nor a farm with the appurtenances of a farm such as presses, storage jars, or grain mills; yet it was to be regarded as a villa because it was a country house (for *trapetas* see I. lv. 5, and note).

 9. Quid . . . membra? for Merula's answer to Axius' question and for his two villas, see above, III. ii. 3, and note. *urbana:* the wealthy Romans were already building villas on the outskirts of Rome and other cities, bringing in this way the country and the country house to the city. They were not intended for profitable farming, but were residences surrounded by extensive acres of land designed as pleasure gardens (see the reference to Lucullus and note above, I. ii. 10): those on the outskirts of Rome were called *villae urbanae* (see above, I. ii. 10: xiii. 7).

Num minus . . . asino: for Axius' two villas in the Rosea see above III. ii. 3 and note. Cicero stayed in one of these, probably the more elegant one in the Rosea on the occasion of the lawsuit.

 10. significasset: the subject is Axius, who agrees with Merula's argument, that both Axius' estates are villas, although one is only a farm and the other is a luxurious country house with a farm attached.

nihilo . . . urbana: " that that was no less a villa which was only a farm than that which was both, farm and city residence ". *utrumque* is difficult to explain grammatically but the meaning is clear.

pastiones: " husbandry ". *pastio* is used for any kind of grazing or feeding of animals either at pasture or at home around the farm. A further explanation is given below (13), showing that it applies to birds and even bees.

probandus sit: the subjunctive has no significance: *est* would have been expected.

 11. oves, aves: a play is made on the sound of words after Varro's jocular manner; an English equivalent might be " beasts ", " bees ".

unde apes nascuntur: see below xvi. 4, for the belief that bees could be engendered from the decaying carcass of an ox.

villam Sei: see below, xvi. 10, and note, for Seius' apiaries and success as a bee-keeper.

pluris: this is a genitive of value associated with *vendis* in the next line.

natos: the boars sold by Axius are reared (*natos*) on his farm, in contrast with the wild boars which Seius sends to market. Both are equally profitable. If Varro refers to Seius' villa at Ostia which Appius is thinking of buying, or another owned by Seius in the same district, the boars which he sells are probably those from the maritime forest, the *silva Laurentina* which grew all along the coast and remains to this day in places. Boars are still preserved in a stretch of forest near Ostia for hunting.

12. Qui minus ... habere possum ?: " what is there to prevent my being able to have ... ". Axius argues that he could have wild boars on his estate and sell them for profit, as does Seius and that they would be just as good, although they would be Reatine. *Qui minus = quominus.*

Siculum ... mel: Sicilian honey was considered to be the best (see below, xvi. 14).

Corsicum: Corsican honey was bitter because the bees fed on wormwood and was therefore inferior in taste and quality.

gratuita: " to be had for nothing ", describing the mast from oak trees in the woods where the boars could be left to feed at will. Axius is asking whether honey and boars from Reate are to be considered inferior just because they are the product of Reate.

Seianas pastiones: husbandry such as Seius practises.

13. Duo ... fundo: here is a clear instance of anacoluthon; the *cum* clause which comes first is not rounded off with a main verb as might be expected, but has been lost sight of in the long trickling sentence which follows: it is in reality out of construction.

agreste: " on pasture land ", lit. " in the field ". This describes the open country where the sheep are grazed (see above II, i, 16-17 and note). There is both the husbandry of the field and, as Varro goes on to say, of the farm itself.

villaticum: " close to the dwelling house ", or " around the home farm ". Here small animals, including birds for the table, could be reared.

Poenus Mago et Cassius Dionysius: for Mago's writings on agriculture, translated by Cassius, see above note on I. i. 10 (text omitted).

ex toto fundo: " from the whole estate ". Seius was a specialist in the produce of the home farm: the implication is that he relies on the productivity of this for profitable returns and does not need, or perhaps does not wish, to raise cattle and sheep as well.

14. qui apparuit Varroni ... accipiebat: " who waited on Varro and used to give me hospitality when his master was away ". Both Varro and Merula had stayed at Seius' villa in Seius' absence. It seems as if owners of villas, especially those near the main roads, gave open house to their friends whether they themselves were there or not. Many of these rich landed proprietors possessed more than one villa; it would therefore be convenient for their friends to be able to stay even if the owner were not there; see also below (15) the mention of Varro's aunt's estate and how Axius used to stay there on his way to and from Reate and Rome.

quinquagena milia: " fifty thousand sesterces ". The ablative of comparison after *plus* would have been expected.

Axio admiranti ... inquam: " I remarked to Axius as he showed surprise ... "

via Salaria: in Roman times the Sabine country was crossed by one of the main highways and at the same time the oldest road in Italy. In origin the *via Salaria* was said to be the route by which the Sabines went down to the marshes at the mouth of the Tiber to obtain salt: the name is therefore thought to be derived from *sal* (" salt ") and to mean " the Salt Way ". After leaving Rome the road hugged the edge of the Tiber valley and reached Eretum at the eighteenth milestone.   Then turning sharply eastward away from the river, it passed near Cures and Trebula Mutusca.   It is somewhere in this region a few miles further on that we should seek to locate the estate of Varro's aunt. Cures (now Passo Corese) is thirty-four kilometres distant from Rome.   Reate itself is eighty-five kilometres from Rome, and her estate was twenty-four Roman miles away.   At a rough estimate we may conclude that this estate, so welcome to Axius as a resting place was situated about half-way on the journey between Reate and Rome; this would place it in the foothill country just before the road mounted the steep hills to enter the Rieti gap and the rough passes on the way to the upland plain on which ancient Reate stood.   The Salaria is an important metalled highway today.

15. Quidni . . . soleam: " how could I not " (lit., " Of course since I make it my habit to divide the day at noon there in summer "").   Axius breaks his journey at Varro's aunt's villa in the heat of the day.

cum eo Reate: " whenever I go to Reate from the city " (*i.e.*, Rome).   A halt at the villa would give a welcome rest before the climb into the hills.   It seems as if Axius spends the summer in Reate to escape the heat of Rome.

noctu ponere castra: " I pitch my camp at night."   Axius often spends the night there in winter on the journey from Reate.   He uses a military term.

Atque . . . ex eo: take in this order: *atque ex eo qui ornithon est in hac villa: ornithon* has been placed in the relative clause, but more naturally would have been in the ablative case in the main clause.

uno . . . villae: " I know that five thousand fieldfares were sold for three denarii each, with the result that that department brought in sixty thousand (sesterces) in that year".   *venisse* is the perfect infinitive of *venere*, " be sold ", shortened from *venum-ire* and used as the passive of *vendere*, " sell ".   There were four *sestertii* to one *denarius*.

16. Scipionis Metelli: the name in this form presents a difficulty.   The reference can only be to that Metellus who celebrated a triumph in 71 B.C. His full name was *Quintus Caecilius Metellus Pius*, and he was consul in 80 B.C. with Sulla.   In 79 B.C. as proconsul he campaigned in Spain against Sertorius, a supporter of Marius.   After long drawn-out fighting he was ultimately successful and triumphed in 71 B.C. for his share in defeating Sertorius.   His adopted son who was consul in 52 B.C. and became Pompey's father-in-law is usually called Metellus Scipio, but far from celebrating a triumph, experienced the defeat of the Pompeians at the battle of Thapsus in 46 B.C.   71 B.C. is seventeen years before the dramatic date of the dialogue.   The word *tunc* may suggest " some years ago ", but perhaps the triumph was particularly memorable for the lavish banquet given to the populace.

collegiorum cenae: clubs of many kinds played an important part in Roman society.   The *sodalitia* or *sodalitates* were the priestly colleges such as the *pontifices* and the *augures* and others which might have a religious, or partly religious, significance: the *collegia* were comprised of men practising the same craft or trade, or sometimes were not more than burial clubs.   During the republic the various clubs increased greatly in number but sometimes took on a political flavour to such an extent that they came to be regarded as dangerous to the state.   Many were suppressed by a decree of the Senate in 64 B.C.   Varro's remarks that they are now countless in number must be related to the date when his conversation was envisaged as taking place.   There is a clear indication in III. ii. 3 to the year 54 B.C. when the lawsuit between the people of Interamna and Reate was conducted by Cicero.   This is ten years after the decree of 64 and Varro's dating seems to show that in spite of the suppression the clubs were once again on the increase: they had been revived again by Clodius in 58 B.C., four years before, but were suppressed again by Julius Caesar.   The reference here undoubtedly belongs to the time when the clubs were again on the increase. It was customary for the members of a *sodalitium* or *collegium* to feast together on special occasions.

excandefaciunt: " cause to flare up ", speaking metaphorically of a rise in the price of provisions which like a fire suddenly bursts into flame.

decoquet: *decoquere* is literally " boil down " and from that it comes to mean " diminish ": here it implies " go bankrupt ".   There is a warning note that the demand for birds will not be consistent because it depends on public demand.   In consequence a steady income must not be relied on.

neque . . . decipiaris: " nor does it happen as the fashion is at the moment that you often have a loss ".   The demand at the present time with so many dinners and public banquets taking place means that a continuous profit can be fairly certain.

Quotus quisque . . . annus: lit. " How many a year is there ? ".   The question implies that there is never a year . . .   It is capped in the next sentence where Axius replies that hardly a day passes without a feast of some kind.

17. cuius . . . libelli: " whose little books are after the manner of Lucilius ". Lucilius lived from about 180 to 102 B.C.   He was the first Roman to write satire.

in Albano: " on his Alban estate ".   An adjective formed from the name of a place and in the neuter gender is often used for the name of an estate.   Lucius Abuccius' estate was situated in the Alban hills to the south-east of Rome. Somewhere in this district was Alba Longa, perhaps on the ridge overlooking the Alban lake, the city which was once as strong as Rome but was conquered by her in the sixth century B.C.   The district is only some fifteen miles from Rome.

Idem . . . recepturum: " He also asserted that if he had obtained possession of a villa by the sea in the place where he wanted one, he would have taken more than a hundred thousand (sesterces) from the home farm."

secundum mare: the coast of Latium is probably meant, which lies between Ostia and Antium (modern Anzio).   Lucius Abuccius was thinking of the yield from fish in addition to the husbandry around the farm, as the next sentence shows.   The Romans used to build fish-ponds out into the sea in which the fish could be kept and fed until wanted.

M. Cato . . . Luculli . . . tutelam: this was the son of Lucullus, famous for his wealth (see I. i. 10 and note).   He left his son, only a young boy, under the guardianship of Cato, the boy's uncle.   It seems that Cato considered that fish-ponds were an unnecessary luxury.

18. Quin . . . minerval: a *minerval* was a fee paid to a teacher in return for instruction: payment was apparently made at the end of the course.   Merula is saying in so many words, that he wants to know what fee he can expect before he begins.

ex ista pastione . . . crebro: Axius implies that his income will be so large after he has acted on Merula's instruction that he will be able to pay the fee over and over again.   If Merula prefers to wait, he will receive more money.

Credo . . . pavones: Appius following up Axius' last word *crebro* makes a sarcastic remark to the effect that Axius will pay his fee as soon as he has some dead birds to sell off.

estis: this shortened form of *editis* (from *edere*, "eat") is frequent but it is remarkable to find it in the same sentence as the longer form: *editis*. In the prose of Cicero and other polished writers, the subjunctive would be used for *editis* in the indirect question introduced by *interest*.

v. 1-7.　*The Aviary for Profit*

1. unde, non ubi: " (the place) from which, not (the place) where . . . ". Merula intends to describe the aviary in which the birds are kept and fed until they were needed, and not the place where they are trapped. There can be little doubt that the fieldfares, blackbirds, quails and ortolans, all wild birds, were netted and not reared on the farm. The very large numbers and the remarks below (6) on migratory habits make this clear. In III. iii. 4 (text omitted) the fowler's net is mentioned. Merula does not intend to give instructions about the catching of the birds.

turdi: see below (6-7) and note.

testudo: lit. " tortoise ", is used for a vaulted shed in shape like a tortoise-shell.

peristylum: a piece of ground surrounded by a colonnade: to make an aviary this could be covered over with a roof and netting, or with netting alone. The word comes from the Greek περίστυλον. It is used for the colonnaded gardens found in Pompeian and other houses.

veneunt: the present passive tense of *vendere*.

miliariae: "fed on millet", *i.e.* fattened for market: these were probably ortolans.

coturnices: see below (6-7) and note.

2. In hoc tectum . . . laborent: the directions given here are yet one more example of the care over detail which the farmer must exercise towards even the smallest creature. A supply of water must be so arranged that it runs in narrow channels in which the birds do not trample and which in this way is not muddied.

ne . . . laborent: " to prevent the birds from being troubled by mud ". The floor must be kept dry.

3. cocliam: this word means literally " a snail shell with spirals ": it is used only in this one passage, and by no other writer, for a door. The inference is perhaps logical that it was an entrance with a winding passage so constructed that there was no straight way in and out: it was intended to prevent any birds from flying out.

cavea: the meaning here is doubtful, but " amphitheatre " makes good sense. Then it would imply an entrance to the amphitheatre which was winding to prevent an animal from escaping: others take the word to mean a " cage " but then the full implication of the *spiral* shell is lost.

marcescere: the question arises as to how far Varro has any real sympathy with the birds, or whether he is only thinking of profit-making.

Tectorio . . . levi: " with smooth stucco " or " plaster " to make a surface in which no mice or other vermin could find any foothold by which to crawl inside.

nequa = *ne qua*.

aliave = *alia ve:* " or any other ".

4. palos: poles set horizontally, as in a hen-coop.

in eis traversis . . . theatri: " with transverse rods fixed to them in steps at moderate intervals so that the appearance is like that of the balustrades of the theatres or the amphitheatres ". The poles for the birds to perch form a kind of lattice: perhaps Varro describes a construction which is like the tiers of seats in the auditorium of a theatre which were often guarded by low stone lattice walls, *cancelli.*

Deorsum: the birds perched all around on the tiers of poles and came down to the floor for water and food.

cibatui: this is a rare use of the predicative dative.

In hoc . . . supplementum: alcoves with shelves set in the wall are described: they are needed as well as the perches.

5. ut domino . . . reddat: this precaution illustrates the careful oversight which the owner of the farm should exercise even to the extent of knowing exactly how many birds are lost out of something like five thousand in number: (see below, 8). The dead birds have to be shown to his master by the slave (*aviarius*) in charge of the aviary.

idoneae: birds which are fit to be sold, those in fact which have been fattened up and are in good condition.

quod . . . ostio: " which is connected with the larger building by a door ".

lumine illustriore: this is a descriptive ablative.

6. Hoc . . . moriantur: there is straight-forward utility in Varro's caution; but we may wonder whether there is at the same time some compassion for the birds.

alieno . . . venditoris: " at a time inconvenient for the seller ", because he aims at having as large a supply as possible ready when a banquet or a triumph (see 8 below) brings a demand.

Non ut advenae . . . turdi: " Not as other migratory birds rear their young, storks in the open, swallows under the eaves, do the fieldfares rear them in any particular place. "

7. Praeterea . . . immani numero: there can be no doubt that the fieldfares and other migratory birds kept in the aviary were netted when they reached Italy. In view of the large numbers described it does not seem possible that they were bred in captivity.

de illo . . . vernum: the large species of thrush, the fieldfare, *turdus pilaris,* is probably the one to which Varro refers. It is a migrant which winters in Italy, but nests in spring in Scandinavia and Russia. It seems from this passage that the Romans did not know where its natural breeding places were. It appears in Italy in autumn and leaves again in spring.

Pontiis, Palmariae, Pandateriae: " on the Pontian islands, on Palmaria and Pandateria ". The names are all in the locative case. A group of small rocky islands now called the Isole Pontine lies off the west coast of central Italy, in the Gulf of Gaeta. They were known to the Romans as the *Insulae Pontiae*: Palmaria is now Palmarola and Pandateria is the modern Ventotene. Thousands of migratory birds come to rest on these islands on first reaching Italy and when they leave again and are still a ready prey to the nets as they were in Roman times.

v. 8-18. *Varro's Aviary at Casinum*

It is not feasible, within the compass of this work to attempt a reconstruction of Varro's aviary, nor to present detailed discussions of the problems involved:

in places, however, a variant, or a simpler explanation of certain difficult phrases and words, has been put forward. The mechanisms of the revolving dining table with hot and cold water taps, the water clock (*clepsydra*) and the weather-vane have not been discussed in detail. The most recent reconstructions of the aviary are those of A. W. Van Buren and R. M. Kennedy, *Journal of Roman Studies*, IX, (1919), p. 59, of C. des Anges and G. Seure, *Revue de Philologie, de Littérature et d'Histoire Ancienne*, IIIᵉ série VI (1932), 217-90 and F. Fuchs, *Mitteilungen des Deutschen Archäologischen Instituts Römische Abteilung*, Band 69 (1962), pp. 96-105. The mechanisms mentioned above are fully described and reconstructed with diagrams by J. M. Gesner, *Scriptores Rei Rusticae* (Leipzig), 1773 and 1774. The aviary which Varro had purely for enjoyment takes a place in the development of the Roman villa from utility to pleasure. Just as the simple farm (the *villa rustica*) evolved in time into a luxurious country estate as a response to various influences from Hellenistic Greece which directed and modified the Romans' innate love of nature, so the aviary which might be within it and in which birds at first were kept for the market and the table became an amenity which added to the pleasures of the owners. The netted peristyle full of many kinds of birds, the fish-ponds, the running water, the round domed *tholos* surrounded by a shady wood, this too netted and full of singing birds, brought natural wild life within the confines of the villa. The diners sat in artificially created but delightful surroundings, around the pond where ducks swam about and fishes darted in the clear water; they could experience, albeit factitiously, the calm, freshness and tranquility which they associated with the idea of nature and which the Romans sought in their country estates just as they did in their pastoral poetry. There was fresh green foliage, shady trees, streams of water, the sweetness of birdsong, the illusion of natural country life around them. The aviary was planted with the flowers and plants loved in a garden: perhaps low hedges of box and sweet-smelling herbs, bushes of laurel, oleander, with trails of ivy, rosemary, violets, roses, and myrtle, such as are seen on the painted walls of Livia's garden room from Prima Porta (now in the Museo Nazionale in Rome).

8. coieceris: the use of this word makes it clear that the birds were caught, not reared: *turdorum* is to be understood with *quinque milia* (see above III. ii. 15).

epulum ac triumphus: see above III. ii. 16: the celebration of a triumph usually ended with a banquet for state officials and the Senate who could amount to several hundred people in all.

sexaginta milia . . . multum: " you will be able immediately to put out at high interest the sixty thousand sesterces which you want ". Axius expresses incredulity and envy when hearing that Varro's aunt had on occasion sold five thousand fieldfares for three denarii each, which amounted to a return of sixty thousand sesterces (see above III. ii. 16).

Tum mihi: *inquit* is to be supplied. Appius then asks Varro to describe the aviary which he has at his villa at Casinum.

alterum genus ornithonis: there are two kinds of aviary: the conversation has first been about profits to be had from birds; there is another to be considered, that for pleasure. Perhaps Varro and his contemporaries came to have some love for birds, especially for their song, quite apart from the profit to be made from them.

animi causa: " for the sake of enjoyment ", " for your pleasure ".

sub Casino: " under Casinum ", implying " at the foot of the hill on which the town of Casinum stands ". Varro's villa was in the valley along the river (see below 9).

in quo . . . Luculli: " in which you are said to have far excelled not only the original aviary of our friend the inventor, Marcus Laenius Strabo, who was our host at Brundisium, (and) was the first to keep birds caged in an enclosure in his peristyle which he fed through a net thrown over it, but also of Lucullus' large aviaries on his Tusculan estate ". It is not known for certain who Strabo was: there is a man of that name mentioned by Cicero in a letter to Atticus (*Ad Att:* XII. 17) but there is no evidence to suggest that it is the same person.

exhedra: the original meaning of *exhedra* is a hall or arcade furnished with recesses and seats, such as was usual in the *gymnasia:* here it is to be understood either as a part of the colonnade of the peristyle shut off by a net to make it into an aviary, or as a pillared cage standing free inside the garden. The former would compare very well with Varro's own more elaborate aviary described below, which was in essence a netted peristyle.

Luculli: see above I. ii. 10 and note. The estate at Tusculum was only one of many which he owned.

9. Cum habeam: There is an unexpressed thought, (I shall willingly tell you about my other kind of aviary at Casinum). " since I have . . . ", the construction of the whole sentence is loose, after Varro's conversational style. The subjunctives in *fluat* and *transeatur* followed by indicatives in *est,* and *confluit* seem unnecessary.

sub oppido Casino: the ancient town of Casinum (modern Cassino) was situated in the southern part of Latium, known to the Romans as *Latium Adiectum*, on the Via Casilina at the foot of the hill now called Monastero. There only remain of the the Roman town an amphitheatre of the first century A.D., a theatre, and a gateway: from the site there is a magnificent view over the valley of the Liris. On the height above is the great monastery of Monte Cassino, the headquarters of the Benedictine Order. The town itself stood on the banks of the river Rapido, a tributary of the Liris, in a wide valley encircled by mountains of great scenic beauty.

marginibus lapideis: " with stone facings "; the stream was faced with stone on both sides and presumably had stone curbs. It was not inconsiderable in size, being fifty-seven feet across.

e villa . . . transeatur: " it is crossed by bridges from (one part of) the villa to the other ". *transeatur* is an impersonal passive.

longum . . . museum: " the length is nine hundred and fifty feet in a straight line from the island which is in the lower reach of the stream, where another stream runs into it, to the upper reach where is the Museum ", a study where Varro could go to read and write and probably had his library. The word is connected with the *Musae* (the Muses), the goddesses of literary arts (see above, I. i. 4).

10. circum . . . conclusus: " around the banks of this (stream) is an uncovered walk ten feet wide on either side: off from this, facing the open country is the site of the aviary enclosed on both sides, on the right and the left, by high walls ". An *ambulatio* was an open walk bordered with hedges, perhaps box or sweet-smelling herbs. This one, because it was along the waterside would be particularly pleasant. Other kinds of walks often had pergolas or colonnades for shade in hot weather.

denos: lit. " ten each ", suggests that there was a walk on both sides of the stream.

Inter ... XXVII: " Between these the site which is that of the aviary is planned in shape like a writing tablet with a head-piece: where the shape is rectangular it extends forty-eight feet in width, in length forty-seven. Where it is round at the head-piece, it extends for twenty-seven feet." *quas* refers to the walls which enclose the aviary on two sides. *deformatus* describes the ground-plan of the aviary: it is in shape like a schoolboy's writing tablet, equivalent to our " slate " but made of wood. This was rectangular with a rounded head, *capitulum,* pierced so that a ring or loop of string could be attached by which it could be carried. The circular building, the *tholos,* described below (12-13) relates to the head-piece.

    11. Ad haec ... aream est: " to this the walk (leads), ruled off as it were from the lower margin of the tablet with a small wing extending from the aviary, in the middle of which are cages through which is the entrance into the quadrangle ". The text of this difficult passage is thought to be corrupt, but the general meaning is that the *ambulatio,* the walk which runs alongside the stream reaches the entrance to the aviary. Perhaps there is an arched entrance in both sides corresponding to the interior colonnades (see the plan). The entrance into the aviary is in the middle of this part of the walk and runs between the netted colonnades. Doubt is felt among scholars about the meaning of *plumula,* a word which is only known in this passage. Some would substitute another word, but there is clearly in its use a pun after Varro's manner and for want of better evidence it should be preserved. A possible meaning is " a small passageway ", describing here the entrance between the colonnades, which is like a small wing attached to the large quadrangle. Another suggestion is that *plumula* describes the façade of the aviary which abuts on the *ambulatio.* This does not alter the interpretation in which the *ambulatio* is understood to pass the aviary, and that leading from it is the entrance. *caveae* is also difficult to explain but is generally taken to indicate that there were bird-cages at the entrance, perhaps of domestic fowls (see below, ix. 6, where it is used of hen coops) but another possible interpretation is to take *caveae* as referring to the netted colonnades which are then seen for the first time by a person who has walked along the *ambulatio.*

In limine ... stylobaten: " On the threshold, on the right and left sides there are colonnades set out with stone columns in the front rows, and instead of stone columns in the inner rows, small trees, while from the top of the wall to the architrave the colonnade is covered with a net of hemp, and also from the architrave to the stylobate. " This describes the colonnading around the quadrangle. There are two rows of columns, the outer one of stone as would be expected in the normal Roman garden (*peristylum*) but what is unusual and original, small trees take the place of a second row between the stone columns and the wall. There is, however, no roof, because the whole is covered with netting which extends from the wall, to the tops of the columns, which are joined together by a continuous piece of stone masonry the *epistylum* (from the Greek ἐπίστυλον, lit. " what rests on the columns ") or architrave. The netting then continues from the top of the colonnading to the base of the columns (the *stylobates*). The colonnading is open to the air and becomes one

large netted area with trees for the birds for perching and shelter.   To judge
from the Roman gardens of those times there is no doubt that the peristyle
would be planted with small shrubs and bushes and flowers both growing among
the trees and making a border.

aqua . . . affluit: there is running water inside the netted colonnades, no
doubt drawn off from the adjacent stream.

12. Secundum . . . versus: " Along the inner side of the base of the columns
on the right and the left (extending) to the upper end of the quadrangle from
the middle on either side are two oblong fish-ponds, not (very) wide, facing the
colonnades."   These ponds are probably to be visualised as extending along
perhaps half the length of the quadrangle.   The Romans enjoyed having
stretches of water within the grounds of their villas.   In them they often
bred ornamental fish.   The contribution to scenic beauty was considerable,
especially when they acted as reflecting pools.

tholum: as Varro explains a *tholus* from the Greek θόλος, was a round, domed
building, an architectural, type Hellenistic in conception, often serving as a
small shrine sacred to some deity of the garden.   It was usually enclosed with
a wall but Varro's is exceptional in that there is no wall.   A domed building
was becoming fashionable as an adornment in the pleasure gardens of the
wealthy, but used for this purpose it had lost its religious significance.

in aede Catuli: there is uncertainty about the reference.   Quintus Lutatius
Catulus had a house in the Palatine Hill in Rome to which he added a portico
after his victory over the Cimbri in 101 B.C. and enlarged it during Cicero's
exile at the expense of Cicero's house which was adjacent.   It is not known
however whether Catulus had a *tholus* on the site as well and the use of the
singular (*aede*) makes a reference to a house difficult (but perhaps not in Varro).
A temple dedicated by Catulus is known from an inscription but no details of
it have come to light.

Extra . . . perluceat: " Outside these columns is a wood planted by hand with
large trees so that the light comes in at the lower part and the whole is enclosed
with high walls."   The trees which surround the *tholus* provide shade, very
necessary during the hot months in Italy, but they have been trimmed or
trained so that there are no lower branches to exclude light from the *tholus*.

13. Intra . . . latus: " Between the outer stone columns of the *tholus* and the
slender inner ones of fir of the same number there is a space five feet wide."
The *tholus* is surrounded by two rows of columns: the outer ones of stone
serve as supports for the domed roof.   The wooden columns perhaps add to
the illusion of an encircling wood.

Inter . . . transire: " Between the exterior columns instead of a wall, there are
nets made of gut so that there is a view into the wood, and whatever is there
can be seen, but no bird can get through."   Nets made of gut were particularly
fine and were used so as to increase the illusion of being in a wood.

Intra . . . obiectum: a net is also stretched between the interior columns;
aviarium is an adjective.

Inter has . . . avium: " Between these and the outer columns is built up in
steps a kind of miniature bird-theatre, with brackets fixed at frequent intervals
on all the columns to serve as perches for the birds."   *mutuli* is the technical
term for the droplets below the triglyphs in the frieze in Doric architecture.
The use of this word in Varro's sense is not known elsewhere.

sedilia: is the usual term for the spectators' seats in a theatre.   Varro seems to describe an arrangement of perches attached to the columns in such a way that the appearances is that of the auditorium of a theatre. A walled-in structure is not to be understood because it would not give the birds freedom of movement nor allow those who observe them to see them with full advantage. The intention is always to give an illusion of being in a wood and being able to see the birds in their natural setting. The seats rise as it were step by step, but they are perches attached to the columns.

14. Subter . . . possint: " Beneath the base of the columns is a stone course rising one foot nine inches above a platform: the platform itself rises about two feet above a pond, and is five feet wide, so that the guests can walk round to their cushions at the foot of the columns." *falere* is only known from this passage; it is thought to be in origin an Etruscean word; in this context it means a platform and here one of stone. It provides a space by the edge of the water just wide enough for a person to walk around, and to find his place at the table which is fixed in and extends from the middle of the pond. The meaning of *columellas* has been thought to be difficult, but these could be the smaller slender columns of fir already described, as some scholars suggest.

Infimo . . . parva: " At the foot of the platform inside is a pond with a border a foot wide, and a little island in the middle."

navalia: this is the technical term for sheds for ships but Varro perhaps uses it here in a jocular fashion: there are shelters for the ducks hollowed out of the stone facing of the platform two feet high where they swim in and out at will.

15. In insula . . . palmum: " On the island is a small column and inside it an axle which supports, instead of a table, a spoked wheel, in such a way that at the circumference where the felloe usually is, is a board hollowed out like a tambourine two and half feet wide and four inches high." The table on which food is served to the guests is shaped like a wheel with a wide rim which rotates. It is set on spokes so that the guests can see the water and watch the ducks. *axis* in Latin is the axle-tree of a wheel, around which the wheel revolves: in this description the axle itself revolves as it supports the wheel-like table.

Haec . . . convivas: a meal served on this kind of revolving table requires only one servant: *puer* is the usual word for slave. He serves all the dishes at once so that the guests may help themselves as the table comes round.

16. Ex suggesto . . . peripetasmata: " from the side of the platform where the coverlets usually are ". The *peripetasmata* were richly embroidered coverlets which were spread over the couches on which the guests reclined, and hung down to the ground. In this case, however, the guests had cushions only (*culcitae*) and instead of seeing coverlets hanging down they watched the ducks coming out of the duck houses and swimming about.

rivus: a stream runs from the pond into the two fish-ponds. There was a stream of water which ran all through the aviary along the colonnades on the inside, through the fish-ponds, the pond inside the *tholus* and in the bird-theatre. No doubt it is one continuous moving watercourse diverted from the river and separated into rivulets. The result was that the guests could see small fish darting about in the pond which swam in and out from the two ponds in the quadrangle.

cum . . . convivam: " while it is so arranged that both hot water and cold flows for each guest from the wooden wheel and the table which I have said is at the ends of the spokes, by turning taps ". This is a refinement of the first order: the spokes of the wheels serve as water-pipes and the taps are presumably set along the edge of the table. The water is drawn up from the pond through the axle (*axis*) which is a hollow tube connected with other hollow tubes branching from it which are the spokes of the wheel which constitutes the dining table.

17. Intrinsecus ... horae: " Inside under the dome, the morning-star by day, the evening star by night so circle around the lower part of the hemisphere and move that they show what the hour is." *Lucifer* and *Hesperus* are different names for the star Venus. It is not possible here to enquire into the working of what Varro describes but it may be assumed that there was a water-clock (*clepsydra*) which caused a pointer with representations of the two stars to move round a face on which the hours were marked. Such clocks which showed the hour by night as well as by day were known in Rome from about the middle of the second century B.C. They were therefore no new thing to Varro. Such a clock could be seen by the public in the Tower of the Winds at Athens (see below 18).

In eodem ... Cyrrestes: " In the middle of the same hemisphere around the axis is a compass of the eight winds as in Athens, in the horologium (clock-tower) which the Cyrrestrian built." This is the so-called Tower of the Winds which still stands in Athens, built by Andronicus of Cyrrhus in the first century B.C. It is an octagonal tower made of marble containing a *clepsydra* on one outside wall. On each of its eight sides at the top is a relief showing a figurative representation of one of the eight winds and the name inscribed in Greek. On the top of the dome on the outside was a bronze triton holding a wand. This was attached to an axle so that it rotated in the wind and indicated with the wand which wind was blowing. The exact date of the building is not known but there is perhaps something of an anachronism in its mention in the dialogue. In 54 B.C., the dramatic date of the dialogue, it would either be very new or not yet built: but by the date of writing (37 B.C.) it would have become well known: certainly for either date it would be a " new thing ", one which clearly caught Varro's imagination.

ibique ... possis: Varro returns to his own *tholus*: a pointer projecting from the revolving axis shows on the inside as well as the outside which wind is blowing. There were probably figures representing the eight winds painted on the ceiling of the *tholus* and there may have been a figure on the dome of Varro's *tholus* similar to the triton on the tower in Athens.

18. in campo: in the Campus Martius where the voting had taken place (see above III. i. 5 and note).

athletae: *athleta* from the Greek means in the first place, a wrestler: from that it comes to mean " a champion " or one who attained special skill: here it implies " old hands ".

Parra: another bird-name adds flavour to the dialogue. *Parra* is a bird of ill-omen, usually a barn owl. The bearer brings bad news.

ad tabulam ... loculum: " a man had been caught in the act of throwing ballots into the ballot-box while the ballots were being sorted at the table ". The offender had probably been bribed to sneak in some extra votes into the ballot box while the tellers' attention was diverted. It is known that in 54 B.C. the dramatic date of the dialogue, Cato made a determined effort to check bribery which was rife at elections. Plutarch in his Life of Cato tells of a Favonius who stood for the office of aedile and was unsuccessful but Cato on examining the votes, found several in the same writing and caused the election to be discounted. Varro may be alluding to this very fraud. Cicero in a letter to Atticus also tells how Cato embarrassed the candidates so much that they decided to deposit 500 sestertia each; if any one were then convicted of bribery he would forfeit his deposit (*Ad Att.* IV. 15).

ix. 1-16.　*Poultry Farming*

1. Ego . . . sodes: "Tell us, please, about the division of fattening in which I am interested, concerning poultry." *sodes* is a colloquialism, shortened from *si audes* (for *audies*) and equal to a polite request. It is found in the plays of Plautus and Terence.

tum . . . licebit: " then, if there is anything worth taking into account in the (other) divisions, we shall be sure (to discuss it)". *ratiocinari* is a technical term from bookkeeping, " reckon ".

Igitur: Merula begins speaking here. *Igitur* is equal to " well ", used merely as an introductory word.

sunt . . . trium: " those that are called poultry are of three kinds ". The case of *generum* is a defining genitive.

villaticae: these are the homely farmyard fowl, reared around the home-farm and not of any particular breed.

rusticae: lit. " wild fowl ". Columella (VIII, ii, 1-2) speaks of fowl he calls *rusticae* as those which live wild on the rocky island of Gallinaria, off the Ligurian coast, facing Albenga (the Roman *Album Ingaunum*). These were netted and then reared as domestic fowl. The island was named after them (*gallinae*) and still keeps the name today: these wild hens were however probably caught in other places as well.

Africanae: a description given below (18, text omitted) in which they are said to be large, speckled, with rounded back, suggests that they were similar to guinea fowl.

deinceps: " continuously ", *i.e.*, they are always to be found on the farm.

2. ornithoboscion: this word is taken directly from the Greek ὀρνιθοβόσκιον " a breeding-place for birds ". The hen was originally a native of Asia: from there its use as a farm animal spread to Greece at the time of the Persian wars, in the fifth century B.C., and eventually to Italy and all Europe. Like bee-keeping (see below xvi. 12) poultry farming seems to have been an art learnt from the Greeks. The hen is not mentioned in Homer or Hesiod, but occurs frequently in Aristophanes.

Deliaci: the inhabitants of the island of Delos in the Aegean Sea excelled in rearing poultry, but they bred mainly fighting-cocks: Columella (VIII. ii. 5) reproaches them for this because ˌof the heavy losses incurred in gambling on the fights.

animadvertant: grammatical accuracy would demand a singular verb to correspond with *vult* above.

pariant: " lay ".

quem ad modem . . . excudant: " how they are set under the brood hen and how they are hatched ".

appendix: " as an addition ", in opposition to *pars*.

3. capi semimares: the half-males which have been castrated are called capons.

Gallos . . . rumpatur: " They castrate cocks so that they may become capons by cauterising them at the lower part of the leg until the skin is broken." This practice is also vouched for by Columella (VIII, ii. 3) and Pliny (*N.H.* X. xxv. 59).

　4. Qui . . . gallinas: a poultry-farm was not complete in Varro's opinion if it did not contain all three varieties, with a preponderance of the ordinary farmyard hen.

E quis . . . aptiores: Varro describes the native Italian breed: the *gallina villatica*. The points which he stresses in such a bird make for good egg production.

　5. qui animadvertunt: the text is difficult but the meaning appears to be that those who wish to obtain cocks which are good for breeding should look for the points that follow. They characterise a strong well set up bird with some fighting qualities.

animalia . . . gallinis: animals which might injure the hens could be dogs, foxes, wolves, eagles, and others.

pertimescant, propugnent: these are generic subjunctives describing " the kind of cocks which would . . . "

　6. seminio: *seminium* from *semen*, " seed ", means " breed ".

Tanagricos: this was a breed called after *Tanagra*, a town in Greece in eastern Boeotia, and undoubtedly originating from there in the first place.

Melicos: Varro explains below (19, text omitted) that *Melicus* is an alternative spelling to *Medicus*, and adds that this breed was originally imported from Media because they were prized for their size, but that later on all large hens were called by that name because of their similarity.

Chalcidicos: another breed originating in Greece came from Chalcis, in Euboea.

steriliores: although Varro does not recommend these Greek strains for breeding purposes, Columella recommends that they should be crossed with Italian hens (VIII. ii. 13).

ducentos: the implication is " for example ".

locus . . . adtribuendus: " a piece of fenced ground is to be assigned to them ". Varro's description of the hen-run and hen-house is to be compared with that of Columella (VIII. iii. 1-7) where greater detail is given.

caveae: " coops "; *cavea*, in origin a hollowed place, is used for any kind of house for animals.

quae . . . versus: *versus* is redundant. The hen coops should face east: this would ensure coolness in summer which is essential in the Italian climate.

viminibus . . . raris: the window-spaces (*fenestrae*) are to be filled in (*factae*) with wattle set in thinly (*raris*), to permit light to enter, but to keep out anything which might be harmful to the birds.

　7. duas: supply *caveas*.

Contra . . . cubilia: " Facing each perch separate nests should be built for them in the wall." The *cubilia* are for laying.

vestibulum saeptum: " hen-run ": this corresponds to the *locus saeptus* mentioned above.

volutari: the passive verb is used with a " middle " meaning lit. " they roll themselves ".

Praeterea . . . firmiter: the large shed is described in which the *gallinarius* has to live. It becomes clear that especial care is taken of the sitting hens:

their nests are all round the walls so that they are under his constant supervision.
ita ut . . . firmiter: the grammar is difficult: a genitive in *cubilia* governed by *plena* would have been expected, but the meaning is not open to doubt. " Built in the walls around or firmly attached. All is filled with hens' nests." **Motus . . . nocet:** *firmiter* is further explained, giving the reason why the nests must be firmly attached.
cum parturient: " at laying time ": the future is used idiomatically.

8. substramen . . . aliut: *aliut = aliud:* the constant laying of fresh straw which must have been a daily task for the *gallinarius* reflects the care and cleanliness which the good farmer looks for.
Quae velis . . . plura: " They say that when you want a hen to sit, she should not sit on more than twenty-five eggs, although she may have laid more." *Quae* agrees with *gallina*, to be supplied. *pepererit* should according to accepted grammatical rules be in the perfect subjunctive, but in Varro it could equally well be the future indicative. Twenty-five is an unusually large number. Columella (VIII. v. 7-8) recommends different amounts for different months: fifteen in January, nineteen in March, twenty-one in April and in the months up to October.

9. aequinoctio . . . autumnale: the spring and summer months are the best for sitting: the hens are more likely to become broody and the newly-hatched chicks will be kept warm.
non supponenda: " are not to be put under the hen ".
pullitris: *pullitra* (a pullet) is a word not found elsewhere, but should be retained in the text since the meaning is quite clear.
quae . . . acutos: this clause describes *vetulis*, the hens most suitable for sitting, those whose claws and beaks are not so sharp that they can damage the eggs.
adpositissimae ad partum: " They are best fitted for laying . . . "

10. si ova . . . coepit: " If you intend to put pea-hens' eggs under hens, you should put the hens' eggs under them only when a hen has already begun to sit for ten days." The singular in *coepit* is to be noticed, following closely on the plural *gallinis*. *subicere* is either the infinitive used to give a command, or governed by *oportet* to be supplied.
Gallinaciis . . . noveni: according to his calculation, the pea-hens' eggs take seven days longer than do the hens' eggs, but Varro shows some margin for eventualities and the ten days mentioned above are according to modern reckoning only nine because the Romans used to include the first and last days in any count of this kind.

11. circumeat . . . vertere: the variety in construction with *oportet* is remarkable in view of the proximity of the two verbs.
animadverti . . . posse: " they say it can be discovered . . . "; this governs the previous clause *ova . . . necne.*
quod . . . desidit: " because an empty one floats, a full one sinks ".
vitales venas: the ancients believed that there was a drop in the egg which developed into the chick. Aristotle (*Hist. Anim.* VI. 3, 10.) believed that this was in the white of the egg and that nourishment was supplied to the embryo from the yolk. Pliny [*N.H.* X. lxxiv (148)] says that eggs have in the centre of the yolk a kind of blood-drop which some authorities think is the heart of the embryo chick.
Idem . . . inane: " The same authorities maintain that when you hold up an

egg to the light, the one through which the light shines is infertile." Pliny [*N.H.* X. lxxiv (151).]

12. haec . . . : i.e., *ova.*

perfricant . . . horas: the instruction must mean " and *leave* for three or four hours", in the salt or brine.

abluta: " after washing them ".

In supponendo ova: the use of an accusative with the ablative of the impersonal gerundive (or possibly here, a gerund) is most remarkable. " In setting eggs."

semen pulli: " the embryo of a chick ".

quadriduo: " four days after ": this is by Roman reckoning but would be equal to three days by ours.

purum unius modi: lit. " clear of one way ", " uniformly clear ", that is without any dark patch. The genitive is to be explained as descriptive.

13. subiciendum ei . . . paucos: " and should be put under (a hen) which has few chicks ". For this and the following directions the passage in Columella VIII. 5. 7-8 should be consulted. He recommends that two or three newly-hatched broods, up to thirty chicks, should be put under one hen on the day they are hatched.

ab eaque . . . pullos: the meaning becomes clear if *ab eaque* is taken as referring to the hen from whom the hatched chickens have been removed: some eggs will be later than others in incubating especially since they may not all have been laid on the same day, or even at the same time: " from this hen, if there are few (unhatched) eggs left, these should be taken away and put under other (hens) which have not yet hatched their eggs and have less than thirty chicks (*i.e.*, eggs still to be hatched) ". The process seems to mean that clutches of eggs about to be hatched are put together: the reason could be twofold, to free some hens so that they may start laying again as soon as possible (see below 15) and to protect the newly-hatched chicks.

subiecto pulvere: " on a layer of dust ". This protects their beaks which are at first very tender. Here is yet another example of the minute attention to detail needed of the careful farmer.

polentam . . . semine: " barley-meal mixed with cress-seed ". *polenta* was a kind of porridge made of pearl-barley: softened meal is desirable for the chickens' first food but it must be thoroughly soaked before it is fed to them. It is not known which plant is meant by *nasturtiun*: there is however no connection with the English garden plant.

14. eligendi . . . consenescunt: the removal of lice must have been a tedious task for the *gallinarius*. The Romans had no insecticides and had only this method of removal by hand as a means of protection against vermin. One removal would probably not be enough but constant supervision would be needed: it must be done frequently (*crebro*).

ex odore: it is perhaps more likely that the chickens might die of fright on seeing a snake.

in stercilinum: it is not clear whether this manure heap is within the hen-run, or in the farmyard. If the former, the *saepta* in 15 below describes the fence round the hen-run adjacent to the coops: it appears now for the first time that the hen-run was covered with a net.

15. apricum: here is another of Varro's inexplicable agreements: perhaps it is best understood as " it is sunny ".

evitantem: this present participle agrees with the imaginary person to whom the directions are addressed: the accusative is governed by some such thought as might be expressed by *oportet* understood.

ad pariendum . . . expeditae: to help towards the profits of the poultry-farm it is necessary to get the mother birds back to laying as soon as possible.

16. Incubare . . . succedunt: this injunction seems to be based on pure superstition. The underlying belief might be that the waxing moon since it was increasing in size had some effect on the growth of the embryo in the egg. **Diebus . . . excudunt:** a statement to this effect has already been made. There have been several repetitions in this chapter which seem to point to hasty writing.

xvi. 1-9. *The Bees: Their Natural History*

The affinities between Vergil and Varro are so close on the subject of bees and both passages so long that it is not within the scope of this work either to comment on them or to make comparisons. It is therefore recommended that *Georgic IV*, 1-280 be read alongside Varro and that the reader draw some conclusions for himself and enjoy Vergil's delicate touch and exquisite language, which is perhaps not surpassed in any other passage in the *Georgics* or elsewhere in his works. See also Chapter III.

1. tertius actus: the conversation in Book III is about the rearing of animals on the home-farm (*pastio villatica*) and is in three parts; the aviary, the preserve, and the fish-pond (III. iii. 1).

ornithones, leporaria, piscinae: these can be regarded as three acts of a play of which the first two have been performed, and the third still remains.

Quid tertius?: Axius argued that was it was not yet time to proceed to the third act (fish-ponds) because bee-keeping should not be overlooked? As honey was the only sweetening agent known not only to the Romans but throughout the ancient and early modern world, it was a profitably agricultural product of the highest importance. The subject of bees is placed here as a kind of interlude between Acts II and III.

mulsum: *vinum* is to be understood, although *mulsum* is often used as equivalent to a noun. Wine mixed with honey was a delicacy (see below 5).

propter parsimoniam: although Appius had been poor in his youth, he is represented in the conversation as a wealthy landowner. He is thinking of buying a villa at Osta [(see above, III. ii. 7) and has a bailiff (III. vii. 1) (text omitted)], therefore he must already have had other landed property.

2. dedi Lucullo: Appius was left as head of the family: when a woman's father died, if she was unmarried she was, by Roman law, in the care of her nearest male relative: it was part of his duties to arrange a suitable marriage. Lucullus had fabulous wealth (see above, I. ii. 10 and note). The fact that Appius could not provide her with a dowry, as was the usual Roman practice, emphasises his straightened circumstances at that time. Appius had in fact three sisters but one might have been married before his father's death.

cessa = concessa: Lucullus relinquished an inheritance which was rightly his, in Appius' favour. The remark seems to suggest that it was in return for receiving his sister for his wife.

me: this form apparently is equal here to *mihi*. The reading may be doubtful: see Lewis and Short, *S.V.* **Ego.**

daretur mulsum: the custom was general in Varro's time of serving honeyed wine to guests at dinner parties, but the inference from what Appius says is that his circumstances were so much improved that he could afford to drink it regularly at his own table (see 5 below).

3. meum erat: " it was my right . . . "; these words on the lips of Appius, suggest a pun on his name from *apis*, " bee ". In the conversation which follows Appius discourses on the nature and character of bees, Merula following him on practical bee-keeping.

volucres: *volucer*, which means a winged creature, can be applied as easily to bees as to any other insect or bird.

Itaque . . . scias: to be taken in this order: *ut scias me eas nosse melius quam te.*

audi: Appius declares his intention to discourse on the character and nature of bees, from his own close knowledge of them.

Merula . . . demonstrabit: after Appius' account of their nature, Merula will follow with the practical aspect of the subject. *historicos = historicus:* the word usually means a writer of history, but in a wider sense, an expert. Merula is a practical bee-master so can speak with knowledge and authority and describe the subject in detail.

quae . . . soleant: " what the bee-keepers are accustomed to follow ", *i.e.*, " their usual practices ".

melitturgoe: the word is Greek, meaning " workers in honey ", and retains the Greek form of the plural. The use of the Greek words suggests that the Romans learnt bee-keeping from the Greeks (see also below, 12, and above, 9).

4. ex bubulo . . . putrefacto: the myth that bees could be generated from the rotting carcass of an ox seems to have been widely accepted in the ancient world; see Vergil's *Georgic IV* where the tale is told at length (317-558), and also the Biblical riddle, *Judges* xiv. 5-18.

Archelaus: was King of Macedonia from 413-399 B.C.; he is to be ranked among the earliest of Greek writers on agriculture and is one of several kings who wrote on the subject: he is coupled by Pliny [*N.H.* XVIII. v. 1. (22)] with Hiero of Sicily, and Attalus Philometor [see above, caption and note I. i. 8 (text omitted) and II. iii. 5]. In the epigram quoted by Varro he describes bees as " the wandering children of a dead cow ", and also in another place he says " wasps are generated from horses, bees from calves ".

sed ut homines: bees have a community life like that of men.

graculi: although rooks like to keep together in flocks, they do not live together in organised communities as do the bees.

hic: *i.e.*, among the bees, contrasted with *illic* in the next line which refers to *graculi.*

facere discunt: the subject of *discunt* is " mankind in general ".

ab his . . . condere: just as the bees work, build, and store food to make possible their social life, so has mankind learned to do. Work, shelter, and food are bare essentials for life, alike for bees as for man.

5. harum: the agreement is not easy to determine, " they have three functions ": *harum* could agree with *apium*, but the Latin is very terse.

Tria . . . domus: three sets of workers carry on the three different occupations; among the bees there is delegation of labour and each has its allotted task.

Quod . . . includatur: every cell in a honeycomb has six sides. Varro's theory that the six-sided shape is related to the number of legs which a bee has, is perhaps sound; the shape would assist movements inside the cell. At the same time we may wonder at the economy of space so observed and the instinct which the bee follows in the building of a comb. Varro seems to attribute this to wisdom and skill (*ratio atque ars*) with which the bees are gifted.

Foris . . . acceptum: " they seek food outside, and within manufacture that which is the sweetest of all things, loved both by gods and men ".

in altaria: honey was often included in offerings to the gods, especially in the more ancient rites.

mel . . . administratur: wine sweetened with honey (*mulsum*) was served at a Roman dinner both at the beginning of the meal, with the *hors d'oeuvres* and at the end of the meal with the dessert (*mensa secunda*).

6. Haec . . . societas: their city states are like those of men, for among them is a king, government, and community.

Secuntur omnia pura: " They seek only the pure." Honey is the most wholesome food known.

Itaque . . . unguenta: so fastidious are bees that not only do they avoid anything impure, but they dislike strong smells even to the extent of avoiding a strong perfume: *unguenta* is an internal accusative with *olet*.

unctus: from *unguere* (or *ungere*) " smear (with fat or oil) "; hence " anoint "; perfumes were in the form of oil among the Romans, with the result that *unctus* means " smelling of perfume ", " heavily scented ".

ut illas: supply *videt*: the negative is not carried on from *nemo*.

Ideo . . . dulcis: " so they alight only on those things the taste of which is sweet ", " which have a sweet taste ".

7. Minime . . . imbecillitatis: "(the bee) is quite harmless in that it destroys no man's work (*i.e.*, the fruits of his toil) by gnawing (or tearing) at it, nor is it (so) cowardly that it does not fight anyone who tries to break up its own work: and yet after all it is not unaware of its own weakness", *i.e.*, " it knows fully its own weakness ", shown by the ways in which it guards the hive.

opus: refers to the farmer's work.

vellicans: *vellicare*, " pluck, pull, tear away ", is a frequentative form of *vellere*. Varro is thinking of the depradations of insect pests on the fruits of orchards, vineyards, and olive groves. Wasps, hornets, and bumble bees, for instance, have strong mandibles with serrated edges with which they can cut into the skin of fruit; the bee's mouth on the other hand is naturally formed for the gathering of nectar from flowers. It has a long, smooth, and rounded proboscis with which it can suck, but cannot bite or tear. It only sucks from fruit which has already been damaged by other insects and for this reason is beneficial to the farmers because it limits the damage by taking in the sugary sap which oozes out of the lesions and so prevents the fruit from becoming mouldy.

ut . . . resistat: the clause is one of consequence.

conetur: the subjunctive is generic.

Musarum . . . volucres: the author of the work from which this quotation comes is not known, although Aristophanes in one of his plays calls a girl " honey bee of the Muses ". The saying however may be proverbial.

si . . . unum: " whenever they are scattered about in flight (the beekeeper) brings them together to one spot by the rhythmic beating of cymbals and clapping of hands ". The gathering of a swarm is accurately described in *displicatae:* at first the bees fly about at random darting and whirling in a thick cloud until they find the queen who is their leader and around whom they cling and so follow to their new home. Rhythmic metallic sounds, sometimes by the beating of cymbals, have been used from time immemorial in the belief that they could prevent a swarm from flying too far away and at the same time cause it to settle in a particular place—probably a hive which had been smeared in readiness with sweet-smelling herbs (see below, 30). The practice is described by Vergil, *Georgic IV*, 64-66. English bee-keepers used to observe this same custom, called " ringing the bees ", by beating on a metal basin or kettle, and even believed that the flight of the swarm could be made faster or slower, higher or lower in accordance with the noise made. The origin of this custom may perhaps be found in the desire of the beemaster to warn his neighbours that his bees are swarming and to enable him to claim them as his own. Modern authorities agree that the " ringing " has in reality little or no effect on the swarm and even doubt whether bees have hearing.

Helicona: the accusative is Greek in form, as is also *Olympon:* for the Muses and Helicon, see above I. i. 4 and note.

sic . . . montes: this sentence seems to carry something more than mere utility; it cannot be denied that Varro shows some inner feelings here and a certain sense of wild beauty.

    8. Regem . . . volunt: the behaviour of a swarm is described. When a swarm of bees leaves the hive a new queen has been nursed to maturity in the hive and the old queen goes away. Those bees which follow her cling closely together around her and seem to be supporting her and carrying her along in their flight.

Regem: in spite of the close observation of the communal life of the bees among both the Greeks and the Romans the function of the queen was not discovered by the ancient bee-keepers and scientists. The importance of the larger bee which led the swarm and to which the other bees clung was realised only in so far as it seemed like a king. It was not until the seventeenth century that the facts were discovered by the Dutch scientist, Swammerdam. His discovery that the so-called " king " is in reality a female and the mother of all the bees in the hive revolutionised our understanding of the communal life of the bees. Now it became known that apart from the time of swarming the queen leaves the hive only once " on her nuptial flight " in the course of which she is fertilised by only one of the many drones which are always hovering around the entrance, waiting for her appearance. The various bees of the hive, workers, drones, and queens are all produced from this one union. Those grubs destined to be queens are housed in larger cells and fed on especially rich pollen-food called by bee-keepers " royal-jelly ".

Neque . . . inertes: five negatives in this sentence are noticeable: *nec non* is equal to *et.*

fucos: the drones do no work and yet eat the honey but the hive cannot feed them during the winter. After the fertilisation of the queen they are a drag on the community, consuming precious food. They are therefore killed off at the end of the summer.

vocificantes: this participle could agree either with *quos* [" in spite of their cries (for mercy) "] or with *plures* (" with threatening cries ").

erithacen: " (with a substance) which the Greeks called *erithace* ". The accusative is in the Greek form. Bees collect a reddish-coloured resinous substance probably the same as *propolis* (see below 23 and note) especially from poplar buds or pine-trees which they use for filling up crevices and glueing the hive, if of straw, to the base-board. They can dilute it into varnish with which they paint the honeycombs in order to cover them with a preservative surface.

9. alternis . . . pariter: they sleep and work regularly in turn. The bees work in shifts so that industry within the hive is never still.

ut colonias mittunt: " they send out as it were colonists ", *i.e.*, when swarming takes place.

iique . . . tubae: " their leaders give out certain orders with the voice, as if in imitation of a trumpet ".

duces: are the queens which lead the swarms.

Merula: Appius has now come to the end of his part of the discussion which has had something of a scientific and philosophical flavour, and turning to Merula invites him in metaphorical terms to talk about the practical and profitable side of bee-keeping.

dum haec . . . physica: " while listening to this (disquisition) of mine on natural history ".

cursu . . . trado: " I hand over to you the torch in the race ", a picturesque manner of saying " now it is your turn ". A form of ritual common to many Greek cults was a torch-race in which sacred fire was carried by rival teams from one altar to another. It formed part of certain important cults especially at Athens. The sacred character of this kind of relay race lasted throughout antiquity and when Varro makes this metaphorical reference he speaks of a practice which is still observed in his own day: *lampada* is in the Greek form of the accusative: *tibe* = *tibi*.

xvi. 10-19.   *The Bees: Profits, Apiary, Hive, Varieties*

10. Seium: see above, III. ii. 7 and 11 and note. Seius seems to have been a very successful farmer.

qui . . . mellis: " who has his apiaries let out annually for five thousand pounds of honey ". Seius seems to have kept bees on a large scale. He sells his honey under contract, he himself taking five thousand pounds of honey from the contractor who then may keep the rest. This means that nearly two tons of honey are allotted to Seius before the contractor can take his share. *quinis milibus* is in the ablative of price: *pondo* is an adverb connected with *pondus*, meaning " by weight ".

in Hispania: Varro had two brothers among his men during his period of active service in Spain. It would seem that Varro saw service in Spain before 54 B.C. (the date of this conversation) and before his disastrous campaign when he fled the country and joined Caesar's forces in Epirus: (see Varro's Life).

ex agro Falisco: " from the Faliscan country ". The *ager Faliscus* was the district on the right bank of the Tiber around the hill Soracte; it extended from Veii which was only nine miles from Rome, northwards to the Ciminian forest. To judge from their names Varro's two brothers were inhabitants of the city. They seem to have come from Veii or the neighbourhood (although *Veianius* is not the usual form of the adjective) and to have had here the villa left them by their father. In early times Veii had been a wealthy and important Faliscan city but it was conquered by Rome in 396 B.C. By Varro's time the site and surrounding country were given over to villas and agriculture. The layout of the ancient city seven miles in circumference can still be traced today and remains of sixth century terracotta statues and revetments in the Villa Julia Museum in Rome testify to her early wealth as an independent Faliscan settlement.

locupletes: their riches were derived from their apiaries.

agellus: a diminutive of *ager:* the small parcel of land rather less than an acre was probably in addition to that surrounding the villa.

hortum: *hortus* is a plot in which vegetables and flowers are grown: the produce would be a source of income in addition to that from the bees. Varro's description of the brothers' small villa and plot of land (*agellus*) may have suggested to Vergil to write about the old Corycian's market garden (*Georgic IV*, 125-48.)

relicum: all the land that remained after the surrounds of the villa and the garden were planted with bee-plants: this is a necessity for bee-keeping on a large scale, to ensure that there are plentiful sources of nectar at hand and enough to sustain a number of hives.

apiastro: the derivation from *apis* is clear: "a plant beloved by the bees". The identification is not certain: the other names given by Varro are all from the Greek and mean "honey-leaf", "bee-leaf", and "bee-herb". We have here another instance of Greek names connected with bee-keeping. See Pliny (*N.H.* XXI. 9. 29), who identifies it with *melissophyllon*.

11. Hos . . . alieno: the accusative and infinitive construction continues: "these (two) never received less, according to their estimated average, than ten thousand sesterces from their honey, since they said that they preferred to wait until they could get a buyer at their own time, instead of selling too soon at an unfavourable time". Differently from Seius who contracted out his honey-gathering, the Veianii stored theirs and sold it when the price was good. This prudent policy would bring in good returns: it is still as much the essence of profitable farming today as in the ancient world.

dena milia sestertia: *sestertium* would have been expected: Varro seems to use *sestertia* as an adjective (see above II. i. 14).

Dic: Axius now changes the subject by asking Merula to give him advice about building apiaries so that he may make handsome profits as did the *Veianii*.

12. Ille: Merula answers Axius' query: a verb of saying should be supplied.

ita: looks on to the advice Merula is about to give: "in the following way".

melittonas . . . mellaria: *melitton* is the Greek μελιττών, or μελισσών, "a bee-house", or "hive": *melitrophion* is the Greek μελιτροφεῖον or a place for raising bees: *mellarium* is the corresponding Latin word. Already Varro has given us the Greek names for the bee-plant (10 above) and before then the Greek name for bee-keepers (3 above). It becomes increasingly clear that the technical terms of bee-keeping came from the Greeks.

quidam: supply *appellant*.

ubi non resonent imagines: "where echoes are not heard". The subjunctive is generic and closely akin to one of consequence: "the kind of place where . . .".

(hic . . . protelum): the meaning of this parenthesis is obscure but appears to mean that echoing sounds might cause the bees to fly away.

ut . . . ortus: if the hive faces the direction of the winter sunrise, it will benefit from the warmth from early morning through the day.

ubi pabulum . . . frequens: the hive should be placed where wild flowers suitable for nectar gathering grow in plenty.

13. Si . . . apes: if there are not enough wild flowers for the bees, then the master must grow the plants which they love best.

dominum: is the bee-master.

Ea . . . est: in his list of bee-plants we have a mixture of garden flowers, wild plants, and several legumes which have already been mentioned as crop plants. The rose (*rosa*) was grown in gardens for garlands and for perfume, and wild thyme (*serpyllum* or *serpillon*) was especially prized for the good honey produced from it; the bean (*faba*), the lentil (*lens*), the pea (*pisum*) are to be found as much in the vegetable garden (*hortus*) as in the field.

ocimum: was probably a kind of clover; *medice* (*medica*) which is alfalfa or lucerne and *cytisum*, shrub trefoil are all among the valuable leguminous crop plants which were grown also for food (see above, I. xxiii. 1 and notes). The frugal bee-master could not fail to make these crops serve a double purpose in providing nectar for the bees as well as feeding the cattle. A modern Italian handbook on bee-keeping puts leguminous crop plants first in the list recommended for honey-making.

cyperum: this is thought to be a species of sweet rush, described by Pliny, *N.H.*, XXI. lxx. (117-18).

quod . . . est: " which is very beneficial for those bees which are ailing ". *minus* is equal in meaning to a negative.

Etenim . . . aequinoctium: this plant begins to flower in early spring in March at the time of the vernal equinox and continues until the beginning of October, at the time of the " other " or autumn equinox. For bees which gather nectar all through the spring and summer months, a plant which flowers continuously is especially valuable.

14. thymum: this is only another name for *serpyllon*.

fert palmam: the metaphor is taken from the games and athletic contests at which the victor received a palm branch: " carries off the prize ".

Itaque . . . causa: this seems to have been done to give a flavour of thyme to the honey: *seminaria* are the plots in which bee-plants were grown.

15. non . . . collocarint: the negatives are superfluous and only combine to make an affirmative, but could be translated: " not but what . . . ". The subjunctive in *collocarint* (shortened from *collocaverint*) is difficult, but could be regarded as expressing consequence: the general meaning of the sentence would therefore be that the best position for the apiary is close to the house with the result that some people have been known to put it actually in the portico. A colonnade surrounded the garden (*peristylum*) of the house, but might be found elsewhere among the farm buildings perhaps around a pond (see above I. xiii. 3). Hives placed in the enclosed garden attached to the house would be safe from marauders. Varro remarks in another passage [III. iii. 5 (text omitted)] how from the first the bees took advantage of the roof of the villa, under the eaves.

Ubi . . . sint: " where the bees may make their home ".

rutundas: the agreement is uncertain, although the neuter plural would have been expected with reference to *alvarium* in the previous sentence: it perhaps looks on to *alvos* below in the next sentence. Round hives made of withes would be similar to the old English skep.

ex arbore cava: a rustic type of hive was made out of part of a hollow tree trunk closed at the upper end with a flat piece of wood.

ex ferulis: hives are still made in Sicily from this same material which is similar in texture to bark, and of this shape. They are placed horizontally with the entrance in one of the short sides.

sed ita . . . animum: " but with the exception that where there are too few to fill the hives, they make them narrower to prevent (the bees) losing heart in a large empty space ". We have here another example of Varro's understanding of the smallest creatures and their difficulties. Although behind these expressed sentiments there is above all a desire for profit it is difficult not to see a sympathy that goes beyond it. In this passage he speaks of the bees as if they have human feelings. *ut* with the subjunctive *conangustent* is used in a limiting sense.

Haec . . . alvos: all kinds of hives are called *alvi*, " bellies ". *alvus* is connected with *alere*, " nourish " which makes Varro's suggested derivation logical.

quas . . . earum: his remark is obscure. It appears to mean that the hives are made very narrow in the middle but this is not feasible, a more reasonable explanation is that the shape of the hives is like that of the belly, but this hardly explains *medias*.

16. easque . . . positae: " and they place these hives on shelves (or brackets) on the wall in such a way that they cannot be disturbed nor touch one another, when they are set in a row ". These hives are presumably placed on the back wall of the colonnade or portico (see above, 15) on shelves or brackets attached to the wall in rows with suitable intervals between them. It is important that the hives are not disturbed in any way, but it is clear that they can be placed on a wall in rows in the same manner as they can be set out in a garden or field. Placed in a portico they would be sheltered from the sun and rain, wind and cold and could be under constant supervision of the bee-master. They would be protected too, to a certain degree, from changes of temperature.

aiunt . . . quartum: " they say that it is better for a further row to be taken away, rather than added ". This is a roundabout way of saying that there should not be more than three rows of hives on the wall.

17. Ad extremam . . . imponunt: " at the back from which the bee-masters may remove the comb, they place (movable) covers ". These were probably in the form of sliding doors which when pulled aside gave access to the combs. Movable frames placed within the hive on which the bees could build the combs were not invented until the nineteenth century nor was it until then that bee-keepers discovered how to exclude the queen bee and consequently the grubs from the combs and honey intended for human food. When the Romans gathered honey they had to smoke out the bees and to strain out grubs and wax from the honey. All modern hives are fitted with movable frames which the bees fill with honey but which the queen cannot reach to deposit her eggs; when honeycomb is bought in the frame it has come straight from the hive in that form and is free from grubs.

corticeae: rustic hives are still made of bark in Italy. They are in the shape of a hollow tree trunk and are closed at the top with another piece of bark.

fictiles: this type is still found in country districts in Italy.

fumigans leniter: the slightest puff of smoke from perhaps a piece of burning wood would be enough to quieten the bees because when they are frightened they gorge themselves with honey, in readiness to leave the hive: so they become lethargic and easy to handle.

vermiculos: "vermin"; among various insect pests such as beetles and ants the most to be feared is the wax moth which creeps into the hive and lays its eggs in the cells of the honeycomb. The larvae which emerge feed on the wax with resultant damage or even complete destruction to the combs. It will attack combs in store as well as those in the hive. Vergil describes the moth as

dirum tineae genus:
" the destructive race of moth ". (*Georgic IV*, 246).

18. ut animadvertat . . . existant: " (the bee-keeper) should see that not too many kinglings are reared ". To be correct according to modern knowledge *reguli* are the young queens which are reared in the hive in spring. There can only be one reigning monarch within the hive. The old queen will not allow another to usurp her authority but will lead out a swarm from the hive when new queens are to be born from their royal cells. When she has left, the first young queen to emerge from her cell asserts her royal prerogative by stinging to death in their cells any others which have not yet emerged. So it is seen how the presence of several queens, old and young, in a hive causes restlessness and swarming. It is not unknown for the first or even second young queen to lead out a swarm in addition to that led out by the old queen.

ducum: these are Varro's *reges*, the modern bee-keepers' queens.

niger ruber varius: there are in reality only two kinds of bees, the black and the striped. It is thought that Varro misquotes Aristotle (*Hist. Anim.* IX. 40) who states that there are two varieties (not three) the black, and the red-and-striped.

ut Menecrates . . . varius: this is a parenthesis "[(or) as Menecrates writes]" and stands outside the construction of the sentence: Menecrates was a poet of Ephesus who wrote on agriculture: he is included in the list of writers recommended by Varro for Fundania's instruction I. i. 9 (text omitted).

varius: the striped bee is to be preferred. The Italian honey-bee, *Apis Mellifica Ligustica Spin.* is probably closely similar to the one described by Varro. Its body is covered with gray or tawny or light-coloured hairs, and the abdomen which is also hairy is striped with three bands of golden and orange yellow. It is very industrious and productive, docile and therefore easy to manage, and resistant to disease. Because of its good qualities the Italian bee has become prized throughout the modern world to the extent that queens are reared especially for export. Many English bee-keepers prefer it to the English black bee. Vergil describes this bee as " shining with markings rough with golden scales ",

maculis auro squalentibus ardens (Georgic IV, 91).

qui ita melior: this refers to the latter.

interficere nigrum: in describing the two varieties of bees, Varro speaks especially of the so-called kings and recommends the bee-keeper to kill the black one. Vergil gives the same advice: " the one which seems inferior give him over to death ".

deterior qui visus . . . dede neci (Georgic IV, 89-90).

It is perhaps reasonable to wonder whether both Varro and Vergil saw some connection between the variety of the so-called " king " and the other bees in the hive.   If so they came near to realising " his " main function, but did not extend this surmise.

19. De reliquis apibus: the other bees in the hives (*i.e.*, the workers) should be of the striped variety and take after the better kind of " king ". English bee-keepers would agree.

Fur . . . fucus: the drone is called a thief because it does no work but yet lives on the honey which the worker bee collects so industriously: the drones' function is to fertilise the queen and thus the less drones there are to a hive the better, because although the drones' sole function is to fertilise the queen only one out of all the hive can have the chance of copulation with her.   Yet the hive must support and feed a large number throughout the spring and summer with consequent drain on the stored honey.

socia . . . operis: " partner in the work ".   Although the wasp is like the bee in appearance it plays no part in honey-making.

Hae: *i.e.*, the bees.

Nunc . . . in cultis: in contrast with the wild bees (*ferae*) which seek their livelihood in the wild, the bees which live in hives are called *cicures*, " tame ".   It does not seem that Varro makes a contrast between uncultivated and cultivated places, but rather between the bees which live in wild places (not necessarily woodland) and the tame bees who live in hives.   They need not always feed on cultivated ground as some translators would have it; *silvester*, although derived from *silva*, can carry the wider meaning of " wild ".

xvi. 20-28.   *The Bees: Purchase, Food, and Drink*

20. opus quod faciunt . . . leve: the honeycomb is described: in a good hive the cells must be evenly shaped and the surface of the wax covering the comb smooth.

Minus valentium signa: " the symptoms of those who are not well (are) . . ." *valentium* is a present participle used as a noun.

nisi opificii . . . tempus: " unless the season of work is driving them on ". Although a shaggy appearance is a sign of ill-health, it can also be a result of overwork.   During the spring and summer months the worker bees do not cease from their nectar-gathering and keep at it from early morning until dusk, never sparing themselves while the sun is warm and there are yet flowers to be visited.   It is only because the bees produce more honey than is sufficient for the upkeep and survival of the hive that there is a surplus for mankind. In this sense they can be said to work blindly in that they extend their work beyond the needs of their community.

21. transferendae: this is speaking of hives which have been bought and are to be moved to a new home; the agreement is with *apes* understood.

tempora: according to modern Italian practice, the best time for moving a hive is in early spring when the combs are not yet filled with honey: (when they are filled the operation would be difficult and dangerous to the bees and especially to the queen who might become drowned in it) and when the number of bees in the hive still remains limited after the winter.   See below, *ut verno potius quam hiberno:* the removal should take place on a sunny day in the morning.

Si e bono . . . fiunt: the bee-master is enjoined to move the hive to a place where there is good food within the bees' reach, perhaps plots of thyme, clover, and other leguminous plants. Otherwise they may try to seek out their old haunts and so desert the hive (see above 13 where the bee-keeper is told to sow plots of bee-plants).

Nec . . . traicias: " if you move your stock from one hive to another in the same aviary . . . ". Merula has spoken above of changing the actual hives from one place to another, now he goes on to talk of transferring the bees from one hive to another.

22. in quam: the relative precedes the antecedent which is *ea*: *quam* agrees with *alvum* from the previous line.

apiastro: this bee-plant has already been mentioned among those which should be grown especially for the hives (see above 13).

ne . . . dicit: the text is corrupt and no satisfactory reading has been suggested. The meaning as far as it can be translated appears to be " for fear that when they have noticed either a lack (of food) . . . "; the subject of *dicit* is probably Menecrates and the argument is undoubtedly that the bees transferred to another hive must find some combs full of honey already there so that they will be encouraged to continue to build on what they find.

Is ait: *i.e.*, Menecrates.

morbidae . . . vernos: bees sometimes suffer from diarrhoea in the early days of spring when they first leave the hive. During the winter they remain in a tight cluster within the hive feeding on the honey stored within the combs. They are unable to excrete unless on the wing but on warm sunny days will go out on so-called " cleansing flights " for this purpose. If the winter has been long and cold so that they have not been able to emerge out of doors they may become affected with this disease. It probably has no connection with the almond and cornel blossom, but it is probable that Varro attributed it to these trees because they are among the first to come into bloom (this is usually in February in Italy).

23. propolim: " bee-glue ", probably the same as the substance which Varro calls *erithace* (see 8 above and note), although he enumerates the two separately below (24) in the list of the four substances which the bees produce. *propolis* is a dark reddish brown aromatic glue gathered from trees which are sticky, especially from pine. It is used for filling in cracks and chinks and in a wild state to secure the ends of the combs.

e quo . . . aestate: " from which they make a roof over the opening of the entrance at the front of the hive especially in summer ". Varro is punning on the Greek derivation from πρό and πόλις (literally " before the city ") in using the word *protectum*.

medici untuntur: *propolis* has antiseptic qualities useful in surgery and was used as a remedy for cuts and burns, tumours, swelling, rheumatic pains, and ulcers. It is still recommended for slow-healing wounds.

carius . . . venit: " it is sold at an even higher price than honey on the Sacra Via ". For shops on the Sacra Via see above, I. ii. 10, where the same expression is made about the sale of fruit.

Erithacen: (see above, 8 and note). It is difficult to understand how Varro came to consider *erithace* as different from *propolis*, except that he separates their uses: according to him *propolis* is for protecting the entrance to the hive and *erithace* for securing the combs in position. It is clear nevertheless that in both cases he is describing a strong glutinous substance.

favos: in modern bee-keeping the oblong movable wooden frame is employed which may even be filled with waxen comb already prepared for the bees to fill with honey. This kind of comb needs no fixing by the bees. In the ancient world on the other hand, whether in a wild state or in a hive the total manufacture of the combs depended on the industry of the bee colony. Such combs were in the shape of flat semi-circular plates hanging closely side by side: these had to be securely fastened to the top of the hive.

melle et propoli: the ablatives are ones of comparison governed by *aliut* which is equal to *aliud*.

itaque . . . illiciendi: this statement looks on to the following sentence: *propolis* (or *erithace*) is aromatic and so is attractive to bees: it can be used to entice a swarm to settle in a particular place as is explained in the statement which follows: " and so in this there is the power of attracting them " (see the use of *inlicium* in 22 above).

24. quot . . . natura: (see above, 5 and note) " as many feet as nature has given to each bee ".

Neque . . . dicunt: Varro's erroneous statements about the substances made by bees and the sources of them which he goes on to enumerate (24-7) are worth considering. There are several misunderstandings and lack of knowledge of their ways, surprising in one who has apparently observed them so closely. In the first place since there is no difference between *propolis* and *erithace* it is clear that Varro himself thinks the bees only manufacture three substances. The wax from which the comb is made is secreted in the form of scales from the bee's abdomen, and honey is the nectar gathered from flowers and digested by the bees before being stored in the combs. He does not seem to know that bees also gather pollen which they manufacture into a food called bee-bread for the grubs, nor about the royal jelly, a kind of milky paste prepared for the grub which will become a queen. It is true, however, that the nectar from some plants makes better honey than others. In addition to bee-resin for *propolis*, bees gather from flowers only nectar and pollen and these it is true may differ in goodness as Varro states below.

cibum: " honey " is meant, but it is to be noted that this would be in the form of nectar which would have to be worked on by the bee before it was converted into real honey (*mel*).

ceram: wax is not gathered from plants but is a secretion (see above).

25. Duplex ministerium: the plants listed are mistakenly said to perform a double service, but ignorance of the source of wax causes this error.

nec non aliter: this phrase is equal to a positive statement: " just so in the same way ".

cibum et mel: " food and honey ", it is difficult to decide what is meant by *cibum* since Varro does not appear to know that pollen in addition to nectar is gathered for food: otherwise pollen would make good sense although it is obtained from most plants and not confined to those mentioned.

lapsano: charlock, or mustard, is a crop highly favoured by bee-keepers. It is mentioned by Columella (IX. iv. 5) as a bee-plant.

Item . . . sequatur: " From other flowers they gather in such a way that they take some ingredients for only one of the substances, others for more than one. They also follow another method of selection in their gathering (or rather it follows them . . .)." This is a summary of what has already been stated, how from some flowers only one substance is gathered, but from others, two or even three. The statement is of course erroneous. The " selection " refers to what follows, how in spite of themselves, the bees produce different qualities of honey because the products of different flowers are different.

26. ut in melle: " so in the case of honey ". *ut* answers to *item* at the beginning of the sentence. The argument is that just as different substances are gathered from different flowers, so different flowers yield different kinds of honey. In the act of gathering, the worker bees when this is possible keep to the same species of flower which accounts for the different qualities and flavours of honey: nowadays, for instance, we can obtain clover honey, heather honey, lime honey, etc.

liquidum mel: " watery " or " thin honey ".

siserae: there is uncertainty about the identity of this plant.

cytiso ... thymo: the brothers *Veianii* (see above, 10) planted shrub trefoil and thyme for their bees: these two plants were grown for bees although the former was also a fodder plant and the wise farmer would put his hives near the fields thereby benefiting the crop by fertilisation as well as the hive: thyme has already been praised as a bee-plant (above 14) (Col. IX. iv. 6).

27. Cibi ... oportet: " Because a part of diet is drink and this for bees is clear water, there must be a place from which they may drink." Water is so necessary for bees that some of the worker bees have the duty of carrying water into the hive where it is fed to the young bees. If there is no natural water at hand modern bee-keepers arrange a supply near the hive, perhaps in an inverted jar standing in a saucer.

ut . . . digitos: the accusative is one of extent: " provided that it does not rise in height more than two or three fingers." The measurement is the thickness of a finger. The water provided near the hive must be quite shallow, not more than three inches in depth, so that as is explained in what follows, pieces of tiles (or potsherds) or small stones can be placed in it for footholds for the bees when they come to drink. Vergil also describes this in one of the most exquisite passages in the *Georgics* (IV. 18-19, 25-29) (the text is that of the Oxford Classical Text, 1969) telling how the bees may rest there and dry their wings if by some mischance they are drenched with rain (they are unable to fly if their wings are wet).

> *at liquidi fontes et stagna virentia musco*
> *adsint et tenuis fugiens per gramina rivus,*
>
> . . . .
>
> *in medium seu stabit iners seu profluet umor,*
> *transversas salices et grandia conice saxa*
> *pontibus ut crebris possint consistere et alas*
> *pandere ad aestiuum solem, si forte morantes*
> *sparserit aut praeceps Neptuno immerserit Eurus.*

" Now there should be running water and standing pools green with weed close by and a shallow stream running through the grass . . . into the middle, whether the water is still or flowing along, throw willow branches and big pebbles so that they may settle on bridges all along and spread their wings to the summer sun if it happens that as they stay to drink, the east wind has spattered them or plunged them headlong in the water."

We have here another case of extreme care for small things, and an understanding of insect behaviour learned from observation. None can fail to see in Varro's care the humanitarianism already perceived in his attitude towards animals, especially lambs, working oxen, and birds.

bibere possint: bees will drink from the water they find out of doors while foraging, but a supply must also be taken into the hive for the young bees who have not yet emerged from the hive.

ad mellificium bonum: honey can very easily become tainted.

28. praeparandus his cibus: if the bees were allowed to live wholly on their own honey which they store in the combs the bee-keeper would be deprived of a good part of his harvest (see below 33). Although some honey is always left for their winter feeding and only a certain proportion removed, there may come a shortage of food for the bees in the hives before the winter is out and in early spring when the honey flow has not yet started: bad weather too may bring the honey in short supply. To counter such shortages food must be provided close to the hives by the bee-keeper.

exinanitas: " leave them empty ", when they have eaten up all the stored honey.

pondo: by weight: the word is an adverb but regularly used to express pound weight.

decoquont = *decoquunt.*

Alii . . . aquam: " Others see that water mixed with honey is placed near (the hive) in basins into which they put pieces of wool through which they can suck, at the same time to prevent them from surfeiting themselves with the drink and from falling into the water." The modern bee-keeper usually provides sugar and water and honey, or sugar and water boiled into syrup. Supplementing the bees' food is perhaps more needful in a cold northern climate such as ours than in Italy.

mulsam: " mixed " with honey, (see above *mulsum,* " wine sweetened with honey " III. xvi. 1 and note).

uvam passam: " raisins ".

passam: from *pandere* means " spread out to dry ".

apponunt . . . possint: " they put (them) in a place where they can come out to feed even in winter ". *tamen* goes closely in meaning with *hieme* and implies the thought that although the bees do not as a general rule emerge from the hive during the winter yet they will after all come out when attracted by the food which they cannot resist; they would be likely to venture out on calm sunny days. Being mixed with wine the scent entices them out.

xvi. 29-31. *The Bees: Swarming*

29. cum adnatae . . . multae: " when the new-born brood is strong and numerous ". *adnatus* derived from *adnasci*: " be born in addition to " is used to mean " a relation by birth ", " a blood relation ". All the bees in a hive are related and so can be spoken of in terms of human beings.

progeniem . . . liberorum: " they wish to send out their young like a colony just as the Sabines used to do: making a habit of doing so repeatedly because of an overplus of children ". The Sabines were a hardy race inhabiting the country to the north-east of Rome, including the mountains of the central Apennines. Together with other early Italian communities they followed the practice called the *ver sacrum,* " the sacred spring ". When the population became too large, they would consecrate to a god all children that should be born in a particular spring. The children on becoming twenty years old were sent in a body out of the land and allowed to go wherever they could find a new settlement. In the same way the bees swarm in spring and go off to a new home.

huius: this is in the neuter and refers back to the statement *cum examen exitura est,* and is associated with *signa.*

scitur: the passive could be explained as impersonal: " you may know that . . . " (lit. = " it is known that . . . ") or the clause " *quod . . . signa* " may be regarded as the subject. In either case the use of *quod* almost equal to the conjunction " that " in English is worthy of comment.

superioribus diebus: on the days immediately before the swarm leaves the hive. This happens when a new queen is about to be hatched and the old queen is preparing to lead out a swarm.

30. Quae primum exierunt: before the swarm emerges from the hive in spring a few bees precede it whose task is to make reconnaissance: they prospect and find the new home to which they will guide the swarm. The good bee-keeper will have a hive ready at this time into which to entice them (see below).

respectantes dum conveniant: " looking back until (they) gather together ". The subjunctive is used in a " final " sense, with the notion of something expected, and counted upon.

30-1. A mellario . . . delectantur: supply *sunt* with *perterritae,* and *eas* with *perducere.* The change from the singular of *mellario* to the plural in *volunt* and *oblinunt* is surprising but after Varro's manner. One way of quietening a swarm so that it can be enticed into a prepared hive is exactly similar to the practice of nowadays: a modern Italian handbook states this but at the same time extends little faith to beating on metal: " to stop the swarm from flying the noises which it used to be customary to make, for example, beating repeatedly on empty tin cans have little effect, but rapid success may be obtained by throwing handfuls of fine earth or sand ", and to this is added " or by a slight sprinkle of water from a hose ".

iaciundo . . . pulvere: Vergil (*Georgic IV*, 67-87) recommends this for separating kings which are fighting each other.

circumtinniendo: see above (7) and note.

31. erithace: another name for *propolis* which is aromatic and therefore likely to attract the bees (see above 23). The swarm is first induced to alight on a place (perhaps a tree trunk) near where a prepared hive is ready. It will not fly straight into the hive but will first gather in a dense mass surrounding the queen. The hive will then be brought up to the swarm and if possible placed beneath it.

Ubi consederunt: when the swarming bees have settled, when, that is, they are no longer flying about, but are hanging in a mass and are quiet. They have yet to be persuaded to enter the new hive.

alvum . . . litam: this is the new hive, prepared to receive the swarm: *litam* is from *linere:* it has been smeared with the substances and sweet-smelling herbs already recommended for the place where the swarm will first alight: the bees' own product, *propolis,* with its aromatic scent will have been applied to it, which will make them feel that the hive is no unfamiliar place, together with the " balm " they love so well and other herbs the scent of which entices them.

apposita: the grammar can only be explained as an ablative absolute, but the normal usages are ignored since the accusative would have been more natural, as the object of *intrare*.

fumo leni: the smoke would only be slight—perhaps a puff or so of wood smoke: this would serve to quieten them although swarming bees are usually half-stupefied through having gorged themselves with honey.

Quae . . . contentae: the compelling reason why the swarming bees remain in the new hive and do not attempt to return to their former homes is the presence of the queen among them.   Such a hold has she upon them that they will never desert her.   Varro of course was unaware of this.

xvi. 32-38.   *The Bees: The Honey Yield*

32. Quod . . . dicam: " Since I have spoken about what I consider has to do with bee-keeping, now at this point I shall talk about the returns, for the sake of which this care is exercised."   Merula has finished his instructions regarding the actual keeping of bees; now he is ready to give the rules for taking the honey for trading purposes.

quoius: this is an archaic form of *cuius*: the relative is placed before the antecedent (*fructu*).

Eximendorum favorum: in Cicero's polished style two heavy genitives plural in a gerundive construction are always avoided: sound is not always considered in Varro's rough Latin.

uiris . . . congerminarit: the text is corrupt, defying even an attempt at translation.

ex ipsis . . . capiunt: the bee-keepers judge from the bees themselves, *i.e.*, the behaviour of the bees, which Varro goes on to explain, indicates when the honey should be taken.

trepidant: " they become agitated ".

opercula . . . remoris: *opercula* are movable covers at the back of the hive: when these are removed the combs are exposed to view.   If the cells have been sealed up with wax, it follows that they are full of honey.   The combs in the ancient type of hive would be in the form of flat half-plates suspended inside from the top of the hive.

remoris: = *removeris:* the verb may be either future perfect indicative, or the perfect subjunctive of the generalising second person singular.   The change of person from the third plural in the same sentence is noteworthy.

membranis: these are membranes of wax with which the ends of the cells are sealed when they are full.

33. decumam relinquere: enough honey must be left in the hive to feed the bees in case of scarcity if for some reason there is shortage, and for winter feeding.   During the winter they gather into a close mass inside the hive and feed on the stored honey, gradually moving upwards.

Ut in aratis . . . fructuosas: " Just as in arable farming those who let their ground lie fallow, reap more grain from interrupted harvests, so in bee-keeping if you do not take the honey every year or not the same amount, you will have your bees by those means both more industrious and more profitable."

eximas, habeas: are subjunctives in a conditional clause with the meaning " if you were to . . . you would . . . ", but at the same time they are also subjunctives of the generalising second person.   There is a sudden change of person in this sentence as in the one before.   The advice which Merula gives represents a principle which is at the root of all good farming: the source of production whether land or bees or anything else must not be impoverished. Bees will be all the more industrious if they have plenty of food: it is no use depriving them of the honey on which they feed and which enables them to work, just for present gain.   The careful farmer must take the long view in every department of the farm.

34. Eximendorum . . . occasum: the Italian climate with the short winter of some three months and the warmth and bountiful sunshine of the rest of the year allows three yields of honey: Varro gives the dates according to the stars. The first comes in spring on May 10th, the second in late summer, at the beginning of September, and the third in early November.   The dating by the constellations should be compared with that of the farmer's calendar, I. xxix-xxxvi.   In England the honey is only taken once in the year, usually in July; the nectar flow begins in May and may last for six weeks.   In Italy the month of May is as hot as an average English summer and in that month the grain harvest is reaped on the plains.   Four months follow of increasing sun into great heat in July and August.

hiemationi: "for their winter feeding".   This may be supplemented with other food, but the careful bee-keeper would see that the hive contains ample honey to last the winter.

ne quid eximatur: if there is not a good supply of honey none is to be taken. The farmer must forego his yield until a better season.   It will not pay him to let his bees suffer from malnutrition.

Exemptio . . . animum: "When the amount taken is large, it should not be taken all at once or openly in case the bees become discouraged."   If they were to find that they had been deprived of a large amount of the honey which they had gathered with so much toil, they might abandon their work altogether. The comparative in *maior* does not carry any particular significance.

Favi . . . praesicatur: "When the combs are being removed from the hive if any part has not been filled or is fouled, it should be cut away".   This probably implies that any such comb found adhering to the walls of the hive should be scraped off so that nothing unclean or infected is left.   Hive and comb must always be kept in as clean a state as possible.

35. Providendum . . . regem: hives are liable to be attacked by robber bees who steal the honey, especially if the stock is not strong.   Sometimes the marauders are individuals who find a way through some aperture which has not been properly sealed.   A whole colony, however, may attack and pillage another weaker one, until the whole hive has been decimated.   Such an assault is described in a modern Italian handbook: " on the step of the hive which is attacked, numerous bees are seen which attempt to approach: others more eager fight furiously with the guard bees, who have been increased in number to keep off the attack.   In the proximity of the doorway the fight is particularly vicious: from the step many bees closely clasped together roll in a confused mass to the ground; others all around the hive try to enter it, especially from above, while others still take up a stand on the front wall of the hive; the buzzing is very loud.   Watching the doorway every so often one can see bees with swelling abdomens which fly out at great speed.   They are plunderers who in the general confusion have succeeded in penetrating into the hive and who, having drunk deep of honey, swiftly transport it to their own hive with the effect of summoning other bees to the fight.   After a short while, the hive—

especially if it is weak or abnormal—surrenders to the violence of the assault and a veritable sack ensues: the conquerors, in their ardour to carry off the honey, spare absolutely nothing. When they have destroyed one hive, they attempt to attack others." The bee-master can strengthen the weaker hives by joining two or three together under one queen and so make a strong colony. Varro's injunction to put the weaker ones under another king may imply this remedy.

36. subfumigandum . . . thymum: the act of smoking lightly would cause the bees to leave the hive including those that were inclined to be idle: the presence of sweet-smelling plants nearby would encourage them to their tasks of nectar gathering.

herbarum: the genitive is partitive.

37. antequam . . . decipiantur: " before they have foreseen that this would happen (although it happens rarely that they are taken unawares)". The subjunctive in *providerint* seems superfluous, though the underlying meaning may be " before they can have . . . ". *fore = futurum esse.*

imbris guttis uberibus: showers of rain in Italy are often sudden torrential downpours like the heavy rain and huge drops we experience during thunderstorms in England. They do not as a rule last very long.

quam . . . tempestate bona: " when the weather is at its best ".

tempestas: can be used for any kind of weather.

ficulneis lignis: the smell of burning fig wood is very pungent: in itself it could help to revive the afflicted bees.

ut . . . tangas: the use of *ut* is limiting, " though you should not touch them with your hand ".

Select Vocabulary

The vocabulary is based on Krumpiegel's word list of Keil's Teubner edition of 1894: *M. Porci Catonis de Agri-Cultura Liber, M. Terenti Varronis Rerum Rusticarum Tres ex recensione Henrici Keilii,* vol. III fasc. II: *Index Verborum in Varronis Rerum Rusticarum Libros Tres composuit* Richardus Krumpiegel (1902 Leipzig.) To keep this list as short as possible more familiar words have been omitted: in doubtful instances Lewis and Short's Latin Dictionary has been followed. Where a reference to the text only and no meaning is given, the word is either doubtful or is only known in that particular context: the corresponding note should be consulted. References to the text are given for more important words and agricultural uses.

A

abduco, 3, *lead* or *take away.*

abicio, -ieci, -iectum, 3, *throw away* or *down.*

abies, -etis, *f., fir-tree, wood of the fir-tree.*

abigo, -egi, -actum, 3, *drive away.*

ablacuo, 1, *turn up the soil around a tree, especially the vine* (I. xxix. 1).

ablego, 1, *send off* or *away.*

abluo, -ui, -utum, 3, *wash off* or *away, cleanse, purify.*

aborto, 1, *give birth prematurely.*

abrodo, -si, -sum, 3, *gnaw off.*

abrumpo, -rupi, -ruptum, 3, *break, break off, tear, sever.*

absolvo, -vi, -utum, 3, *make loose, set free;* pensum absolvere, *complete a task* (II. ii. 1).

absterreo, 2, *drive away by terrifying, frighten away.*

abstineo, -tinui, -tentum, 2, *hold back, keep away from.*

absumo, -sumpsi, -sumptum, 3, *take away, lessen, destroy, consume.*

abundo, 1, *overflow, be rich in, grow in abundance.*

accedo, -cessi, -cessum, 3, *approach, come near.*

accerso, -ivi, -itum, 3, *call, summon, invite;* a more usual form is **arcesso.**

accessus, -us, *m., an approach to, access, entrance.*

accido, -cidi, 3, *fall down, happen.*

accipio, -cepi, -ceptum, 3, *take, receive.*

accipiter, -tris, *m., hawk.*

acerbus, *bitter, unripe, sour.*

acervatim, *adv., in heaps.*

acervos, -i, *m., heap;* more usually **acervus.**

acinus, -i, *m., berry, grape.*

acratophorum, -i, *m., vessel for unmixed wine* (I. viii. 5), *pitcher.*

actio, -onis, *f., action, action in a court of justice, legal formula* (II. v. 11).

actus, -us, *m., motion, action, act* (theatrical performance).

acus, -us, *f., needle, bodkin.*

acus, -eris, *n.* (also us, *f.*), *chaff.*

acutus, -a, -um, *sharpened, pointed, shrill, sharp, keen.*

adcognosco, 3, *recognise* or *know perfectly.*

adduco, -duxi, -ductum, 3,

313

lead to, draw to, accustom, persuade.

adeo, adv., to that point, so far, followed by ut with the subjunctive, to such an extent that.

adeo, -ii, -itum, 4, come to, approach; adire hereditatem, enter on an inheritance (legal term) (II. x. 4).

adeps, -ipis, c., soft fat of animals.

adfero, adferre, adtuli, adlatum, carry, bring to.

adfinis, -e, neighbouring, akin (I. ii. 15).

adfinitas, -atis, f., neighbourhood, relationship.

adflatus, -us, m., blowing or breathing on.

adflo, 1, blow on, blow (of wind).

adfluo, -fluxi, -fluctum, 3, flow into or to.

adfundo, -fudi, -fusum, 3, pour into.

adgero, -gessi, -gestum, 3, carry or bring to.

adgnatus: see **adnatus.**

adhibeo, 2, bring one thing to another, lay hands on, apply, employ, exercise.

adicio, -ieci, -iectum, 3, throw into, add to.

adigo, -egi, -actum, 3, drive to (of lambs, II, v. 16).

adimo, -emi, -emptum, 3, take away, with dative (I. ii. 17).

aditialis, -e, pertaining to an entrance; aditialis cena, inaugural feast given by a magistrate when entering upon office.

adiungo, -nxi, -nctum, 3, join to, bind, unite.

adiuto, 1, help, be of service.

adiuvo, -iuvi, -iutum, 1, help, give help, assist, support.

adminiculum, -i, n., prop, support, the pole on which a vine is trained and supported.

administro, 1, help, manage, direct, serve (at table).

admiror, dep., 1, admire, wonder at, be astonished or surprised at.

admisceo, -miscui, -mix-

tum, 2, mix with, add to by mixing.

admissarius, -i, m., stallion, breeding animal.

admissio, -onis, f., mating.

admissura, -ae, f., breeding.

admitto, -misi, -missum, 3, send to, mate, allow in, assemble (III. ii. 4).

admoveo, 2, move, bring to or close, place near.

adn, see also **agn** and **ann.**

adnatus, -a, -um, relation descended from a common ancestor, cousin (III. xvi, 29).

adnecto, -nexui, -nexum, 3, bind or fasten to, connect with.

adnumero, 1, count out, pay.

adp, see also **app.**

adprime, adv., by far, above all, exceedingly.

adruo, 3, scrape up, heap up.

ads, see also **ass.**

adsero, -sevi, -situm, 3, plant at or near (I. xvi. 6).

adsignifico, 1, show, make evident, point out.

adsisto, 3, place oneself, take up a position, stand, help.

adsitus, p.p.p. of adsero.

adsuesco, -suevi, -suetum, 3, accustom, accustom oneself, become accustomed.

adsumo, -sumpsi, -sumptum, 3, take to oneself, take in addition to, take for use, choose.

adtexo, -texui, -textum, 3, weave on or to, add.

adtribuo, -ui, -utum, 3, allot, assign to.

adulterinus, -a, -um, adulterous, not genuine, mixed (of seeds, I. xl. 2), false.

adveho, -vexi, -vectum, 3, carry or bring to a place.

advena, -ae, c., stranger, foreigner, migratory (of birds, III. v. 6).

adveneror, 1, dep., reverence, venerate, honour, beseech.

advenio, -veni, -ventum, 4, come to, approach.

adventicius, -a, -um, coming from outside or from a foreign country, casual (of water, I, viii. 4; xli, 3).

adventus, -us, *m., arrival.*

adversarius, -a, -um, *turned towards, opposed, contrary, harmful.*

adversus, adversum, *adv., opposed to, against,* prep. *with accus. towards, facing, opposite to.*

adverto, -verti (-vorti), -ver- sum (-vorsum), 3, *turn towards, direct* or *give attention to, take precautions.*

advoco, 1, *summon, call.*

advolo, 1, *fly to, hasten towards.*

aedes and **aedis,** is, *f., building, temple, pl. house.*

aedificatio, -onis, *f., build- ing, act of building.*

aedificium, -i, *n., building.*

aedilicius, -a, -um, *relating to an aedile.*

aedilicius, -i, *m., one who has been an aedile.*

aedilis, is, *m., aedile* (I. ii. 2).

aeditumus, -i, *m.,* old form of **aedituus** (I. ii. 1).

aedituus, -i, *m., keeper* or *sacristan of a temple* (I. ii. 1).

aeger, -gra, -grum, *sick, ill.*

aegre, *adv., with difficulty, scarcely, hardly.*

aegrotus, -a, -um, *sick, ill.*

aequabilis, -e, *like, similar, equal.*

aequinoctialis, -e, *equinoc- tional, relating to the equinox* (I. xii. 1).

aequinoctium, -ii, *n., equinox, the time of equal days and nights* (III. v. 7).

aequus, -a, -um (old nom. masc. sing. -os, I. vi. 6), *equal, level.*

aer, *m., air, atmosphere.*

aes, aeris, *n., copper, bronze* (III. xvi. 30), *money.*

aestas, -atis, *f., summer.*

aestivus, -a, -um, *relating to summer* (archaic accus. sing. masc., aestivom).

aestus, -us, *m., heat, seeth- ing, raging.*

aetas, -atis, *f., the period of life, life, age.*

aff, see also **adf.**

agellus, -i, *m.* dim. (ager), *small piece of field, little field, plot.*

aggero, 1, (agger), *heap up like an* agger *or earth work, pile* or *heap up.*

agg, see also **adg.**

agnatus (adg), -a, -um, *born to, connected by birth, uncle, cousin.*

agninus, -a, -um, *pertaining to a lamb, of a lamb.*

agnosco, -novi, -nitum, 3, *know well, recognise.*

agnus, -i, *m., lamb.*

agrestis, -e, *pertaining to land, county, rustic, rural.*

aio *defective verb, say, state.*

ala, -ae, *f., wing.*

alacer, -cris, -e, *lively, brisk, quick.*

alea, -ae, *f., game with dice, game of chance* (I. iv. 3).

ales, alitis, *winged, c., bird, fowl.*

alesco, 3, *grow up, increase.*

algor, -oris, *m., cold.*

alias, *adv., at a time other than the present, at another time.*

alibilis, -e, *giving nourish- ment, nourishing.*

alieno, 1, *make one person or thing another, separate out, discard.*

alienus, -a, -um, *that belongs to another person, another's.*

alimonium, -ii, *n., nourish- ment, support.*

aliquanto, *adv., somewhat, in some degree.*

aliqui, aliqua, aliquod, *indef. adj., some, any.*

aliquis, aliquid, *indef. pro- noun, some, any.*

aliquot, *indec. a few, some.*

aliquotfariam, *adv., in some* or *several places.*

allatus, *p.p.p.* of adfero (aff-).

alligo, 1, *bind to something.*

alnus, -i, *f., alder.*

alo, alui, altum, and alitum, 3, *feed, nourish.*

alsiosus, -a, -um, *easily freez- ing, susceptible to cold.*

altaria, -ium, *n.*, only in *plur., that which was placed upon the altar proper* (ara) *for the burning of the victim, high altar, altar.*

alternus, -a, -um, *one after another, by turns, alternate.*

alteruter, -tra, -trum, *one of two, the one or the other.*

altilis, -e, *fattened, fat.*

altitudo, -inis, *f., height, depth.*

alvarium, -ii, *n.* 2, (more usually **alvearium**), *beehive, apiary* (III. xvi. 11).

alvus, -i, *f., belly, womb, beehive (archaic nom. sing.* alvos, *and accus. sing.* alvom I. xxxi. 4).

amarus, -a, -um, *bitter.*

ambulatio, -onis, *f., walk, place for walking, promenade.*

amitto, -misi, -missum, 3, *send away, dismiss, let go, lose.*

amnis, is, *m., stream.*

amphora, ae, *f., clay vessel with two handles, often for wine, pitcher, measure for liquids.*

amplus, -a, -um, *great, large, strong.*

amurca, -ae, *f., the watery part that flows out when olives are pressed, the lees or dregs of oil.*

anas, anatis, *f., duck.*

angustus, -a, -um, *narrow, small in space.*

anima, -ae, *f., air, breath, life.*

animadverto, -ti, -sum, 3, *direct the mind* or *attention to a thing, punish* (I. xvi. 5).

animosus, -a, -um, *full of courage, spirited.*

animus, -i, *m., soul, reason, intellect.*

anniculus, -a, -um, *a year old, yearling.*

annona, -ae, *f., yearly produce, corn, grain, provisions.*

annuo, -ui, 3, *nod to, give approval by nodding.*

anser, -eris, *m., goose.*

antelucanus, -a, -um, *before light, before day.*

antiquus, -a, -um, (also **anticuus**), *old, ancient.*

aper, -pri, *m., wild boar.*

apiastrum, -i, *n., a plant of which bees are fond.*

apica, -ae, *f., a sheep that has no wool on the belly* (II. ii. 3).

apis, is, *f., bee.*

apodyterion, *n., undressing room in the baths* (II praef. 2).

appareo, 2, *appear, become visible.*

appello, -puli, -pulsum, 3, *drive to a thing, or toward.*

appendix, -icis, *f., that which hangs to anything, appendage, addition.*

applico, 1, *join to, fasten.*

appono, -posui, -positum, 3, *place at, unite.*

apricus, -a, -um, *warm, sunny.*

aptus, -a, -um, *fitted, suitable.*

apud, *prep.* (also **aput**), *near, at the house of.*

aquarius, -a, -um, *of,* or *relating to, water;* -i, *m., water-carrier.*

aquatilis, -e, *found in or near water, aquatic.*

aquila, -ae, *f., eagle.*

aquilo, -onis, *m., north-wind.*

aquosus, -a, -um, *rainy, moist, abounding in water.*

aratio, -onis, *f., ploughing, agriculture.*

arator, -oris, *m., ploughman, farmer.*

aratrum, -i, *n., plough.*

arbitror, 1, *dep. observe, make a decision.*

arbuscula, -ae, *f., small tree, shrub.*

arbustum, -i, *n., place where trees are planted, orchard, plantation, vineyard planted with trees.*

arbutus, -i, *f., wild strawberry tree, arbutus.*

archetypon, *n., original, archetype* (III. v. 8).

arcturus, -i, *m., the brightest star in the constellation Bootes* (I. xxxv. 2, II. i. 18).

arcuo, 1, *make in the form of a bow, bend.*

arduitas, -atis, *f., steepness.*

area, -ae, *f., piece of level ground, vacant space.*

areo, *be dry.*

aresco, *become dry.*

argentum, -i, *n., silver.*

argilla, -ae, *f., white clay, potter's earth.*

argillosus, -a, -um, *full of clay.*

aridus, -a, -um, *dry, parched.*

aries, -ietis, *m., ram.*

arituado, -inis, *f., dryness, drought.*

armenium, -ii, *n., ultra-marine, fine blue colour.*

armentarius, -i, *m., herds-man.*

armenticius, -a, -um, *relating to a herd of cattle.*

armentum, -i, *n., cattle for ploughing, herd.*

aro, I, *plough, till.*

articulus, -i, *m., dim.* (artus), *joint, knuckle, limb.*

artifex, -ficis, *m., artisan, worker.*

arvum, -i, *n., arable field, ploughed land.*

arx, arcis, *f., stronghold, fortress, citadel.*

as, assis, *m., copper coin.*

ascendo, -scendi, -scensum, 3, *ascend, climb.*

asellus, -i, *m., dim.* (asinus) *little ass, ass's colt.*

asinarius, -a, -um, *pertaining, or belonging to an ass.*

asininus, -a, -um, *of or produced by an ass.*

asparagus, -i, *m., asparagus.*

aspargo, -ersi, -ersum, 3, *scatter, strew* (sometimes **aspergo),**

aspectus, -us, *m., look, sight.*

aspergo, see **aspargo.**

asperitas, -atis, *f., uneven-ness, roughness, harshness, severity.*

aspero, I, *make rough, un-even.*

aspicio, -spexi, -spectum, 3, *look at, see.*

ass, see also **ads.**

assentior, 4, *dep., agree with, approve.*

assido, -sedi, 3, *sit down, seat one's self somewhere.*

assiduus, -a, -um, *busy, occupied with, continual.*

assuefacio, -feci, -factum, 3, *accustom to something.*

astrologia, -ae, *f., know-ledge of the stars, astronomy.*

astrologus, -i, *m., astrono-mer, astrologer.*

astrum, -i, *n., star, con-stellation.*

atavus, -i, *m., father of a great-great-grandfather or great-great-grandmother, ancestor.*

ater, -tra, -trum, *black, dark, gloomy.*

athleta, ae, *c., athlete.*

attineo, 2, *hold to, keep, belong to.*

auceps, -cupis, *c., bird catcher.*

auctor, -oris, *c., founder, originator.*

auctoritas, -atis, *f., author-ity, reputation.*

aucupor, I, *dep., go bird-catching, watch for.*

augur, -uris, *c., augur, diviner, priest.*

auratus, -a, -um, *orna-mented with gold, gilded.*

aureolus, -a, -um, *golden, dim.* (aureus).

aureus, -a, -um, *of gold, golden.*

auricula, -ae, *f., external ear, ear-lap.*

auriga, -ae, *c., charioteer, driver* (also **origa,** an old form).

auris, is, *f., ear* (the organ of hearing).

aurum, -i, *gold, a thing made of gold.*

auspicium, -ii, *n., divina-tion from the flight of birds, sign, omen.*

avaritia, -ae, *f., greed, avarice.*

aversus, *p.p.p.* averto.

averto, -ti, -sum, 3, *turn away.*

aviarius, -a, -um, *pertaining to birds, of birds.*

avidus, -a, -um, *eager, ear-nest, greedy.*

avis, is, *f., bird.*

avus, -i, *m., grandfather, ancestor.*

axis, is, *m., axle-tree around which a round body, such as a wheel, turns.*

B

baca, -ae, *f., berry.*

bacillum, -i, *n., small staff, wand, handle.*

balatro, -onis, *m., babbler, buffoon.*

balneae, -arum, *f. pl., public baths.*

balo, 1, *bleat.*

benevolentia, -ae, *f., good-will, kindness.*

bestia, -ae, *f., beast.*

bestiola, -ae, *f., dim.* (bestia), *small beast.*

betaceus, -a, -um, *of the beet,* **betaceus,** -i, *m., beet-root.*

bibo, bibi, 3, *drink.*

biduum, -ii, *n., period of two days, two days.*

biennium, -ii, *n., period of two years.*

bifariam, *adv., twofold, double.*

bifer, -era, -erum, *bearing fruit twice a year.*

bimestris, -e, *of two months' duration.*

bimus, -a, -um, *two years old.*

bini, -ae, a, *two each, two at a time.*

bipedalis, -e, *two feet long broad or thick.*

bivium, -ii, *n., place where two ways meet.*

bolus, -i, *m., throw,* in game of dice, *gain, haul, winning piece of fortune.*

bombus, i, *m., hollow, deep sound, buzzing.*

bonitas, -atis, *f., goodness.*

bos, bovis, *c., ox, bull, cow.*

brassica, -ae, *f., cabbage.*

brevis, -e, *short, narrow, of small extent, in space and time.*

brocchus, -a, -um, *projecting,* of the teeth of animals

bruma, -ae, *f., shortest day* of the year, winter solstice, *winter,* from brevissimus.

brumalis, -e, *pertaining to the shortest day, wintry, of winter.*

bubile, -is, *n., stall for oxen.*

bubulcus, -i, *m., ploughman, herdsman.*

bubulus, -a, -um, *pertaining to cattle or oxen.*

bulla, -ae, *f., bubble, trifle, amulet.*

bumammus, -a, -um, *of the vine, with large clusters.*

bura, see **buris.**

buris, is, *m., curved hinder part of the plough, plough-beam* (I, xix. 2).

buturum, -i, *n., butter.*

C

cacumen, -inis, *n., peak, top, utmost point.*

cado, cecidi, casum, 3, *fall, drop off.*

caducus, -a, -um, *falling, fallen* (II. ii. 12 *fallen in mowing*).

caedo, cecidi, caesum, 3, *cut, hew, cut off.*

caeduus, -a, -um, *that can be cut, meant for cutting.*

calco, 1, *tread under foot, trample on.*

caldor, -oris, *m., warmth.*

calidus, -a, -um, *warm, hot.*

calix, -icis, *m., cup, goblet.*

callis, is, *m., foot path, cattle-drove, mountain pasture, defile* (II. ii. ix).

calor, -oris, *m., heat, glow.*

calx, calcis, *f., heel.*

campester, -tris, -tre, *pertaining to a level field, flat.*

canaliculus, -i, *m., dim.* (canalis), *small channel, pipe* or *gutter.*

canalis, -is, *m., pipe, groove, channel.*

cancelli, -orum, *m., dim.,* (cancer) *lattice, enclosure, railings.*

candeo, 2, *shine, glitter.*

candidus, -a, -um, *shining white, clear, bright.*

canicula, -ae, *f., dim.* (canis), *small dog or bitch, the lesser dog-star in the constellation* canis.

cannabinus, -a, -um, *of hemp.*

cannabis, is, *f., hemp.*

cantherius, -ii, *m., gelding, ass, mule.*

cantrix, -icis, *f., female singer.*

canus, -a, -um, *white, hoary.*

capella, -ae, *f., dim.* (caper), *she-goat.*

capillus, -i, *m., hair of the head, hair.*

capitatus, -a, -um, *having a head* (II. ix. 15).

capitulum, -i, *dim.* (caput), *small head* (III. v. 10).

capra, -ae, *f., she-goat.*

caprea, -ae, *f., a kind of wild she-goat.*

capreolus, -i, *m., a kind of wild goat,* of vines, *the small tendril which grasps the prop* (I. xxxi. 4).

caprile, -is, *n., enclosure* or *stall for goats.*

caprilis, -e, *of* or *pertaining to the goat.*

caprinus, -a, -um, *of,* or *pertaining to, goats.*

capus, -i, *m., capon.*

caput, -itis, *n., head.*

carcer, -eris, *m., enclosed space, prison, jail.*

cardo, -inis, *m., pivot and socket upon which a door swung, hinge of a door.*

care, *adv., at a high price, clearly.*

careo, 2, *be without, want.*

careota, -ae, *f., kind of nut-shaped date.*

caritas, -atis, *f., dearness, costliness.*

carnarium, -ii, *larder.*

caro, carnis, *f., flesh.*

carpo, -psi, -ptum, 3, *pick, gather.*

carptura, -ae, *f., of bees, sucking* or *gathering from flowers.*

casa, ae, *f., cottage, cabin, shed.*

caseus, -i, *m., cheese.*

castro, 1, *castrate.*

catulus, -i, *m., dim.* (catus), *young of animals, puppy.*

cauda, -ae, *f., tail.*

cauliculus, -i, *m., dim.* (caulis), *small stalk* or *stem.*

caulis, is, *m., stalk* or *stem of plant.*

causa, -ae, *f., cause, reason, opportunity.*

cavea, -ae, *f., excavated place, hollow.*

caveo, 2, *be on one's guard, take care.*

cavo, 1, *make hollow, hollow out.*

cavus, -a, -um, *hollow, excavated.*

-ce, *strengthening demonstrative particle.*

cedo, cessi, cessum, 3, *go, happen, depart, yield in rank, be inferior to.*

celeber, -ebris, -ebre, *much frequented, populous, renowned.*

celebriter, *adv., frequently.*

cella, ae, *f., store-room, place for keeping grain* or *fruit, granary.*

cena, ae, *f., dinner, supper.*

cenalis, -e, *pertaining to dinner* (I. ii. 11).

ceno, 1, *dine, eat.*

censorius, -a, -um, *pertaining to the censor, censorial.*

census, -us, *m., registering and rating of Roman citizens, census.*

centenarius, -a, -um, *consisting of a hundred.*

centeni, -ae, -a, *hundred each, hundred.*

cera, -ae, *f., wax.*

certamen, -inis, *n., contest, struggle, rivalry.*

certus, -a, -um, *determined, settled, certain.*

cervinus, -a, -um, *pertaining to a deer.*

cervix, -icis, *f., neck, nape of the neck.*

cervus, -i, *m., stag, deer.*

cestus, -i, *m., girdle* (I. viii. 6).

ceterus, -a, -um, *the other, what is left.*

character, -eris, *n., instrument for branding, character, style.*

chorion, *membrane surrounding the* foetus *in the womb* (II. i. 19).

chors, see **cohors.**

cibarius, -a, -um, *suitable for food,* **cibaria,** -orum, *n. pl., food, provisions.*

cibatus, -us, *m., food, nourishment.*

cibus, -i, *m., food, fodder.*

cicer, -eris, *n., chickpea.*

cicercula, -ae, *f., dim.,* (cicera; al. cicer) *kind of small chickpea.*

ciconia, *f., stork.*

cicur, -uris, *tame.*

cimex, -icis, *m., bug.*

cinerarius, -a -um, *pertaining to ashes.*

cingulum, -i, *n., girdle.*

cinis, -eris, *m., ashes.*

circensis, -e, *adj., pertaining to the Circus* (I. ii. 11).

circiter, *adv. and prep., round about, about, near.*

circuitus, -us, *m., going round, circling, way around.*

circulus, -i, *m., circle.*

circum, *adv. and prep., around, all around.*

circumago, -egi, -actum, 3, *drive in a circle, turn round.*

circumcido, -cidi, -cisum, 3, *cut around, trim.*

circumdo, -dedi, -datum, *put around, wrap around, enclose.*

circumeo, -ivi or -ii, -itum, *go around.*

circumfluo, -xi, 3, *flow around.*

circumitus, -us, *m., a going round, revolving.*

circumspicio, -exi, ectum, 3, *look about one's self, observe, be cautious.*

circumtinnio, -ire, *ring or tinkle around.*

circus, -i, *m., circle, the oval circus* (Circus Maximus), *between the Palatine and Aventine hills in Rome* (II. praef 3).

cis, *prep., on this side.*

cisterna, -ae, *f., subterranean reservoir for water, cistern.*

citer, -tra, -trum, *on this side.*

citerior, *comparative of* citer.

cithara, -ae, *f., cithara, guitar, lute.*

cito, I, *put into quick motion, excite, stimulate.*

cito, *adv., quickly, soon.*

citro, *adv. in the connection and position* ultro citroque, ultro et citro, *hither and thither* (III. v. 16).

citrum, -i, *n., wood of the* citrus, *citrus-wood.*

civilis, -e, *pertaining to citizens, civil, civic.*

civitas, -atis, *f., citizenship, state, body-politic.*

clam, *adv., secretly, prep., without knowledge of, unknown to.*

classis, is, *f., class, fleet.*

claudus, -a, -um, *limping, lame.*

clavis, is, *f., key.*

clavola, -ae, *f., dim.* (clava), *graft* (I. xl. 41).

clavulus, -i, *m. dim.* (clavus), *small nail, stud.*

cludo (claudo), -si, -sum, *shut.*

clunis, is, *m. and f., haunch.*

coacesco, -acui, 3, *become acid or sour.*

coalesco, -alui, -alitum, *grow together, unite.*

coda, see **cauda.**

coemo, -emi, -emptum, 3, *purchase together.*

coeliacus, -a, -um, *relating to the stomach, afflicted with a disease of the bowels.*

coepio, coepi, coeptum, 3, *begin.*

coerceo, -cui, -citum, 2, *surround, encompass, control.*

cogito, I, *ponder, think.*

cognatus, -a, -um, *related by blood, kinsman, relation.*

cognitio, -onis, f., *knowledge, knowing.*

cognomen, -inis, *n., name added to the* nomen (*or name of the* gens), *usually the third word in order in the full name of each citizen, Roman surname.*

cognosco, -gnovi, -gnitum, 3, *become thoroughly acquainted with, understand, learn.*

cogo, coegi, coactum, 3, *collect, drive together, assemble.*

cohors, or **cors,** -rtis, *f., place enclosed around, courtyard, company of soldiers.*

coicio, see **conicio.**

colesco, see **coalesco.**

coliculus, see **cauliculus.**

colis, is, *m.,* see **caulis.**

collare, -is, *n., band for the neck, collar.*

collega, -ae, *m., colleague, partner in office.*

colligo, -legi, -lectum, 3, *gather together, assemble.*

collinus, -a, -um, *hilly, found* or *growing on a hill.*

collis, is, *m., high ground, hill.*

colloco, 1, *place together, put, set.*

collum, -i, *n., neck.*

colo, colui, cultum, 3, *cultivate, till, inhabit.*

colonia, -ae, *f., farm, colonial town, settlement, colony, planters.*

colonicus, -a, -um, *pertaining to agriculture, to a colony.*

colonus, -i, *m., husbandman, farmer, colonist.*

columba, -ae, *f., dove, pigeon.*

columbinus, -a, -um, *pertaining to a dove or pigeon.*

columellaris, -e, *pillar-formed,* referring to grinders of horses (II. vii. 2).

columna, -ae, *f., column, pillar, post.*

comburo, -ussi, -ustum, *burn up, consume.*

comedo, -edi, -esum, *eat, eat up.*

comes, -itis, *c., companion, comrade.*

comitium, -ii, *n., the place where the Romans assembled to vote by the* curiae (I. ii. 9).

comitor, 1, *dep., accompany, attend.*

commeto, *go frequently.*

comminuo, -ui, -utum, 3, *make small, break up, crumble.*

committo, -misi, -missum, 3, *join, entrust.*

commodus, -a, -um, *complete, suitable, convenient.*

commoveo, -movi, -motum, 2, *put in violent motion, displace.*

communico, 1, *divide a thing with one, share.*

communis, -e, *common, general.*

communiter, *adv., together, in common.*

commutatio, -onis, *f., change, alteration.*

commuto, 1, *change entirely.*

comoedia, -ae, *f., comedy.*

comp, see also **conp.**

conangusto, 1, *make narrow* (III. xvi. 15).

conb, see also **comb.**

concalefacio, -feci, -factum, *warm thoroughly.*

concalesco, -lui, 3, *become thoroughly warm.*

concedo, -cessi, -cessum, *go, walk, depart.*

conceptus, -us, *m., collecting, conceiving, pregnancy.*

concessio, -onis, *f., allowing, granting, yielding.*

concido, -cidi, 3, *fall down, fall together.*

concido, -cidi, cisum, 3, *cut up, kill.*

concipio, -cepi, -ceptum, 3, *take* or *take hold of, receive.*

concito, 1, *put in violent motion, rouse, influence.*

concludo, -si, -sum, 3, *shut up closely, confine.*

concurro, -curri, -cursum, 3, *assemble together in crowds.*

concutio, -cussi, -cussum, 3, *shake, agitate, disturb.*

condenso, 1, *press close together.*

condicio, -onis, *f., agreement, stipulation.*

condisco, -didici, 3, *learn carefully.*

condo, -didi, -ditum, 3, *bring, put together, build, store.*

conducticius, -a, -um, *hired, rented* (I. xvii. 2).

confercio, -fertum, 4, *cram together, press close together.*

confero, contuli, collatum, *bring or carry together, collect.*

conficio, -feci, -fectum, 3, *complete, accomplish, weaken an object.*

confinis, -e, *adjoining, neighbouring.*

confinium, -ii, *n., boundary.*

confirmo, 1, *make firm, strengthen.*

confluo, -xi, 3, *flow, or run together.*

confracuit (I. xiii. 4), see **fracesco.**

confundo, -fudi, -fusum, 3, *pour together, mingle, confound, put in disorder.*

congelo, 1, *freeze up, congeal, curdle.*

congermino, 1, *shoot forth at same time.*

congero, -gessi, -gestum, 3, *bear or carry together, heap up.*

congius, -ii, *m., liquid measure, containing the eighth part of an* amphora (nearly six pints).

conglobo, 1, *gather into a ball or sphere.*

conglutino, 1, *glue, join together, cement.*

congrego, 1, *collect into a flock.*

congruo, -ui, 3, *run* or *come together.*

conicio (coniicio, coicio), -ieci, -iectum, 3, *throw together, unite.*

coniecto, 1, *throw or bring together, guess.*

coniectura, -ae, *f., guess, conjecture.*

coniicio, see **conicio.**

coniunctio, -onis, *f., joining together, union.*

coniungo, -nxi, -nctum, 3, *bind together, unite.*

conl, see also **coll.**

conopium (conopeum), -ei, *n., gauze bed curtain* (II. x. 8).

conor, 1, *try, attempt.*

conparo, 1, *compare,* 2, *prepare.*

conpasco, -pastum, *feed together, pasture.*

conpello, -puli, -pulsum, 3, *assemble, drive together.*

conpenso, 1, *poise, balance, make good.*

competitor, -oris, *m., rival, competitor.*

conpleo, -evi, -etum, 2, *fill up, complete.*

conplico, 1, *fold together.*

conplures, *more than one, several, very many.*

conpluviatus, -a, -um, *fashioned like a* conpluvium (I. viii. 2).

conpluvium, -ii, *n., open space in the roof of a Roman house through which rain-water fell and was collected in a basin* (impluvium) *below* (I. xiii. 3).

conpono, -posui, -positum, 3, *gather, put together.*

conporto, 1, *carry or bring together.*

conprehendo, -di, -sum, 3, *take hold of, seize, perceive.*

conprendo, see **comprehendo.**

conprimo, -pressi, -pressum, 3, *press together, restrain.*

conquiesco, -quievi, quietum, 3, *rest, be quiet.*

conr, see also **corr.**

consaepto, 1, *fence round.*

conseco, -cui, -ctum, 1, *cut up, lop, prune.*

consector, 1, *dep., pursue, follow.*

consenesco, -ui, 3, *grow old together.*

consentes, *pl. adj., twelve superior deities* (I. i. 4).

consentio, -sensi, -sensum, 4, *feel together, agree.*

consequor, -secutus, *dep.,* 3, *follow, press upon.*

consero, -sevi, -situm, or -satum, 3, *sow* or *plant with something.*

conservo, 1, *retain, preserve.*

conservus, -i, *m.,* **conserva,** -ae, *f., fellow-slave.*

considero, 1, *look at closely, inspect, meditate.*

consido, -sedi, -sessum, 3, *take one's place, sit.*

consilium, -ii, *n., consultation, counsel, intention, council.*

consisto, -stiti, -stitum, 3, *stand still, stop.*

consolor, I, *dep., console, cheer.*

consono, I, *sound together, resound.*

conspargo (also **conspergo),** -si, -sum, 3, *sprinkle, moisten, strew.*

conspectus, -us, *m., seeing, sight, view, appearance.*

constituo, -ui, -utum, 3, *set, put, erect.*

consto, -stiti, -statum, *stand together, agree with, cost* (II. i. 14).

consudo, I, *sweat profusely.*

consuefacio, -feci, -factum, 3, *accustom one to a thing, habituate.*

consuesco, -suevi, -suetum, 3, *accustom.*

consuetudo, -inis, *f., custom, habit.*

consulo, -lui, -ltum, 3, *consider, have regard to, consult.*

consultor, -oris, *m., adviser, counsellor.*

consumo, -sumpsi, -sumptum, 3, *eat, consume.*

consurgo, -surrexi, -surrectum, 3, *rise up together, stand up.*

contemno, -tempsi, -temptum, 3, *despise, value little.*

contendo, -di, -tum, 3, *stretch, draw tight.*

contenebro, I, *darken.*

contentus, -a, -um, *contented, pleased with.*

contero, -trivi, -tritum, 3, *grind, pound, waste.*

contineo, -tinui, -tentum, 2, *hold* or *keep together.*

contingo, -tigi, -tactum, *touch, take hold of, happen.*

continuatus, see **continuo.**

continuo, *adv., immediately, speedily.*

continuo, I, *make continuous, connect* (II. ix. 2).

continuus, -a, -um, *uninterrupted.*

contra, *adv., opposite, fronting, contrary to.*

contrarius, -a,-um., *opposite.*

contrarium, -ii, *m., opposite, reverse.*

controversia, -ae, *f., turning against, quarrel, dispute* (III. ii. 3).

contrudo, -si, -sum, 3, *thrust* or *crowd together.*

contueor, -uitus, 2 *dep., gaze on, survey, observe.*

contundo, -tudi, -tusum, 3, *beat, pound, bruise.*

convalesco, -lui, 3, *regain health, gain strength.*

convallis, is, *f., valley enclosed on all sides.*

conveho, -vexi, -vectum, 3, *carry* or *bring together.*

convenio, -veni, -ventum, 4, *come together.*

converto, -ti, -sum, 3, *turn* or *whirl round, cause to turn, reverse.*

convicior, I, *dep., revile, taunt.*

conviva, -ae, *c., one who feasts with another, table companion.*

convivium, -ii, *n., meal in company, banquet.*

cooperio, -rui, -rtum, 4, *cover over.*

copiosus, -a, -um, *well supplied, rich.*

coquo, -xi, -ctum, 3, *cook.*

cor, cordis, *n., heart;* cordi est alicui, *it pleases.*

corbis, is, *c., basket* (I. l. 1).

corbula, -ae, *dim.* (corbis), *f., little basket.*

cordus, -a, -um, *late-born,* or *produced late in season* (II. i. 19).

corium, -ii, *n., rind, skin, hide.*

corneus, -a, -um, *of horn, horny.*

cornu, -us, *n., horn.*

cornus, -i, *f., cornel cherry tree.*

corona, -ae, *f., garland, wreath, crown,* sub corona vendere, *sell captives as slaves* (II. x. 4).

corpus, -oris, *n., body, person.*

correpo, 3, -psi, *creep.*

corrigo, -rexi, -rectum, 3, *make straight, set right.*

corruda, -ae, *f., wild asparagus.*

corrumpo, -rupi, -ruptum, 3, *break to pieces, waste, spoil.*

cortex, -icis, *m.,* rarely *f., bark, shell, rind, cork.*

corticeus, -a, -um, of *bark* or *cork.*

corvus, -i, *m., raven.*

costatus, -a, -um, *ribbed.*

cotidianus, -a, -um, *daily.*

cotidie, *adv., daily.*

coturnix, -icis, *f., quail.*

coxendix, -icis, *f., hip* (I. xx. 1).

crassitudo, -inis, *f., thickness, density.*

crassus, -a, -um, *solid, thick, fat.*

cratis, is, *f., wicker-work, hurdle, harrow.*

creber, -bra, -brum, *thick, frequent, repeated.*

credo, -didi, -ditum, 3, *trust, believe in.*

creo, I, *produce, create.*

crepitus, -us, *m., noise.*

cresco, -crevi, -cretum, 3, *grow, arise, increase.*

creta, -ae, *f., chalk, white earth.*

crista, -ae, *f., crest, comb of a cock.*

crocum, -i, *n.* and crocus, -i, *m., saffron, saffron-crocus* (I. xxxv. 1).

crudus, -a, -um, *bloody, raw, unripe.*

crus, -uris, *n., leg, shin.*

cubile, -is, *couch, bed.*

cubo, -ui, -itum, I, *recline, sleep.*

cucumis, -eris, *m., cucumber.*

cucurbita, -ae, *f., gourd.*

culina, -ae, *f., kitchen.*

culleus, -i, *m., leather bag, sack for holding liquids* [*nom. pl. neut.* cullea (I. ii. 7)].

cultellus, -i, *m., small knife.*

culter, -tri, *m., plough-share, knife.*

cultura, -ae, *f., cultivation, agriculture, husbandry.*

cum, also **quom** and **quum,** *prep., with.*

cum, also **quom,** *conj., when, although.*

cupiditas, -atis, *f., desire, wish.*

cupio, -ivi, or ii, -itum, 3, *long for, wish, desire.*

cupressus, -i, *f., cypress.*

cura, -ae, *f., care, concern.*

curatio, -onis, *f., taking care of, management.*

curator, -oris, *m., manager, overseer.*

curculio, -onis, *m., corn-worm, weevil.*

curiosus, -a, -um, *careful, curious, inquisitive.*

curo, I, *care for, administer, cause something to be done.*

curriculum, -i, *n., running race* (I. ii. 11), *race-course.*

curso, I, *freq., run to and fro.*

cursura, -ae, *f., running.*

cursus, -us, *m., running* (*on foot, on a horse, chariot,* etc.), *way, passage.*

cuspis, -idis, *f., point, pointed end.*

custodia, -ae, *f., guard, protection.*

custos, -odis, *c., guard, attendant.*

cymbalum, -i, *n., cymbal* (III. xvi. 7).

cyperum, -i, *n., species of rush?* (III. xvi. 13), *galingale?*

cytisum, -i, *n., shrub trefoil* (I. xxiii. 1).

D

damno, I, *harm, condemn, sentence to punishment* (II. ii. 6).

datio, -onis, *f., giving.*

debeo, 2, *owe, ought, must.*

decarpo, see **decerpo.**

decerpo, -psi, -ptum, 3, *pluck off, gather.*

decet, -cuit, 2, *impers., it is seemly, fitting.*

decimus, -a -um, *the tenth.*

decipio, -cepi, -ceptum, 3, *catch, deceive.*

declino, I, *turn aside,* or *away, deviate.*

decolo, I, *to come to nought, fail* (I. ii. 8).

decoquo, -xi, -ctum, 3, *boil down, consume, waste.*

decresco, -crevi, -cretum, 3, *grow less, decrease, wane.*

decumus, see decimus.

decurro, -cucurri, or -curri, 3, *run down, race.*

deduco, 3, *turn from a straight course, draw off.*

defatigo, I, *weary out, tire.*

defendo, -di, -sum, 3, *ward off, defend, guard.*

defensor, -oris, *m.*, rarely, *f.*, *defender, protector.*

defero, -tuli, -latum, *bear or bring away, carry down.*

defervesco, -fervi and -ferbui, 3, *cease boiling, grow cooler.*

deficio, -feci, -fectum, 3, *loosen, remove from, fail.*

defluo, -xi, -xum, 3, *flow down.*

deformo, I, *deform, disfigure.*

defringo, -fregi, -fractum, 3, *break off* or *to pieces.*

degusto, I, *taste.*

deicio, -ieci, -iectum, 3, *throw down.*

dein, see **deinde.**

deinceps, *adj.* and *adv.*, *following, next.*

deinde, *next, thereafter.*

deiungo, 3, *unyoke.*

delabor, -lapsus, 3, *dep.*, *fall, slip down.*

delapsus, *p.p.* of delabor.

delapsus, -us, *m.*, *falling off, descent.*

delectatio, -onis, *f.*, *delight.*

delecto, I, *entice.*

delibo, I, *take off, enjoy, gather, take away from.*

deliciae, -arum, *f. pl.*, *delight, pleasure.*

delineo, also **delinio,** I, *sketch out, delineate.*

delphinus, -i, *m, dolphin,* constellation (II. v. 13).

deminuo, -ui, -utum, 3, *lessen, diminish.*

demitto, -misi, -missum, 3, *send down, drop.*

demo, -mpsi, -mptum, 3, *take off, withdraw.*

demonstro, I, *point, show.*

denarius, *Roman silver coin.*

denego, I, *refuse, reject, deny.*

deni, -ae, -a, *distrib. num.*, *ten each.*

denigro, I, *colour very black.*

denique, *adv.*, *and then, lastly.*

densus, -a, -um, *thick.*

dentatus, -a, -um, *toothed, having teeth.*

denudo, I, *lay bare.*

deorsum, also **deorsus,** *adv.*, *downwards.*

depello, -puli, -pulsum, 3, *drive out or down, remove, expel.*

dependeo, 2, *hang from, on, be dependent on.*

deplanto, I, *take off a twig or shoot.*

derectus, see **dirigo.**

derideo, -si, -sum, 2, *laugh at, scoff at.*

derigo, see **dirigo.**

descendo, -di, -sum, *come down, descend.*

descensus, -us, *m.*, *descent.*

descisco, -ivi, -ii, -itum, 3, *free one's self, depart, be unfaithful to.*

describo, -psi, -ptum, 3, *represent, describe.*

deseco, -cui, -ctum, I, *cut off.*

desiderium, -ii, *n.*, *longing, desire.*

desidero, I, *long for, desire.*

desidiosus, -a, -um, *slothful, lazy.*

desido, -sedi, 3, *fall or settle down, deteriorate.*

desino, -sii, 3, *cease, stop.*

desisto, -stiti, -stitum, 3, *set down, cease.*

despicio, -exi, -ectum, 3, *look down upon, despise.*

despondeo, -spondi, -sponsum, 2, *promise.*

despuo, 3, *spit out, reject.*

destituo, -ui, -utum, 3, *set down, place, forsake.*

desuefacio, -feci, -factum, 3, *disuse, disaccustom* (II. ix. 12).

desum, -fui, -esse, *be absent, fail.*

deterior, -ius, *comp. adj.,* *worse, poorer.*

deterreo, 2, *frighten from,* *prevent.*

deversorius, -a, -um, *belonging to an inn.*

deverto, -ti, -sum, 3, *turn away, turn aside.*

devoro, I, *devour.*

dexter, -tra, -trum, *on the right side, right.*

diduco, -xi, -ctum, 3, *draw apart, divide.*

dies, diei, *m.,* sometimes *f.,* *day.*

differo, distuli, dilatum, *carry different ways, scatter, disperse.*

difficultas, -atis, *f., difficulty, poverty.*

difficulter, *adv., with difficulty.*

diffindo, -fidi, -fissum, 3, *cleave asunder, divide.*

diffundo, -fudi, -fusum, 3, *pour out, spread.*

digitabulum, -i, *n., glove worn for gathering olives* (I. lv. 1).

digitus, -i, *m., finger, finger's breadth.*

dignitas, -atis, *f., worthiness, dignity, rank.*

diiungo, or **disiungo,** -xi, -ctum, 3, *separate, unyoke.*

dilectus, -us, *m., selection, culling.*

diligens, -entis, *careful, attentive.*

diligentia, -ae, *f., carefulness, diligence.*

dilucide, *adv., clearly, evidently.*

diluo, -ui, -utum, 3, *dissolve, wash.*

dimidiatus, -a, -um, *halved, half.*

dimidium, -ii, *n., the half.*

dimitto, -misi, -missum, 3, *send different ways, dismiss.*

directus, *p.p.p.,* from dirigo.

diribeo, -itum, *lay apart, separate, sort* (III. ii. 1).

dirigo, -rexi, -rectum, 3, *set in straight line, arrange.*

discedo, -cessi, -cessum, 3, *divide, forsake, leave.*

discerno, -crevi, -cretum, 3, *separate, discern.*

disciplina, -ae, *f., instruction, teaching.*

disco, didici, 3, *learn.*

discrimen, -inis, *n., interval, distance, decisive point, decision.*

discriminatim, *adv., with a difference.*

discrimino, I, *divide, separate.*

discutio, -cussi, -cussum, 3, *dash to pieces, shatter.*

dispargo, see **dispergo.**

dispereo, -ii, 4, *go completely to ruin.*

dispergo, -si, -sum, 3, *scatter about, disperse.*

disperiliter, *adv., of differing importance* (I. vi. 6).

dispersim, *adv., here and there.*

displicatus, -a, -um, *scattered, dispersed.*

displodo, -sum, 3, *be spread out, extend.*

disseco, -ui, -ctum, I, *cut in pieces, cut up.*

dissero, -sevi, -situm, 3, *scatter seed, sow.*

dissero, -rui, -rtum, *argue, discuss.*

dissimilis, -e, *unlike, different.*

distinguo, -nxi, -nctum, 3, *separate, distinguish.*

disto, *stand apart, be separate distant, differ.*

disturbo, I, *throw into disorder, disturb.*

diu, *adv., by day, a long time.*

diurnus, -a, -um, *daily.*

dius, see **divus.**

diuturnus, *lasting, long.*

divarent, II. v. 8.

diversus, *p.p.p.* from diverto.

diverto, -ti, -sum, 3, *turn or go different ways.*

dives, -itis, and **dis,** dite, *rich.*

divido, -visi, -visum, 3, *part, divide.*

divisio, -onis, *f., division, separation.*

divus, -a, -um, *belonging to a deity, divine.*

doceo, -cui, -ctum, 2, *teach.*

docilis, -e, *easily taught, docile.*

dodrans, -antis, *m., nine twelfths or three fourths of any thing.*

doleo, 2, *feel pain, suffer.*

doleum, see **dolium.**

dolium, -ii, *n., large jar with wide mouth* (I. xiii. 7).

dolor, -oris, *m., pain.*

domesticus, -a, -um, *belonging to the house.*

domicilium, -ii, *n., dwelling.*

dominicus, -a, -um, *belonging to a master.*

dominium, -ii, *n., property, ownership.*

domo, -ui, -itum, I, *tame, break, conquer.*

domus, -us, and i, 2nd and 4th *decl., f., house.*

dorsum, -i, *n., back.*

dos, -otis, *f., dowry, gift.*

dossum, see **dorsum.**

dossuarius, -a, -um, *that carries on its back.*

dubius, -a, -um, *doubting, doubtful, uncertain.*

duceni, -ae, -a, *distrib. num., two hundred each.*

ducenti, -ae, -a, *two hundred.*

dulcis, -e, *sweet.*

dumtaxat, *adv., to this extent, merely, at least.*

duplex, -icis, *twofold, double.*

duplico, I, *double.*

duplus, -a, -um, *double, twice as much* (II. x. 5).

duritia, -ae, *f., hardness, austerity.*

E

ecquis, ecquid, *interrog. pron., Is there anyone who ?*

eculus, see **equulus.**

edo, edi, esum, 3, *eat.*

editus, -a, -um, *elevated, lofty.*

educo, I, *rear.*

educo, -xi, -ctum, 3, *lead out, bring away.*

edulia, -ium, *n. pl., eatables, food.*

effero, extuli, elatum, *carry or bring out, produce.*

efficio, -feci, -fectum, 3, *effect, complete, accomplish.*

efflictus, *p.p.p.* **effligo.**

effligo, -xi, -ctum, 3, *strike dead, destroy.*

efflo, I, *blow or breathe out* (I. xii. 3).

effluo, -xi, 3, *flow out.*

effodio, -fodi, -fossum, 3, *dig out,* or *up.*

egredior, -gressus, 3, *dep., go* or *come out.*

egregius, -a, -um, *distinguished, eminent.*

eicio, -ieci, -iectum, 3, *thrust or drive out.*

eiuncidus, -a, -um, *grown like a rush, slender.*

elatus, *p.p.p.* of effero.

eleganter, *tastefully, gracefully.*

eligo, -legi, -lectum, 3, *pick out, choose.*

eliquesco, *become liquid (by being pressed)* (I. lv. 4).

emblema, -atis, *n., inlaid work* (III. ii. 4).

emendo, I, *correct, improve.*

emineo, 2, *stand out, project.*

eminulus, -a, -um, *dim.* (emineo), *projecting a little.*

emitto, -misi, -missum, 3, *send out, produce, let go.*

emo, emi, emptum, 3, *buy, purchase.*

emplastrum, -i, *n.,* or **emplastra,** -ae, *f. pl., plaster, poultice.*

emporium, -ii, *n., market-town, market.*

empticius, -a, -um, *bought, purchased.*

emptio, -onsi, *f., purchase.*

emptor, -oris, *m., purchaser.*

empturiens, -entis, *desiring to buy.*

enascor, -natus, 3, *dep., sprout out, be born.*

enitor, -nisus, or nixus, *dep.,* 3, *dep., strive, struggle.*

enixus (*p.p.* of enitor), -a, -um, *strenuous, earnest.*

eo, *adv., there, on that account, by so much, to that place.*

eodem, *adv., to the same place.*

emphippium, -ii, *n., horse-cloth* (II. vii. 15).

epigramma, -atis, *n., inscription, epigram.*

epistylum, (also **epistylium**), -ii, *n., the cross-beam that rests on the columns, architrave* (III. v. 11).

epitonium, -ii, *n., bung, cock in a water pipe* (III. v. 16).

epulor, I, *dep., hold an entertainment, feast.*

epulum, -i, *n., rich food, banquet.*

equa, -ae, *f., mare.*

equarius, -a, -um, *belonging to a horse.*

equile, -is, *n., stable for horses.*

equinus, -a, -um, *belonging to a horse.*

equulus, -i, *dim.* (equus) (also **eculus**), *foal.*

eradico, I, *root out.*

erectus, *p.p.p.* from erigo.

erigo, -rexi, -rectum, 3, *raise, set up, excite.*

erithace, -es, *f., a kind of propolis* (III. xvi. 8).

erro, I, *wander, stray.*

eructo, I, *belch out.*

erudero, I, *clear from rubbish.*

ervila, -ae, *f.,* (also **ervilia**), *kind of pulse, bitter vetch* (I. xxxii. 2).

escendo, -di, -sum, 3, *climb up, ascend.*

esurio, -itum, 4, *wish to eat, be hungry.*

esus, -us, *m., eating.*

etiamnum, *yet, still, even to this time.*

evado, -si, -sum, 3, *go out, escape.*

evanno, -ere, *throw out the chaff of grain from the fan* (vannus), *winnow* (I. lii. 2).

eveho, -xi, -ctum, 3, *carry or convey out, lift up.*

evello, -velli, -vulsi, 3, *tear or pluck out.*

eventus, -us, *m., event, success.*

everro, -verri, -versum, 3, *sweep out.*

everto, -ti, -sum, 3, *overturn, turn upside down.*

evito, I, *shun, avoid.*

evoco, I, *call out, bring out, summon.*

evolo, I, *fly out.*

exacuo, -ui, -utum, 3, *sharpen.*

exaequo, I, *make even* or *level.*

exalatus, see **exhalo.**

examen, -inis, *n., swarm.*

exaresco, -rui, 3, *dry up.*

exceptus, *p.p.p.* from excipio.

excipio, -cepi, -ceptum, *take out, make an exception, catch, receive.*

excito, I, *call out, wake, rouse up.*

excludo, -si, -sum, 3, *shut out, prevent.*

excolo, -colui, cultum, 3, *tend, cultivate.*

excudo, -di, -sum, 3, *strike, beat or hammer out.*

exculpo, -psi, -ptum, 3 (also **exsculpo**), *dig out, cut out, carve.*

excurro, -cucurri, -cursum, 3, *run out, hasten out.*

excutio, -cussi, -cussum, 3, *shake out, drive out.*

exedo, -edi, -esum, 3, *eat up, devour.*

exemplum, -i, *n., example.*

exemptio, -onis, *f., taking out, removing.*

exeo, -ii, -itum, *go out, depart.*

exerceo, -ui, -itum, 2, *drive on, work, exercise, practise.*

exercitatio, -onis, *f., setting in motion, practice.*

exercitus, -us, *m., army, host.*

exhedra, *recess* (III. v. 8).

exhalo, I, *breathe out, evaporate.*

exigo, -egi, -actum, 3, *drive out, demand.*

exilis, -e, *small, thin, slender.*

exilium (or **exsilium**), -ii, *n., banishment, exile.*

eximo, -emi, -emptum, 3, *take away, remove.*

exin, see **exinde.**

exinanio, -ivi, or -ii, -itum, 4, *empty, make desolate.*

exinde, *adv., from there, after that, then.*

existimo, I, *value, esteem.*

existo, see **exsisto.**

exitus, -us, *m., going out, way out, end.*

exorior, -ortus, 3 and 4, *spring up, rise, originate, appear.*

exorno, I, *fit out, supply, adorn.*

exortus, -us, *m., coming forth, rising.*

expartus, -a, -um, *past bearing.*

expedio, -ivi, or -ii, -itum, 4, *extricate, set free, prepare.*

experientia, -ae, *f., trial, experiment, practice, experience.*

experior, -pertus, 4, *dep., try, put to test, attempt.*

expleo, -evi, -etum, 2, *fill up, finish.*

explico, I, *unfold, set in order.*

expono, -posui, -positum, 3, *set out, expose, explain.*

exporto, I, *bear or carry out, convey away.*

exprimo, -pressi, -pressum, 3, *press* or *squeeze out.*

expromo, -mpsi, -mptum, 3, *take out, fetch out, exhibit, display.*

expuo, -ui, -utum, 3, *spit out.*

expurgo, I, *cleanse, purify.*

exradico, see **eradico.**

exscribo, -psi, -ptum, 3, *write out, copy.*

exsisto (also **existo**), -stiti, -stitum, 3, *come forth, emerge, arise, exist, to be.*

exsugo, -xi, -ctum, 3, *suck out.*

exsuctus, -a, -um, *sucked out, dried up.*

exta, -orum, *n. pl., internal organs such as heart, lungs, liver.*

extenuo, I, *make thin, reduce, lessen.*

extergeo, -si, -sum, 2, also **extergo,** 3, *wipe.*

exterior, -us, *outer, exterior.*

extero, -trivi, -tritum, 3, *rub out.*

exterreo, 2, *strike with terror.*

exto (also **exsto**), I, *stand out, project, exist.*

extollo, 3, *lift out, raise up, exalt.*

extra, *adv., on the outside.*

extraordinarius, -a, -um, *out of common order.*

extrinsecus, *adv., from without, on the outside.*

extumidus, -a, -um, *raised* (I. li. I).

exuctus, see **exsuctus.**

exulcero, I, *make sore, aggravate.*

F

faba, -ae, *f., bean.*

fabalis (fabul-), -e, *belonging to a bean, bean-.*

fabella, -ae, *f., short history, story.*

faber, -bri, *m., worker in wood, stone, metal, etc., smith.*

facies, -ei, *f., make, figure, shape, appearance, face.*

facio, feci, factum, 3, *make, do:* passive **fio,** fieri, factus sum.

factito, I, *make* or *do frequently.*

facultas, -atis, *f., capability, skill, opportunity.*

faenisicia, -ae, *f.,* and -orum, *n. pl., mown hay.*

faenum, -i, *n., hay.*

falere, -is, *n., pile, pedestal* (III. v. 14 and 16).

fallo, fefelli, falsum, 3, *deceive, be mistaken, escape notice.*

falx, falcis, *f., sickle, reaping-hook, pruning-hook, scythe.*

fames, -is, *f., hunger.*

familia, -ae, *f., the slaves in a household or estate, a house and all belonging to it, family estate.*

familiaricus, -a, -um, *of or belonging to the house-servants.*

familiaris, -e, *m., servant, of,* or *belonging to a household, household, family, friend.*

far, farris, *n., sort of grain* (I. ii. 6) emmere.

farcio, farsi, fartum, 4, *stuff, fill full.*

farina, -ae, *f., ground corn, meal, flour.*

farraceus, or -ius, -a, -um, *of corn* or *grain.*

farrago, -inis, *f., mixed fodder.*

fascis, is, *m., bundle.*

fasti, -orum, *m. pl., calendar, almanac.*

fastidio, -ivi, -ii, -itum, 4, *feel disgust, or nausea, shrink from.*

fastidiosus, -a, -um, *full of disgust.*

fastidium, -ii, *n., loathing, aversion.*

fastigium, -ii, *n., top of a gable, summit.*

fauces, -ium, *f. pl., throat, gullet.*

fautor, -oris, *m., promoter, patron.*

favonius, -ii, *m., west wind, zephyr.*

favus, -i, *m., honey-comb.*

febris, is, *f., fever.*

Februarius, -ii, *m., February.*

fecunditas, -atis, *f., fertility.*

fecundus, -a, -um, *fertile.*

fel, fellis, *n., gall-bladder, gall, bile.*

femen, -inis, *n., upper part of the thigh, thigh.*

femineus, -a, -um, *belonging to a woman, womanly.*

femininus, -a, -um, *of the feminine gender, feminine.*

fenestra, -ae, *f., window.*

fenus (also **faenus**), -oris, *n., interest on a loan.*

ferax, -acis, *fruitful, fertile.*

fere, *adv., almost.*

feriae, -arum, *f. pl., holidays, festivals.*

feriatus, -a, -um, *keeping holiday, at leisure.*

ferio, 4, *strike, kill.*

fero, ferre, tuli, latum, *bear, carry.*

fermento, I, *ferment.*

ferrago, (I. xxxi. 5).

ferramentum, -i, *n., tool made of iron.*

ferreus, -a, -um, *made of iron.*

fertilis, -e, *fertile, fruitful.*

ferula, -ae, *f., the plant fennel-giant* (III. xvi. 15).

ferus, -a, -um, *wild, untamed.*

ferveo, -bui, 2, or **fervo,** -vi, 3, *boil, ferment.*

fervidus, -a, -um, *glowing hot, fiery.*

fervor, -oris, *m., boiling heat.*

festum, -i, *n., holiday, festival.*

festus, -a, -um, *festive, festal.*

fetura, -ae, *f., bearing of young, breeding.*

fetus, -a, -um, *pregnant, breeding.*

fetus, -us, *m., bringing to birth, young, litter.*

ficetum, -i, *n., fig-plantation.*

fictilis, -e, *made of clay.*

ficulneus, -a, -um, *of the fig-tree.*

ficus, -i, and -us, *f., fig-tree, fruit of the fig-tree.*

fides, -ei, *f., trust, confidence.*

figilinus (also **figlinus** and **figulinus**), -a, -um, *belonging to a potter, potter's.*

figura, -ae, *f., form, shape.*

figuro, I, *form, shape.*

fimus, -i, *m., manure, dung.*

fingo, finxi, finctum, 3, *form, fashion.*

finio, -ivi, or -ii, -itum, 4, *limit, enclose, bring to an end, cease.*

finis, -is, *m., bondary, pl., territory.*

finitimus, -a, -um, *neighbouring.*

fio, see **facio.**

firmo, I, *make firm, strengthen.*

fiscella, -ae, *f., small basket* (II. ii. 14).

fistula, -ae, *f., pipe.*

flabellum, -i, *n., dim.* (flabrum), *small fan.*

flaccus, -a, -um, *flabby, hanging down.*

flagellum, -i, *n., dim* (flagrum), *whip.*

flagrum, -i, *n., whip, lash.*
flammeus, -a, -um, *flaming.*
flatus, -us, *m., blowing, breathing.*
flo, flavi, flatum, 1, *blow.*
Floralia, -ium and -orum, *n., pl., festival of Flora* (I. 1. 6).
floreo, -ui, 2, *bloom.*
floresco, -ere, *begin to blossom.*
floridus, -a, -um, *flowery.*
flos, -oris, *m., flower.*
fluo, -xi, -xum, 3, *flow.*
focus, -i, *m., hearth.*
fodina, -ae, *f., pit, mine.*
fodio, fodi, fossum, 3, *dig.*
foetidus (also **fetidus**), *that has an evil smell, stinking.*
folium, -ii, *n., leaf.*
folliculus, -i, *m., dim.* (follis), *shell, husk, skin.*
foramen, -inis, *n., opening, hole.*
foras, *adv., out of doors, out.*
forensis, -e, *belonging to the market or forum.*
fores, -um, *f. pl., two leaves of a door.*
foris, *adv., out of doors, abroad.*
forma, -ae, *f., shape, beauty.*
formica, -ae, *f., ant.*
formidulosus, *fearful, terrible.*
formosus, -a, -um, *beautiful, handsome.*
formula, ae, *f. dim.* (forma), *beauty, mould, rule of evidence on which an enquiry is conducted, form of contract* (II. ii. 5).
fortasse, *adv., perhaps.*
forte, *adv., by chance, accidentally.*
forum, -i, *n., outside space, public place, market-place* (I. liv. 2).
fossa, -ae, *f., ditch, trench.*
fossicius, -a, -um, *dug out.*
foveo, fovi, fotum, 2, *keep warm, cherish.*
fracesco, fracui, 3, *become soft, rot.*
frango, fregi, fractum, 3, *break, break in pieces.*
freni, -orum, *m. pl., bridle, bit.*

frequens, -entis, *frequent, often.*
frigor, -oris, *m., cold.*
frigus, -oris, *n., cold.*
frondeo, -ere, *be in leaf.*
frondosus, -a, -um, *leafy.*
frons, also **fros,** -dis, *f., green bough, foliage.*
frons, frontis, *f., forehead, brow.*
fructuosus, -a, -um, *fruitful, productive.*
fructus, -us, *m., enjoyment, produce, fruit.*
frugalis, -e, *economical, thrifty.*
frugalitas, -tatis, *f., economy thrift.*
frugi, dat. form as *indecl., adj., honest, useful, see* **frux.**
frumentarius, -a, -um, *belonging to corn, corn-, provision.*
frumentum, -i, *n., corn, grain.*
fruor, fructus, 3, *dep., with abl., enjoy.*
frustra, *adv., in vain, to no purpose.*
frustratio, -onis, *f., deceiving, disappointment.*
frutex, -icis, *m., shrub, bush.*
frux, frugis, *f., more frequently, pl.* **fruges,** -um, *fruits of the earth, pulse, legumes.*
fucus, -i, *m., drone* (III. xvi. 19).
fuga, -ae, *f., flight, running away.*
fugio, fugi, fugitum, 3, *flee.*
fugitivus, -a, -um, *fugitive, runaway.*
fugo, 1, *put to flight.*
fullo, -onis, *m., fuller.*
fulmen, -inis, *n., lightning that sets on fire, thunderbolt.*
fumigo, 1, *smoke, fumigate* (III, xvi. 17).
fumus, -i, *m., smoke.*
fundamentum, -i, *n., foundation, ground-work.*
fundo, fudi, fusum, 3, *pour, shed.*
fundus, -i, *m., piece of land, farm, estate.*

funiculus, -i, *m. dim.* (funis), *slender rope, cord.*

fur, furis, *c., thief.*

furca, -ae, *f., two-pronged fork, instrument of punishment* (I. xx. 2).

furcilla, -ae, *f. dim.* (furca), *a small fork.*

furfur, -uris, and **furfures,** -um, *m., bran.*

furtum, -i, *n., theft, robbery.*

fuscus, -a, -um, *dark, dusky.*

G

galbulus, -i, *m., nut of the cypress-tree.*

Gallicani, -orum, *m., the Gallicans* (I. xxxii. 2).

Gallicus, -a, -um, *belonging to the Gauls.*

gallina, -ae, *hen.*

gallinaceus, *belonging to domestic fowls or poultry.*

gallinarius, *belonging to poultry.*

gallus, -i, *m., cock.*

gelum, -i, *n.,* **gelus,** -us, *m.,* **gelu,** *n., frost, cold.*

geminus, -a, -um, *born at the same time.*

gemini, -orum, *m. pl., twins.*

gemmare, I, *put forth buds, bud.*

generalis, -e, *general, relating to all.*

geno, genui, genitum, 3 (old form of **gigno**), *beget, bear, produce.*

gens, gentis, *f., race* or *clan* (embracing several families).

gentilis, -e, *of* or *belonging to the same clan, relative bearing the same name.*

genu, -us, *n., knee.*

genus, -eris, *n., birth, origin, race.*

geometra, -ae, *m.* (also **geometres**), *geometrician.*

gibberosus, -a, -um, *hunchbacked.*

gigno, see **geno.**

glaber, -bra, -brum, *smooth, bald.*

glacies, -ei, *f., ice.*

glaeba, -ae, *lump of earth, clod.*

glandarius, *of acorns* or *mast.*

glans, glandis, *f., acorn.*

glis, gliris, *m., dormouse.*

glomero, I, *form into a ball, gather into a round heap.*

globo, -ere, *remove bark, peel.*

gluma, -ae, *f., hull* or *husk.*

graculus, -i, *m., jackdaw.*

gradatim, *adv., step by step, gradually.*

gradus, -us, *m., step.*

Graece, *adv., in the Greek language.*

granarium, -i, *n., granary.*

grandis, -e, *large, abundant.*

granum, -i, *n., grain, seed.*

gratuitus, -a, -um, *that is done without pay, free.*

gratulor, I, *wish a person joy, congratulate.*

gravis, -e, *heavy, great, burdened.*

gravo, I, *load, burden.*

gregalis, -e, *belonging to the herd or flock.*

gregales, -ium, *m. pl., comrades.*

grex, gregis, *m., flock, herd.*

grus, gruis, *f., crane.*

gurgulio, -onis, *m., gullet.*

gustare, I, *taste.*

gutta, -ae, *f., drop of fluid* (III. xvi. 37).

gymnasium, -ii, *n., place for gymnastic exercises, gymnasium* (II *praef.* 2).

H

habito, I, *dwell in a place.*

haereo, haesi, haesum, 2, *hold fast, stick, cling.*

hara, -ae, *f., pen, coop.*

harena, -ae, *f., sand, place of combat in the amphitheatre, arena.*

harenaria, -ae, *f., sand-pit.*

harundinetum, -i, *n., thicket of reeds.*

harundo, -inis, *f., reed.*

haruspex, -icis, *m., soothsayer.*

helvus, -a, -um, *light bay.*

hemisphaerium, -ii, *n.,* *half-globe, hemisphere.*

herba, -ae, *f., grass, herbage, herb.*

herbidus, -a, -um, *full of grass, grassy.*

herbosus, -a, -um, *full of grass, grassy.*

hercle, vocative, *by Hercules !*

hereditas, -atis, *f., inheritance.*

hesperus, -i, *m., evening star.*

hexagonon (also **hexagonum,** -i, *n.*), *six-sided figure* (III. xvi. 5).

hiatus, -us, *m., opening, cleft.*

hiberno, I, *pass the winter.*

hibernus, -a, -um, *belonging to winter, wintry.*

hiematio, -onis, *f., passing the winter.*

hiems, -emis, *f., winter.*

hircus, -i, *m., he-goat, buck,* also **ircus,**

hirtus, -a, -um, *rough, hairy.*

hirundo, -inis, *f., swallow.*

histon, -onis, *m., place where a loom stands, weaving-room* (I. ii. 21) (-as, Greek accusative pl. form).

historicus, -i, *m., writer of history.*

holus, -eris, *n., kitchen or garden herbs, vegetables, especially cabbage* (I. xvi. 6).

honestus, -a, -um, *full of honour, honourable.*

honor, -oris, *m., honour, repute.*

hora, -ae, *f., hour.*

horda, -ae, *f., cow that is with calf* (II. v. 6).

hordeacius, -a, -um, *of or relating to barley.*

hordeum, -i, *n., barley.*

hordicidia, -ae, *f., sacrifice of a cow that is with calf* (archaic form of **fordicidia** (II. v. 6).

horologium, -ii, *n., clock,* either *sundial* or *water-clock.*

horreum, -i, *n., storehouse, granary.*

horridus, -a, -um, *rough, shaggy, bristly, wild.*

hortus, -i, *m., garden of any kind.*

hospes, -itis, *m., host, guest.*

hospitalis, -e, *relating to a guest or host, hospitable.*

hospitium, -ii, *n., hospitality.*

hostia, -ae, *f., victim for sacrifice.*

humanitas, -atis, *f., humanity, gentleness.*

humilis, -e, *low, humble.*

humus, -i, *f., earth, ground, soil.* **humi,** *on the ground.*

hypenemius, -a, -um, *containing wind, windy,* ova, *wind-eggs* (II. i. 19).

I

iaceo, -cui, -citum, 2, *lie, lie ill.*

iacio, -ieci, -iactum, 3, *throw, fling.*

iacto, I, *freq.* (iacio), *throw, hurl, toss about.*

iaculor, -atus, I, *dep., throw, cast.*

ianua, -ae, *f., door.*

ibidem, *adv., in the same place.*

ictus, -us, *m., blow, stroke, thrust.*

ideo, *adv., for that reason.*

idoneus, -a, -um, *fit, suitable.*

Idus, -uum, *f., one of the three days in each month from which the other days were reckoned in the Roman calendar: the Ides: they fell upon the fifteenth day in March, May, July, and October and on the thirteenth in the remaining months.*

ieiunus, -a, -um, *fasting, hungry.*

ignavus, -a, -um, *lazy, cowardly.*

ignis, -is, *m., fire.*

ignoro, I, *not to know, pay no attention to.*

ignotus, -a, -um, *unknown.*

illinc, *adv., from that place.*

illuc, *adv., to that place.*

ill, see also **inl-.**

imago, -inis, *f., imitation, likeness, statue.*

imbecillitas, -atis, *f., weakness.*

imbecillus, -a, -um, *weak, feeble.*

imber, -bris, *m., rain, rainstorm.*

imitatio, -onis, *f., imitation.*

imbuo, -ui, -utum, 3, *moisten, fill, imbue.*

imbutus, *p.p.p.* from imbuo.

imitor, I, *imitate, copy, portray.*

imm-, see also **inm.**

immanis, -e, *enormous, huge.*

imp, see also **inp.**

imperium, -ii, *command, power, authority.*

impero, I, *command.*

impetus, -us, *m., attack, onset.*

importo, I, *bring or carry in.*

improvisus, -a, -um, *unforeseen, unexpected.*

imprudentia, -ae, *f., want of foresight.*

impune, *adv., without punishment.*

imus, -a, -um, *lowest, last.*

inaequabilis, -e, *uneven, unequal.*

inanis, -e, *empty, useless.*

inaresco, -arui, 3, *dry up.*

inaro, I, *plough in.*

incedo, -cessi, -cessum, 3, *go, step, advance.*

incentivus, *that sets the tune* (I. ii. 15).

incertus, -a, -um, *uncertain, doubtful.*

incido, -cidi, -casum, 3, *fall, light upon, happen.*

incido, -cidi, -cisum, 3, *cut into, through.*

inciens, -entis, *pregnant, with young* (II. ii. 8).

incipio, -cepi, -ceptum, 3, *begin.*

inclinatus, *p.p.p.,* from inclino, *bent down, sunken.*

inclino, I, *cause to lean, turn back, bend.*

includo, -si, -sum, 3, *shut up* or *in.*

incolumis, -e, *uninjured, safe, whole.*

incomitatus, -a, -um, *unaccompanied.*

incommode, *adv., inconveniently, unfortunately.*

incontaminatus, -a, -um, *uncontaminated, pure.*

incubo, -ui, -itum, I, *lie in* or *upon, brood, hatch.*

incultus, -a, -um, *untilled, uncultivated.*

incunatus, *fouled, dirty* (III. xvi. 34).

incurvus, -a, -um, *bent, crooked, curved.*

inde, *adv., from that place, thereafter.*

index, -dicis, *c., pointer, indicator, mark, sign.*

indico, I, *point out, show.*

indigeo, -ui, 2, *need, want.*

indiscretus, -a, -um, *unseparated, closely connected.*

indivisus, -a, -um, *undivided.*

indomitus, -a,-um, *untamed, fierce.*

induco, -xi, -ctum, 3, *lead* or *bring in.*

ineo, -ivi, or -ii, -itum, *go into, enter.*

iners, -ertis, *unskilled, lazy, idle.*

inertia, -ae, *f., want of skill, idleness.*

infans, -fantis, *that cannot speak, young child, babe.*

inferior, -ius, *comp., lower.*

infero, intuli, illatum, *carry* or *bring into.*

infestus, -a, -um, *unquiet, unsafe.*

inficiens, -entis, *inactive* (III. xvi. 8).

inficio, -feci, -fectum, 3, *stain, dye.*

infimus, -a, -um, *lowest, last* (also **imus).**

infirmitas, -atis, *f., weakness.*

infirmus, -a, -um, *weak, feeble.*

influo, -xi, -xum, 3, *flow into.*

infra, *adv.* and *prep., below, underneath.*

infrio, I, *rub into, strew upon.*

infundo, -fudi, -fusum, 3, *pour in* or *upon.*

ingenium, -ii, *n., natural quality, natural disposition, talent.*

ingredior, -gressus, 3, *dep., go into, enter.*

inicio, -ieci, -iectum, 3, *throw in* or *on.*

inigo, -egi, -actum, 3, *drive into* or *to.*

inimicus, -a, -um, *hostile,*
inimicus, -i, *m., enemy.*

initium, -ii, *n., entrance, beginning.*

iniussus, -us, *m., only in abl., without command.*

inlicio, -lexi, -lectum, 3, *allure, entice.*

inlicium, -ii, *n., allurement, inducement.*

inlustris, -e, *clear, bright.*

inlustro, I, *light up, illuminate.*

inmitto, -isi, -issum, 3, *send into, let loose.*

inmolo, I, *sprinkle a victim with sacrificial meal, offer, sacrifice.*

innumerabilis, -e, *countless, innumerable.*

inopia, -ae, *f., want, lack.*

inp, see also **imp-.**

inpello, -puli, -pulsum, 3, *push, drive, strike against.*

inpendo, -di, -sum, 3, *weigh out.*

inpensa, -ae, *f., outlay, cost.*

inpleo, -evi, -etum, 2, *fill up.*

inplico, I, *infold, involve, clasp.*

inpono, -posui, -positum, 3, *put into.*

inquam, 3, *defective verb, I say.*

inquino, I, *foul, stain, pollute.*

inr, see also **irr.**

inrigatio, -onis, *f.* (also **irrigatio,** -onis, *f.*), *watering, irrigation.*

inrigo, I, *conduct water to a place, irrigate.*

inriguus, -a, -um, *watered, irrigated.*

inrito, I, *excite, tease.*

inscribo, -psi, -ptum, 3, *write in* or *upon.*

insector, I, *dep., pursue.*

insero, -sevi, -situm, 3, *sow in, plant in, graft.*

insiticius, -a, -um, *that is inserted,* somnus, *noon-day nap, siesta.*

inspicio, -spexi, -spectum, 3, *look into, consider, inspect.*

instituo, -ui, -utum, 3, *put or place into, establish, instruct.*

instrumentum, -i, *n., implement, apparatus.*

instruo, -xi, -ctum, 3, *build, erect, provide.*

insuavis, -e, *unpleasant.*

insum, -fui, *be in* or *upon, be contained in.*

insuo, -ui, -utum, 3, *sew in,* or *up in.*

integer, -tegra, -tegrum, *untouched, whole, uninjured.*

intego, -texi, -tectum, 3, *cover, protect.*

intellego, -exi, -ectum, 3, *perceive, understand.*

intentus, -a, -um, *strained, intent upon, eager.*

inter, *adv.* and *prep., between, among.*

intercedo, -cessi, -cessum, 3, *come between, intervene.*

interdiu, *adv., during the day, by day* (also **interdius,** II. x. 5).

interdum, *adv., sometimes.*

interea, *adv., in the meantime.*

intereo, -ii, -itum, 4, *go among, perish.*

interficio, -feci, -fectum, 3, *consume, kill.*

intericio, -ieci, -iectum, 3, *throw* or *place between.*

interior, -ius, *inner, interior.*

internodium, -ii, *n., space between two knots* or *joints.*

intero, -trivi, -tritum, 3, *rub into, bruise, crumble.*

interpello, I, *interrupt, disturb.*

interpono, -posui, -positum, 3, *place* or *set between.*

interputo, I, *prune here and there* (I. xxx).

intersum, -fui, *be between, differ.* **interest,** *impers., it makes a difference, it is of interest.*

intervallum, -i, *n., space between, interval of time.*

intorqueo, -torsi, -tortum, 2, *turn round, wrench.*

intra, *adv.* and *prep., on the inside, within.*

intrinsecus, *adv., on the inside, inwardly.*

intro, *adv., on the inside.*

intro, I, *enter.*

introeo, -ivi or -ii, -itum, 4, *go into, enter.*

introitus, -us, *m., going in, entrance.*

introrsum, *adv., inwards, into.*

intus, *adv., within.*

inuro, -ussi, -ustum, 3, *burn, brand.*

inusitatus, -a, -um, *unusual.*

inutilis, -e, *useless.*

invectus, -i, *m., means of transportation* (I. xvi. 2).

inveho, -vexi, -vectum, 3, *carry, bear, pass, ride.*

invenio, -veni, -ventum, 4, *come upon a thing, find, meet with.*

inventor, -oris, *m., discoverer, inventor.*

invideo, -vidi, -visum, 2, *look askance at, envy.*

invoco, I, *call upon, invoke.*

involo, I, *fly into* or *to a place.*

iocosus, -a, -um, *humorous, droll.*

ircus, see **hircus.**

irr, see also **inr-.**

iste, -a, -ud, *that near you, that of yours.*

Italicus, -a, -um, *of* or *belonging to Italy.*

Italicus, -i, *m., an Italian.*

Italus, -a, -um, *Italian.*

Italus, -i, *m., an Italian.*

item, *adv., likewise.*

iter, -itineris, *n., way, journey.*

itero, I, *repeat, do a second time.*

itidem, *adv., in the same way.*

iuba, -ae, *f., mane, crest of an animal.*

iubeo, iussi, iussum, 4, *bid, command.*

iucunditas, -atis, *f., pleasantness, enjoyment.*

iucundus, -a, -um (also **iocundus**), *pleasing, delightful.*

iudicium, -ii, *n., judgment, verdict, opinion* (II. ii. 6).

iugerum, -i, *n., acre,* or a *iuger* of land about two thirds the size of an English acre.

iuglans, -glandis, *f., walnut, walnut-tree.*

iugo, I, *bind to rails, yoke* (of vines) (I. viii. 1).

iugum, -i, *n., yoke for oxen, team of draught-cattle.*

iumentum, -i, *n., beast of burden, draught-animal.*

iuncetum, -i, *n., rush-bed* (I. viii. 3).

iuncus, -i, *m., a rush.*

iungo, -nxi, -nctum, 3, *join together.*

ius, iuris, *n., broth, soup.*

ius, iuris, *n., what is obligatory, right, justice, legal right or authority.*

iustus, -a, -um, *just, righteous.*

iuvencus, -a, -um, *young. bullock.*

iuventus, -utis, *f., youth.*

iuxta, *adv.* and *prep., near together.*

K

Kalendae, -arum, *f. pl., first day of the month.*

L

labor, lapsus, 3, *dep., fall, slide, sink.*

laboro, I, *labour, take pains, suffer.*

labrum, -i, *n., upper lip.*

lac, lactis, *n., milk.*

lacertosus, -a, -um, *muscular, powerful.*

lactarius, -a, -um, *belonging to* or *containing milk.*

lacuna, -ae, *f., ditch, hole, hollow, furrow* (I. xxix. 3).

lacus, -us, *m., basin, lake, reservoir.*

laedo, -si, -sum, 3, *hurt, injure.*

laetus, -a, -um, *joyful, happy.*

lampas, -adis, *f., torch, lamp,* (lampada, Greek accusative singular).

lana, -ae, *f., wool.*

lanaris, -e, *woolly, wool-bearing* (II. ix. 1).

lanio, 1, *tear in pieces, lacerate.*

lapideus, -a, -um, *of stone, stone-.*

lapicidinae, -arum, *f. pl., stone-quarries.*

lapidicinae,see **lapicidinae**

lapillus, -i, *m., little stone.*

lapis, -idis, *m., stone.*

lapsanum, -i, *n., edible plant, charlock* (III. xvi. 25).

largus, -a, -um, *abundant, large, much.*

lascivus, -a, -um, *playful, frisky.*

lateo, -ui, 2, *lurk, lie hid.*

latitudo, -inis, *f., breadth.*

latratus, -us, *m., barking.*

latrocinium, -ii, *n., robbery, fraud.*

latus, -a, -um, *broad, wide.*

latus, -eris, *n., side, flank.*

lavo, 1, *wash, bathe.*

laxus, -a, -um, *wide, loose, roomy.*

legaricus, -a, -um, *gathered by pulling* (I. xxxii. 2).

legitimus, -a, -um, *fixed by law, legitimate.*

lego, legi, lectum, 3, *gather, collect, pick out, read.*

legumen, -inis, *n., pulse, leguminous plant.*

lembus, -i, *m., small boat with sharp prow* (II. iii. 7).

lenis, -e, *soft, smooth.*

leniter, *adv., gently.*

lens, -tis, *f., lentil.*

lentiscus, -i, *f.,* also **lentiscum,** -i, *n., mastic-tree.*

lentus, -a, -um, *tough, sticky.*

leoninus, -a, -um, *belonging to a lion, lion's.*

leporarius, -a, -um, *belonging to a hare, a hare's.*

leporinus, -a, -um, *of a hare.*

lepus, -oris, *m., hare.*

lepusculus, *m., dim.* (lepus), *young hare, leveret.*

levis, -e, *light in weight, quick.*

lēvis, -e, *smooth.*

leviter, *adv., lightly.*

levo, 1, *lift up, relieve.*

libella, -ae, *f., dim.* (libra), *small silver coin, tenth part of a denarius.*

libellus, -i, *dim.* (liber), *small book, pamphlet.*

libenter, *adv.* (also **lubenter**), *willingly.*

liber, -era, -erum, *free, unrestricted,* **liberi,** *m. pl, children.*

liberalis, -e, *befitting a freeman, bountiful.*

libero, 1, *set free.*

libertus, -a, -um, *set free,* **libertus,** -i, *m., freedman.*

libido, -inis, *f.* (also **lubido**), *desire, passion.*

librarius, -a, -um, *of or containing a pound weight.*

libum, -i, *n., cake* of meal, made with milk or oil and spread with honey.

licet, -cuit, *impersonal, it is allowed* or *permitted.*

ligneus, -a, -um, *of wood, wooden.*

lignum, -i, *n., wood.*

ligurrio (also **ligurio**), *lick* (III. xvi. 6).

limes, -itis, *m., boundary.*

limito, 1, *enclose within boundaries.*

linea, -ae, *f., linen thread, string, line.*

lingo, -nxi, -nctum, 3, *lick.*

lingua, -ae, *f., tongue, speech, language.*

lino, levi (livi), litum, 3, and **linio,** -ivi, -itum, 4, *daub, smear, rub over.*

linum, -i, *n., flax.*

liro, 1, *plough* or *harrow in the seed* (I. xxix. 2).

lithostrotum, -i, *n., mosaic-work* (III. ii. 4).

littera, -ae, *f., letter, written sign.*

litterarius, -a, -um, *belonging to reading and writing.*

litus, -oris, *n., sea-shore, river-bank.*

loco, I, *place, set, arrange.*

loculus, -i, *m., dim.* (locus), *a small place, box, ballot-box* (III. v. 18).

locuples, -etis, *rich.*

longinquus, -a, -um, *long, extensive, distant.*

longitudo, -inis, *f., length.*

longurius, -ii *m., long pole* (also **longorius**).

loquor, -cutus, 3, *dep., speak, talk.*

lora, -ae, *f., wine made of the pressed skins of grapes, after-wine.*

lotus, *p.p.p.* from lavo.

lubenter, see **libenter.**

lucifer, -fera, -ferum, *light-bringing* (III. v. 17).

ludus, -i, *m., a play, game:* **ludi,** *m. pl., public games, spectacles, shows* (I. i. 6).

lumbus, -i, *m., loin, flank.*

lumen, -inis, *n., light, lamp.*

lunaris, -e, *of the moon, lunar.*

lupinus, -i, *m., lupin* (I. xiii. 3).

luridus, -a, -um, *pale yellow, sallow.*

lusciniola, -ae, *f., small nightingale.*

luscus, -a, -um, *one-eyed.*

lutulentus, -a, -um, *muddy, dirty.*

lutum, -i, *n., mud.*

luxuria, -ae, *f., luxuriance, extravagance.*

lympha, -ae, *f., water, spring-water.*

lyra, -ae, *f., lute, lyre, a constellation* (II. v. 12).

M

macellarius, -a, -um, *belonging to the meat-market.*

macellum, -i, *n., meat-market.*

macer, -cra, -crum, *lean, thin.*

maceria, -ae, *f., enclosure, wall.*

macero, I, *soak, soften by steeping.*

macesco, *grow thin* (III. xvi. 9).

magis, *adv., more, in a higher degree.*

magnitudo, -inis, *f., greatness, size.*

maialis, -is, *m., gelded boar, barrow, hog.*

maiestas, -atis, *f., greatness, dignity.*

Maius, -i, *m., month of May* (I. xxviii. I).

māla, -ae, *f., cheek-bone, jaw.*

maleficus, -a, -um, *evil-doing, wicked.*

malo, malui, *infin.,* malle, *prefer.*

malus, -a, -um, *evil, bad.*

mālus, -i, *f., apple-tree.*

mālus, -i, *m., mast, pole, beam.*

mamma, -ae, *f., breast, teat.*

mammosus, -a, -um, *full-breasted.*

mammula, -ae, *f. dim.* (mamma), *small breast* or *teat.*

mancipium, -ii, *n., legal, formal purchase of a thing* (II. x. 4), *property, slave.*

mando, I, *entrust, commend.*

mane, *adv., in the morning.*

manipulus, -i, *m., handful, bundle* (I. l. I), *company.*

mansuefacio, -feci, -factum 3, *tame, make softer, make gentle.*

manus, -us, *f., hand.*

marcesco, *wither, decay, pine away.*

mare, -is, *n., sea.*

margo, -inis, *m.* and *f., edge, brink.*

marinus, -a, -um, *of the sea, sea-.*

marisca, -ae, *f., large, inferior kind of fig* (I. vi. 4).

maritimus, -a, -um, *of the sea, sea-.*

marito, I, *marry, give in marriage to a man.*

marmoreus, -a, -um, *consisting of marble.*

mas, maris, *m.*, *male.*

materies, -ei, *f.*, *stuff, materials,* of which anything is composed, *timber.*

matertera, -ae, *f.*, *aunt on the mother's side.*

matrix, -icis, *f.*, *breeding animal.*

maturesco, -rui, 3, *ripen.*

maturus, -a, -um, *ripe.*

medeor, 2, *dep.*, *heal, remedy.*

medica, -ae, *f.*, *medic, alfalfa* (I. xxiii. 1).

medicina, -ae, *f.*, *healing* or *medical art, medicine.*

medicus, -a, -um, *pertaining to healing, medical.*

medicus, -i, *m.*, *medical man, doctor.*

mediocris, -e, *moderate, ordinary, mediocre.*

medius, -a, -um, *in the middle,* **medium,** -ii, *n.*, *the middle, midst.*

medulla, -ae, *f.*, *marrow of bones, pith of plants.*

mel, mellis, *n.*, *honey.*

melicus, -a, -um, *musical, tuneful.*

melior, -ius, *comp.* of bonus.

meliphyllon, -i, *n.*, also **melissophyllon,** -i, *n.*, *a plant of which bees are fond* (III. xvi. 10).

melitrophion, *n.*, *bee-hive* (III. xvi. 12).

melittaena, -ae, *f.*, *a bee-plant* (III. xvi. 10).

melitton, *n.*, *bee-hive* (III. xvi. 12).

melitturgus, -i, *m.*, *bee-keeper* (III. xvi. 3).

melium, -ii, *n.*, *dog's collar.*

mellarium, -ii, *n.*, *bee-hive.*

mellarius, -a, -um, *of honey, honey-.*

mellificium, -ii, *n.*, *making of honey.*

membrana, -ae, *f.*, *skin* or *membrane* that covers parts of the body (III. xvi. 32).

membrum, -i, *n.*, *limb, member.*

memini, -isse, *remember.*

memoria, -ae, *memory.*

mens, mentis, *f.*, *mind, disposition, heart, soul.*

mensis, is, *m.*, *mouth.*

menstruus, -a, -um, *of* or *belonging to a month, monthly.*

mentum, -i, *n.*, *chin.*

mercator, -oris, *m.*, *trader, merchant, dealer.*

mercennarius, -a, -um, *hired* (I. xvii. 2).

mercor, 1, *dep.*, *trade, purchase.*

meridianus, -a, -um, *of mid-day.*

meridies, -ei, *m.*, *mid-day.*

merula, -ae, *f.*, *blackbird.*

merus, -a, -um, *pure, unmixed.*

messio, -onis, *f.*, *reaping.*

messis, is, *f.*, *harvest.*

meta, -ae, *f.*, *mark at a boundary, conical columns at each end of Roman Circus, turning-post* (I. iii. 1).

metallum, -i, *n.*, *mine, quarry, metal.*

metior, mensus, 4, *dep.*, *measure.*

meto, -messui, 3, *reap, mow, crop.*

metuo, -ui, -utum, 3, *fear.*

miliarius, -a, -um, *containing a thousand.*

miliarius, -a, -um, *of millet, fed on millet* (III. v. 2).

militaris, -e, *of a soldier, military.*

militaris, -is, *m.*, *military man, soldier.*

milium, -ii, *n.*, *millet.*

mina, -ae, *f.*, *smooth,* mina ovis, *smooth-bellied, with no wool on the belly* (II. ii. 6).

minerval, -alis, *n.*, *gift in return for instruction* (III. ii. 18).

minister, -tri, *m.*, *servant.*

ministerium, -ii, *n.*, *service.*

ministra, -ae, *f.*, *female servant, handmaid.*

ministro, 1, *serve.*

minium, -ii, *n.*, *vermilion, red-lead, minium* (III. ii. 4).

minor, -us, *less, smaller, comp.* of parvus.

minuo, -ui, -utum, 3, *lessen, diminish.*

minusculus, -a, -um *dim.* (minus), *rather less, rather small.*

minutatim, *adv., piecemeal, gradually.*

miraculum, -i, *n., wonder, marvel, miracle.*

miror, 1, *dep., wonder* or *marvel at.*

mirus, -a, -um, *wonderful, marvellous.*

miscellus, -a, -um, *mixed.*

misceo, miscui, mixtum, 2, *mix, blend.*

miser, -era, -erum, *wretched, unfortunate, pitiable.*

mitto, misi, missum, 3, *send.*

mobilis, -e, *easy to be moved, movable.*

modicus, -a, -um, *keeping a proper measure, modest, temperate.*

modius, -ii, *m.,* also **modium,** -ii, *n.,* Roman corn-measure.

modo, *adv., only, merely.*

modulus, -i, *m., dim.* (modus), *small-measure, measure.*

modus, -i, *m., size, measure, bound, limit.*

mola, -ae, *f., millstone.*

molarius, -a, -um, *belonging to a mill* or *to grinding.*

moles, -is, *f., huge mass, bulk.*

molestus, -a, -um, *troublesome, irksome.*

mollis, -e, *supple, soft, tender.*

mollitia, -ae, *f., softness, tenderness.*

molo, -ui, -itum, 3, *grind in a mill.*

momentum, -i, *n., movement, motion.*

montanus, -a, -um, *of a mountain, mountainous.*

montuosus, -a, -um, *mountainous.*

morbidus, -a, -um, *sickly, diseased.*

morbosus, -a, -um, *ailing, diseased.*

morbus, -i, *m., disease.*

morior, mortuus, 3, *dep., die.*

moror, 1, *dep., delay, linger.*

morsus, -us, *m., biting, bite.*

morticinus, -a, -um, *dead, of an animal that has died of itself* (II. ix. 10).

morus, -i, *f., mulberry-tree.*

mos, moris, *m., manner, practice, custom.*

motus, -us, *m., moving, motion.*

moveo, movi, motum, 2, *move, set in motion.*

mula, -ae, *f., mule.*

mulgeo, -si, -sum, 2, *milk.*

mulier, -eris, *f., woman.*

mulsus, -a, -um, *mixed with honey.*

multa (mulcta), -ae, *f., fine, penalty.*

multicavatus, -a, -um, *with many hollows,* describing honeycomb (III. xvi. 24).

multiplex, -icis, *that has many parts, complicated.*

multitudo, -inis, *f., large number, crowd.*

multus, -a, -um, *many, many a.*

mulus, -i, *m., mule.*

mundus, -i, *m., universe, world.*

munio, 4, *fortify, defend.*

muria, -ae **(muries,** -ei), *f., salt liquor, brine, pickle.*

murus, -i, *m., wall,* especially city wall.

mus, muris, *c., mouse.*

Musa, -ae, *f., Muse* (goddess of poetry and other arts).

musca, -ae, *fly* (III. xvi. 6).

Museum, -i, *n., academy, study.*

mustum, -i, *n., new wine, must.*

muto, 1, *move, change.*

mutulus, -i, *m., mutule bracket* (in architecture) (III. v. 13).

mutus, -a, -um, *dumb, mute.*

N

nares, -ium, *f. pl., nostrils, nose.*

narro, 1, *tell, relate.*

nascor, natus, 3 *dep., be born.*

nasturtium, -ii, *n., kind of cress* (III. ix. 13).

natio, -onis, *f., birth.*

nativus, -a, -um, *born, that is produced by nature.*

nato, I, *swim, float.*

natura, -ae, *f., nature, natural property of a thing, disposition, organ of generation* (II. ii. 14).

naturalis, -e, *natural.*

natus, -us, *m., birth, growth.*

nauta, -ae, *m., sailor, mariner.*

navalia, -ium, *n. pl., dock, dockyard.*

navigo, I, *sail.*

navis, -is, *f., ship.*

nebulosus, -a, -um, *misty, cloudy.*

necdum, *and not yet, not yet.*

necessarius, -a, -um, *needful, necessary.*

necesse, neuter adj., *indispensable, necessary.*

neclego (neglego), -exi, -ectum, 3, *neglect, be indifferent to.*

neco, I, *kill.*

necto, -xui, -xum, 3, *bind, tie, fasten.*

necubi, adv., *that nowhere, lest anywhere.*

neglegenter, adv., *heedlessly, carelessly.*

nego, I, *deny.*

nemus, -oris, *n., wood with open glades for cattle, grove.*

nequa = ne qua.

nequeo, -ivi, -itum, 4, *not to be able.*

nequi, see **nequis.**

nequiquam, adv., *in vain, to no purpose.*

nequis = **ne quis,** *that no one.*

nescio, -ivi or -ii, -itum, *not to know, be ignorant.*

nescius, -a, -um, *ignorant.*

neuter, -tra, -trum, *of the neuter gender, neither, of two.*

nidus, -i, *m., nest.*

niger, -gra, -grum, *black, dark.*

nigrans, -antis, *black, dusky.*

nimbus, -i, *m., rain-storm.*

nimius, -a, -um, *excessive, too much.*

nitidus, -a, -um, *shining, bright.*

no, I, *swim, float.*

nobilis, -e, *noted, celebrated, noble.*

noceo, -cui, -citum, *hurt, injure.*

noctu, adv., *by night.*

nolo, nolui, nolle, *not to wish, be unwilling.*

nomen, -inis, *n., name.*

nomino, I, *call by name.*

Nonae, -arum, *f., the fifth day in every month of the year except March, May, July, and October in which it was the seventh.*

nonagesimus, -a, -um, *the ninetieth.*

nondum, adv., *not yet.*

nonne, adv., interrogative, expecting an affirmative answer, *not.*

nosco, novi, notum, 3, *get to know,* in perfect tense, *have learned, know.*

notus, -a, -um, *known.*

novalis, -is, *f.,* also **novale,** -is, *n., fallow land* (I. xxix. 1).

novellus, -a, -um, *dim.* (novus), *new.*

novenarius, -a, -um, *consisting of nine.*

novenus, -a, -um, *nine each.*

novi, see **nosco.**

novicius, -a, -um, *new* (I. xix. 2).

novies (-iens), *nine times.*

novo, I, *renew.*

novus, -a, -um, *new,* **novissimus,** -a, -um, *last, latest.*

noxa, -ae, *f., hurt, harm, injury.*

nubilarium, -ii, *n., shed* in which corn ready to be threshed was kept from the rain (I. xiii. 5).

nubilo, I, *be cloudy or overcast.*

nuculeus (nucleus), -i, *m., small nut* (I. xli. 6).

nudus, -a, -um, *naked, bare.*

nugatorius, -a, -um, *worthless, useless* (II. 5. 9).

nullus, -a, -um, *not any, none.*

nullus, -ius, *m., no one, nobody.*

num., *adv.*, interrogative particle usually implying that a negative answer is expected; in indirect speech, *whether.*

numero, I, *count, reckon.*

numerus, -i, *m., number.*

nummus, -i, *m., coin, money.*

nunquam or **numquam**, *adv., never.*

nuncubi, *adv., anywhere.*

nuper, *adv., lately, recently.*

nuptiae, -arum, *f. pl., marriage, wedding.*

nutricatus, -us, *m., suckling, nursing.*

nutrico, I, and **nutricor**, I, *dep., suckle, nourish.*

nutrix, -icis, *wet-nurse, nurse.*

nutus, -us, *m., nod.*

nux, nucis, *f., nut.*

O

ob., *prep., on account of, in consideration of.*

obaerarius, -ii, *m., debtor, one who must work off his debt* (I. xvii. 2).

obduco, -xi, -ctum, 3, *lead or draw before, lead against.*

obduresco, -rui, 3, *grow or become hard.*

obicio, -ieci, -iectum, 3, *throw or put before or towards, throw in the way.*

obligo, I, *bind, oblige, make liable.*

oblino, -levi, -litum, 3, *daub or smear over.*

oblitesco, -tui, 3, *hide one's self* (I. li. I).

oblivio, -onis, *f., forgetfulness, oblivion.*

oblongus, -a, -um, *rather long.*

obruo, -rui, -rutum, 3, *overwhelm, overthrow.*

obscurus, -a, -um, *dark, dusky.*

obsero, -sevi, -situm, 3, *sow, plant:* also **opsero.**

observo, I, *watch, take notice of.*

obsitus, *p.p.p.* of obsero.

obstruo, -xi, -ctum, 3, *build, block* or *wall up, hinder, obstruct.*

obstupesco, also **obstipesco**, -pui, 3, *be amazed.*

obsum, obfui, *be against, hinder.*

obtero, -trivi, -tritum, 3, *bruise, crush.*

obtineo, -tinui, -tentum, 2, *take hold of, hold.*

obvenio, -veni, -ventum, 4, *come* or *fall to one's lot.*

obvolvo, -vi, -utum, 3, *wrap around.*

occasus, -us, *m., going down, setting.*

occido, -cidi, -cisum, 3, *strike down, crush* (I. xxxi. I).

occido, -cidi, -casum, 3, *fall down, perish.*

occipitium, -ii, *n., back of the head* (I. viii. 6).

occo, I, *harrow* (I. xxix. 2).

occulco, I, *tread, trample down.*

occupo, I, *take possession of, seize.*

oceanus, -i, *m., the great sea that encompasses the land, the ocean.*

ocimum, -i, *n., fodder-plant, kind of clover* (III. xvi. 13, I. xxxi. 4).

ocinum, -i, *n., a mixed fodder crop* (I. xxiii. I).

octingenarius, -a, -um, *consisting of eight hundred* (II. x. 11).

octogeni, -ae, a, *distrib. num., eighty each.*

octogesimus, -a, -um, *eightieth.*

octoginta, *eighty.*

oculus, -i, *m., eye.*

odi, odisse, *hate.*

odium, -ii, *n., hatred, offence.*

odor, -oris, *m., smell, scent.*

odoratus, -a, -um, *that has a smell, fragrant.*

offa, -ae, *f., morsel, pellet* (III. v. 4).

offendo, -di, -sum, 3, *hit, strike.*

officio, -eci, -ectum, 3, *come in way of, hinder, obstruct.*

offigo, -xi, -xum, 3, *drive in, fasten.*

offringo, -egi, -actum, 3, *plough a second time* (I. xxix. 2).

offuca, -ae, *f. dim.* (offa), *little bit, small piece.*

olea, -ae, *f., olive, olive-berry, olive-tree.*

oleagineus, -a, -um, *of the olive tree.*

olearius, -a, -um, *of oil.*

oleo, -lui, 2, *smell, give out a smell.*

oleum, -i, *n., oil, olive-oil.*

olfacio, -eci, -actum, 3, *smell, scent, cause to smell of* (II. ii. 16).

olim, *adv., some time ago, formerly.*

olivetum, -i, *n., place planted with olive trees, oliveyard.*

ollula, -ae, *f., dim.* (olla), *a small pot.*

omnigenus, -a, -um, *of all kinds.*

omnino, *adv., altogether, entirely.*

onus, -eris, *n., load, burden.*

opera, -ae, *f., service, work, purpose, attention.*

operarius, -i, *m., labourer.*

operculum, -i, *n., cover, lid.*

operio, -ui, -ertum, 4, *cover.*

opifex, -icis, *c., worker, maker.*

opificium, -ii, *n., working, work* (III. xvi. 20).

opilio (also **upilio**), -onis, *m., shepherd.*

opimus, -a, -um, *rich, fat, splendid.*

opinor, 1, *dep., believe, think.*

oporotheca, -ae, also **oporothece,** -es, *f., place for storing fruit* (II *praef.* 2).

oportet, -uit, 2, *impers, it is needful or proper, must.*

oppidum, -i, *n., town.*

oppleo, -evi, -etum, 2, *fill up.*

oppono, -posui, -positum, 3, *set opposite, oppose.*

opportunus, -a, -um, *fit, convenient, suitable.*

opprimo, -essi, -essum, 3, *press against, overwhelm, crush.*

opsero, also **obsero.**

opturo, 1, also **obturo,** *stop up, close up.*

opulus, -i, *f., species of maple-tree* (I. viii. 3).

opus, -eris, *n., work, labour,* opus est mihi, *I have need of.*

ora, -ae, *f., border, rim, edge.*

orbiculus, -i, *m., dim.* (orbis), *small disk, or wheel* (I. lii. 1).

orbile (III. v. 15).

orbis, -is, *m., circle, round surface, world, universe.*

orca, -ae, *f., large-bellied vessel, butt.*

orcus, -i, *m., death* (I. iv. 3).

ordinatus, -a, um, *orderly, appointed.*

ordo, -inis, *m., regular row, order.*

oriens, see **orior.**

origa, see **auriga.**

origo, -inis, *f., earliest beginning, birth, origin.*

orior, ortus, 4 *dep., rise, become visible,* **oriens,** -tis, *m., rising sun, the East.*

ornamentum, -i, *n., equipment, trappings, decoration.*

ornatus, -us, *m., furnishing, adornment.*

ornithoboscion, *poultry-farm* (III. ix. 2), *poultry-yard.*

ornithon, -onis, *m., bird-house, poultry-house.*

ornithotrophion, *aviary* (III. 5. 8).

ortus, -us, *m., rising* of the heavenly bodies.

os, oris, *n., face, mouth.*

os, ossis, *n., bone.*

ostendo, -di, -sum, and -tum, 3, *expose to view, show.*

ostiarius, -ii, *m., door-keeper, janitor.*

ostium, -ii, *n., entrance, door-way.*

otiosus, -a, -um, *at leisure, unemployed, idle.*

otium, -ii, *n., leisure.*

oviarius, also **oviaricus,** -a, -um, *of sheep.*

ovillus, -a, -um, *of sheep.*

ovis, -is, *f., sheep.*

ovum, -i, *n., egg.*

P

pabulatio, -onis, *f.*, *pasture* (III. xvi. 21).

pabulor, I *dep.*, *feed*, *graze*.

pabulum, -i, *n.*, *food*, *fodder*.

paciscor, pactus, 3, *dep.*, *make a bargain* or *contract* (II. x. 5).

paeminosus, -a, -um, *full of chinks*, *uneven* (I. li. 1), also **peminosus**.

pala, -ae, *f.*, *spade*.

palaestra, -ae, *f.*, *wrestling school*, *place for exercise*.

palam, *adv.*, *openly*, *publicly*.

palea, -ae, *f.*, *chaff*.

palea, -ae, *f.*, also **palear,** -aris, *n.*, *dew-lap* of an ox, *wattles* of a cock.

palma, -ae, *f.*, *palm of the hand*, *palm-tree*, *palm-branch*, *victory*.

palmar, -e, *hand's breadth* (I. xxxv. 1).

palmus, -i, *m.*, *palm of the hand*, as measure of length, *a span*, *twelve digits*.

palumbus, -i, *m.*, *wood-pigeon*, *ring-dove*.

palus, -udis, *f.*, *swamp*, *marsh*.

paluster, -tris, -tre, *marshy*.

pampino, I, *prune off superfluous tendrils*, *shoots and leaves of vines*, *trim* (I. xxxi. 2).

pando, pandi, passum, 3, *spread out*.

panicum, -i, *n.*, *panic-grass*.

panis, -is, *m.*, *bread*.

papaver, -eris, *n.*, *poppy*.

papilla, -ae, *f.*, *dim.* (papula), *nipple*, *teat*.

par, paris, *equal*.

parens, -entis, *c.*, *parent*.

pareo, -ui, paritum, 2, *come forth*, *be present*, *obey*.

paries, -etis, *m.*, *wall*.

parilema (I. xxxi. 3).

pario, peperi, paritum, and partum, 4, *bring forth*, *bear*.

pariter, *adv.*, *equally*.

paro, I, *get ready*, *provide*, *get*.

pars, partis, *f.*, *part*.

parsimonia, -ae, *f.*, *frugality*, *thrift*.

particulatim, *adv.*, *piecemeal*, *severally*.

partim, *adv.*, *partly*, *in part*.

partio, -onis, *f.*, *bearing*, *laying of eggs*.

parturio, -ivi, or -ii, 4, *be in travail* or *labour*.

partus, -us, *m.*, *bringing forth*, *birth*.

parum, *adv.*, *too little*.

pascito, I, *frequentative*, *pasture*, *feed*.

pasco, pavi, pastum, 3, *feed*, *pasture*.

passim, *adv.*, *in every direction*, *everywhere*.

passus, -a, -um, *outspread*, *dried*, uva passa, *raisins* (III. xvi. 28) (*p.p.p.* of pando).

pastio, -onis, *f.*, *pasturing*, *grazing*.

pastor, -oris, *m.*, *shepherd*, *herdsman*, *countryman*.

pastoralis, -e, *of herdsmen* or *shepherds*, *pastoral*.

pastoricius, or **-tius,** -a, -um, *of a shepherd*.

pastus, see **pasco**.

pateo, -ui, 2, *lie open*, *be open*.

patior, passus, 3 *dep.*, *bear*, *endure*, *suffer*.

patria, -ae, *f.*, *fatherland*, *native place*.

patronus, -i, *m.*, *defender before a court of justice*, *advocate*.

patulus, -a, -um, *standing open*, *wide*.

pauci, -orum, *m. pl.*, *few*.

paulum, -i, *n.*, *little*, *trifle*.

paulum, *adv.*, *a little*, *somewhat*.

paulo, *adv.*, *by a little*, *somewhat*.

pauper, -peris, *poor*, *of small means*.

pauperculus, -a, -um, *dim.* (pauper), *poor*.

pavimentum, -i, *n.*, *beaten* or *hard floor*, *pavement*.

pavio, -ivi, -itum, 4, *beat*, *strike*, *ram down*.

pavo, -onis, *m.*, *peacock*.

pavoninus, -a, -um, *of a peacock*.

pectus, -oris, *n.*, *breast*, *chest*.

pecuarius, -a, -um, *of cattle*.

pecuarius, -ii, *m., grazier.*
peculiaris, -e, *relating to private property* (I. xvii. 7).
peculium, -ii, *n., property, private property.*
pecunia, -ae, *f., property, wealth.*
pecus, -oris, *n., cattle, herd, flock.*
pecus, -udis, *f., single head of cattle, animal, one of a herd.*
pedalis, -e, *belonging to the foot.*
pedamentum, -i, *n., stake, prop* (I. viii. 4).
pelagius, -a, -um, *belonging to the sea* (III. iii. 10).
pelagus, -i, *n., the sea, open sea.*
pellis, -is, *f., skin, pelt.*
pellitus, -a, -um, *covered with skins, jacketed* (II. ii. 18), *of sheep covered with skins to protect their fine wool.*
pendeo, pependi, 2, *hang, be suspended.*
pendo, pependi, pensum, 3, *suspend, weigh, pay.*
pendulus, -a, -um, *hanging down.*
pensilis, -e, *hanging down.*
penuria, also **paenuria,** -ae, *f., need, scarcity.*
per, *prep., through, on account of.*
peraeque, *adv., quite equally.*
perambulo, I, *ramble, go through.*
peraresco, -arui, 3, *become very dry* (I. xlix. 1).
percalefacio, -feci, -factum, 3, *pass.,* **percalefio,** -factus, *make very warm, heat thoroughly.*
percenseo, -ui, 2, *go through or over, reckon up.*
percurro, -cucurri, or -curri, -cursum, 3, *run or hurry through, pass over.*
percutio, -cussi, -cussum, 3, *strike through and through, kill, beat* (I. lv. 3).
perduco, -xi, -ctum, 3, *lead or bring through, guide.*
perennis, -e, *that lasts the year through, perpetual, perennial.*

pereo, -ii, or -ivi, -itum, *pass away, be lost, perish.*
perfectus, see **perficio,**
perfero, -tuli, -latum, *bear or carry through.*
perferus, -a, -um, *very wild or savage.*
perficio, -feci, -fectum, 3, *achieve, carry out, complete.*
perfluo, I, *blow through.*
perfrico, -cui, -catum, and -ctum, I, *rub all over, scratch.*
perfrigesco, -frixi, 3, *grow very cold, catch cold* (II. ix. 13).
perfundo, -fudi, -fusum, 3, *pour over, wet, moisten.*
perhibeo, -ui, -itum, 2, *grant, give, attribute, assert.*
peripetasmata, -um, *n. pl. coverings, curtains, carpets.*
peripteros, -on, *surrounded with a row of columns on the outside, pergola* (II, *praef.* 2).
peristylon, see **peristylum.**
peristylum, -i, *n., part of the building enclosing the court-yard or garden surrounded by columns on the inside* (III. v.1).
peritus, -a, -um, *experienced, skilled.*
perluceo, -xi, 2, *shine through, appear, be transparent.*
permagnus, -a, -um, *very great, very large.*
permaneo, -mansi, -mansum, 2, *to stay to the end, endure, remain.*
permultus, -a, -um, *very much, very many.*
permuto, I, *change completely, change position of* (I. iv. 5).
perna, -ae, *f., ham, gammon.*
pernicies, -ei, *f., destruction, calamity.*
pernocto, I, *pass the night.*
perpetuus, -a, -um, *continuous, uninterrupted.*
perpolitus, -a, -um, *thoroughly polished.*
perpurgo, I, *cleanse thoroughly.*
perscribo, -psi, -ptum, 3, *write at length, write out.*
persequor, -cutus, 3 *dep., continue to follow.*

persero, -sevi, 3, *sow* or *plant.*

perterreo, -ui, -itum, 2, *frighten thoroughly.*

pertica, -ae, *f., pole, long staff.*

pertimesco, -mui, 3, *be greatly afraid.*

pertinax, -acis, *that holds fast, tenacious.*

pertineo, -ui, 2, *stretch out, reach, concern.*

pertundo, -tudi, -tusum, 3, *bear, bore through.*

pertusus, -a, -um, *p.p.p.* of pertundo, *perforated, that has an opening.*

perungo, or **-unguo,** -unxi, -unctum, 3, *besmear, anoint.*

pervenio, -veni, -ventum, 4, *come to, arrive, attain.*

pes, pedis, *m., foot.*

pestilens, -entis, *infected, unhealthy.*

pestilentia, -ae, *f., contagious disease, plague.*

pestis, -is, *f., contagious disease, pest.*

peto, -ivi, and -ii, -itum, 3, *attack, seek, ask.*

philosophus, -i, *m., philosopher, lover of wisdom and knowledge.*

Phoenicia, -ae, *f., Phoenicia.*

physicus, -a, -um, *belonging to natural philosophy, natural.*

picatus, -a, -um, *p.p.p.* of pico, I, *covered with pitch* (I. liv. 2).

pictor, -oris, *m., painter.*

pila, -ae, *f., ball.*

pilosus, -a, -um, *hairy, shaggy.*

pilus, -i, *m., hair.*

pinacotheca, -ae, also **pinacothece,** -es, *f., picture-gallery.*

pingo, pinxi, pinctum, 3, *paint.*

pinguis, -e, *fat, rich, fertile.*

pinguitudo, -inis, *f., fatness, richness.*

pinna, also **penna,** -ae, *f., feather, pl., wing, plumage.*

pinso, pinsi and pinsui, pinsum, 3, *beat, pound.*

pinus, -us, *f., pine-tree.*

piraticus, -a, -um, *of pirates, piratical.*

pirum, -i, *n., pear.*

pirus, -i, *f., pear-tree.*

piscator, -oris, *m., fisherman.*

pisciculus, -i, *m., dim.* (piscis), *little fish.*

piscina, -ae, *f., fish-pond.*

piscis, is, *m., fish.*

pisum, -i, *n., leguminous plant, pea* (III. xvi. 13).

placenta, -ae, *f., cake.*

placidus, -a, -um, *gentle, quiet, calm.*

plaga, -ae, *f., blow, stripe.*

planta, -ae, *f., sprout, shoot, graft, cutting* (I. xl. 4 and I. lv. 3).

planus, -a, -um, *even, flat.*

platanus, -i, *f., plane-tree.*

plaustrum, -i, *n., wagon, cart.*

plausus, -us, *m., clapping of hands* (III. xvi. 7).

plebs, -is, *f., the common people, the plebians.*

plenus, -a, -um, *full, filled with.*

plerusque, -raque, -rumque, *very many, the most;* generally in plural: **plerumque,** *adv., for the most part.*

plostellum, -i, *n., dim.* (plaustrum), *small wagon or cart* (I. lii. 1).

pluma, -ae, *f., small soft feather, down.*

plumula (III. v. 11).

pluo, plui, 3, used impersonally, *rain.*

plurimus, -a, -um, superlative of multus, *most.*

plus, pluris, comparative of multus, *more.*

plusculum, *adv., rather more, too much* (II. vii. 10).

pluvius, -a, -um, *rainy, bringing rain.*

poena, -ae, *f., punishment, penalty.*

Poenicus (also **Punicus),** -a, -um, *Punic, Carthaginian.*

poeta, -ae, *f., poet.*

polenta, -ae, *f., peeled barley.*

politus, -a, -um (*p.p.p.* of polio), *polished, refined.*

pomarium, -ii, *n., fruit garden, orchard.*

pomerium, also **pomoerium,** -i, *n., the open space within and without the walls of a city, bounded by stones and limiting the city auspices.*

pompa, -ae, *f., solemn or public procession* (I. ii. 11).

pomum, -i, *n., fruit-tree of any kind.*

ponderosus, -a, -um, *weighty, heavy.*

pondo, *adv., by weight, in weight.*

pondus, -eris, *n., weight.*

pono, posui, positum, 3, *put, place, set.*

pons, -ntis, *m., bridge.*

populus, -i, *m., people, the people.*

populus, -i, *f., poplar-tree.*

porca, -ae, *f., sow, ridge between two furrows* (I. xxix. 3).

porcus, -i, *m., hog, pig.*

porricio, -eci, -ectum, 3, *lay before, offer as sacrifice* (I. xxix. 3).

porro, *adv., forward, at a distance, afar off.*

portendo, -di, -tum, 3, *point out, foretell.*

porticus, -us, *f., walk covered by a roof supported on columns, colonnade, arcade.*

portio, -onis, *f., share, portion.*

posco, poposci, 3, *beg, request.*

post, *adv. and prep., behind, after, afterwards.*

postea, *adv., after this or that, afterwards.*

posteaquam, *adv., after, afterwards.*

posterior, posterius (*comp.* of posterus), *that comes or follows after, next in order, later.*

posterus, -a, -um, *coming after, following, future.*

postridie, *adv., on the day after, on the following day.*

postulo, 1, *ask, demand.*

potestas, -atis, *f., ability, power.*

potio, -onis, *f., drink, draught.*

potior, -itus, *dep., with abl., become master of, take possession of, get.*

potis, pote, *able, capable.*

poto, 1, -avi, -atum or potum, *drink.*

potus, *p.p.p.* of poto, *that has drunk, drunk.*

potus, -us, *m., drinking, draught.*

praebeo, -ui, -itum, 2, *hold forth, offer, grant, show.*

praeceptum, -i, *n., rule, precept, command.*

praecido, -cidi, -cisum, 3, *cut off in front, cut off* (I. xl. 6).

praecipio, -cepi, -ceptum, 3, *take or seize beforehand, anticipate, advise, recommend.*

praeclarus, -a, -um, *very clear or bright.*

praecox, -cocis, *ripe before its time, early ripe, premature.*

praeda, -ae, *f., spoil, plunder.*

praedium, -ii, *n., farm, estate.*

praeeo, -ivi, and -ii, -itum, *go before, lead the way.*

praefectus, -i, *m., overseer, director, chief, prefect.*

praegnas, -atis, also **praegnans,** -antis, *pregnant.*

praegnatio, -onis, *f., pregnancy.*

praemium, -ii, *n., profit derived from booty, booty, profit, advantage.*

praeparo, 1, *prepare, equip, provide.*

praepono, -posui, -positum, 3, *put or set before, place at the head of.*

praescribo, -psi, -ptum, 3, *write before, order, direct, prescribe* (II. ii. 5).

praesaepe (praesepe), -is, *n., enclosure, stable, fold.*

praesaepes or **praesaepis,** -is, *f., enclosure, stable, fold.*

praesens, -tis, *present, current* (from praesum).

praesertim, *adv., especially, chiefly.*

praesico, also **praeseco,** -cui, -ctum, or -catum, I, *cut off before, cut off,* or *out.*

praesidium, -ii, *n., protection, help.*

praesto, -iti, -atum, or -itum, I, *stand in front, be superior.*

praesto, *adv., ready, present.*

praesum, -fui, *be before, preside, rule over.*

praeter, *adv.* and *prep., beyond, in addition to.*

praeterea, *adv., beyond this* or *that, besides, thereafter.*

praetereo, -ivi or -ii, -itum, *go past, pass away, perish.*

praeterfluo, 3, *flow by* or *past.*

praetermitto, -misi, -missum, 3, *allow to go by, omit.*

praeterquam, *adv., beyond, except.*

praetor, -oris, *m., leader, Roman magistrate who had charge of administration of justice.*

pratum, -i, *n., meadow.*

precor, I, *dep., ask, entreat, pray.*

prehendo, -di, -sum, 3, *grasp, seize.*

prelum, -i, *n., press-beam for pressing olives, grapes, wine* or *olive press* (I. liv. 2).

premo, -essi, -essum, 3, *press.*

prendo, see **prehendo.**

pretium, -ii, *n., money, worth, price, value.*

primatus, -us, *m., first preference* (I. vii. 10).

primigenius, -a, -um, *first of its kind, primitive.*

primitus, *adv., at first, originally.*

primoris, -e, *first, earliest, foremost.*

primus, -a, -um, *first,* **primum,** *adv., at first,* **primo,** *adv., at first, firstly.*

princeps, -cipis, *m., first man, principal, leader.*

principium, -ii, *n., beginning, origin.*

prior, -oris, *former, first,* **prius,** comparative *adv., before, sooner, first.*

priscus, -a, -um, *belonging to former times, olden, primitive.*

priusquam, *before that, before.*

privatus, -a, -um, *belonging to an individual, private, apart from the state.*

pro, *prep., before, in front of, on behalf of.*

proavos, also **proavus,** -i, *m., great grandfather.*

probatio, -onis, *f., proving, trial, approval, consideration* (I. xx. 1).

probo, I, *try, test, be satisfied with, approve.*

procedo, -cessi, -cessum, 3, *go forth* or *before, advance.*

procer, -eris, *m., chief.*

procerus, -a, -um, *tall, high.*

proclivis, -e, and **proclivus,** -a, -um, *sloping, going downhill.*

procoeton, -onis, *m., anteroom* (II *praef.* 2) (Greek accusative, ona).

procreo, I, *bring forth, create, produce.*

procuratio, -onis, *f., charge, superintendence, management.*

prodeo, -ii, -itum, *go forth, appear.*

prodigo, -egi, -actum, 3, *drive forth,* or *to a place.*

prodo, -didi, -ditum, 3, *give* or *put forth, produce, publish.*

proelior, I, *dep., join battle, fight.*

profectus, see **proficiscor.**

profero, -tuli, -latum, *carry* or *bring out, offer.*

profestus, -a, -um, *not kept as a holiday, working.*

proficio, -feci, -fectum, 3, *go forward, advance, perform, accomplish, be of service.*

proficiscor, -fectus, 3 *dep., set out, march.*

profiteor, -fessus, 2 *dep., declare publicly, acknowledge.*

profundo, -fudi, -fusum, 3, *pour out, shed copiously,* **profusus,** *p.p.p., spread out, extravagant.*

progenies, -ei, *f., descendants, posterity.*

progredior, -gressus, 3 *dep., come* or *go forth, advance.*

prohibeo, -ui, -itum, 2, *hold back, hinder, ward off.*

proinde, *adv., in the same manner, just, even, accordingly.*

prolixus, -a, -um, *stretched far out, long* (II. ii. 3), *prolix.*

promitto, -misi, -missum, 3, *send* or *put forth, foretell, promise, assure.*

promo, -mpsi, -mptum, 3, *bring out, produce.*

promum, -i, *n., store-room.*

promus, -i, *m., steward, butler* (I. xvi. 5).

pronus, -a, -um, *leaning forward, stooping.*

propago, I, *fasten down, propagate by slips* (III. iii. 10).

prope, *adv.* and *prep., near, nearby, about.*

propello, -puli, -pulsum, 3, *drive before one's self, drive forward.*

propemodum, *adv., nearly, almost.*

propero, I, *do with haste, hurry.*

propinquus, -a, -um, *near, neighbouring.*

propitius, -a, -um, *favourable, gracious, kind.*

propolis, -is, *f., a glue-like substance with which the bees close the crevices of their hives, propolis* (III. xvi. 23).

propono, -posui, -positum, 3, *put* or *set forth, display, imagine.*

proprius, -a, -um, *one's own, special.*

propter, *adv.* and *prep., near, close to, on account of* (III. xvi. 14).

propterea, *adv., for that cause, on that account.*

propugno, I, *rush out to fight, contend for.*

propulso, I, *keep or ward off, repulse.*

proruo, -rui, -rutum, 3, *throw down, hurl to the ground, rush forth.*

proscindo, -scidi, -scissum, 3, *rend, split* (I. xxxii. 1).

prospere, *adv., favourably, fortunately.*

prospicio, -exi, -ectum, 3, *look forward, look out, watch.*

prostratus, see **prosterno.**

prosterno, -stravi, -stratum, 3, *strew in front of, throw down,* **prostratus,** *p.p.p. prostrate.*

prosum, -fui, *be of use to, benefit, profit.*

protectum, -i, *n., projecting part of the roof* (III. xvi. 23), *eaves, gable.*

protelum, -i, *n., sign, signal* (III. xvi. 12).

proverbium, -ii, *n., proverb, maxim.*

provideo, -vidi, -visum, 2, *act with foresight, foresee, care for.*

proximus, -a, -um, *nearest, next.*

pruina, -ae, *f., hoar-frost.*

psittacus, -i, *m., parrot.*

publicanus, -a, -um, *farmer-general of the Roman revenues.*

publicus, -a, -um, *of the people, public.*

pulcher, -chra, -chrum, *beautiful, handsome.*

pulex, -icis, *m., flea.*

pullitribus, *dat., pl.* (III. ix. 9).

pullus, -i, *m., young animal, foal, pullet.*

pulverulentus, -a, -um, *full of dust, dusty.*

pulvinus, -i, *m., cushion, pillow, flower-bed* (I. xxxv. 1).

pulvis, -eris, *m., dust, powder.*

pungo, pupugi, punctum, 3, *pierce into, penetrate, sting, grieve, annoy.*

Punicus, -a, -um, *Punic, Carthaginian.*

purgatio, -onis, *f., cleansing, purgation.*

purgo, I, *cleanse, purify.*

purus, -a, -um, *clean, pure, unmixed.*

putamen, -inis, *n., that which falls off in pruning or trimming, clippings, shell of nuts* (I. vii. 3).

349

putatio, -onis, *f., pruning of trees* (I. vi. 5).

puter, also **putris,** -tris, -tre, *rotten, decaying.*

puto, I, *cleanse, trim or prune trees or vines, reckon, consider, reflect.*

putrefacio, -feci, -factum, 3, *pass.* **putrefio,** -factus, *make rotten, become rotten, putrefy.*

putresco, 3, *grow rotten, decay.*

Q

qua, *adv., in which place, where, by what way, by what manner.*

quaad, see **quoad.**

quadragesimus, -a, -um, *fortieth.*

quadraginta, *forty.*

quadrans, -antis, *m., a fourth part, quarter,* as a coin, *fourth part of an as,* as a weight, *quarter of a pound,* as a liquid measure, *the fourth part of* a sextarius, three cyathi.

quadratus, -a, -um, *squared, square.*

quadriduum, -ii, *n., space of four days.*

quadrigae, -arum, *f. pl., team of four, four-horse chariot.*

quadrigarius, -ii, *m., one who drives a four-horse chariot, chariot-racer.*

quadrimestris, -e, *of four months.*

quadrimus, -a, -um, *of four years.*

quadringenti, -ae, -a, *four hundred.*

quadripes (also **quadrupes**), -edis, *m., f.,* and *n., quadruped, four-footed creature.*

quadripertior, -iri, 4, *divide into four parts.*

quaero, -sivi, or -sii, situm, 3, *seek, procure, ask.*

quaeso, see **quaero.**

quaestio, -onis, *f., inquiry, question.*

qualis, -e, *of what sort, of such a kind, such as,* often with correlative *talis.*

quam, *adv., in what manner, how much,* in comparisons, *as, than,* with superlative and *tam, the more, the more* (II. ix. 12).

quamvis, *adv.* and *conj., as you will, very much, however much.*

quando, *adv., at what time? when? conj., when, since.*

quantus, -a, -um, *how great, how many:* corresponding with *tantus.*

quaque, *adv., wherever.*

quare, *adv., by which means, whereby.*

quartus, -a, -um, *fourth.*

quasi, *adv., as if, as it were.*

quatio, quassum, 3, *shake, wield, shatter.*

-que, *enclitic* conj., *and, and so,* repeated, *-que . . . -que, both and.*

quemadmodum or **quem ad modum,** *adv., in what manner, how.*

queo, -ivi, and -ii, -itum, 2, *be able.*

quercus, -us, *f., oak-tree* (I. vi. 4).

queror, questus, 3, *dep., complain.*

querquetum, also **quercetum,** -i, *n., oak-wood, oak-forest.*

quia, *conj., because.*

quicumque, quaecumque, quodcumque, *whoever, whatever, all that.*

quidam, quaedam, quoddam, and noun quidam, *a certain, somebody, something.*

quidem, *adv., indeed, but, yet,* ne . . . quidem, *not even.*

quies, -etis, *f., rest, quiet.*

quietus, -a, -um, *at rest, calm, quiet.*

quin, *conj. interrog. particle, why not?* as *relative, who . . . not, that not,* = Engl. *without* and *participial clause,* especially after verbs of *doubting, hesitating.*

quincunx, -uncis, *m., five twelfths* of a whole, in money, land measurement, weight, liquid measure: *trees planted*

in the form of a quincunx
(I. vii. 2).

quini, -ae, -a, *five each.*

quinquagenarius, -a, -um,
consisting of, or *containing,*
fifty.

quinquageni, -ae, a, *fifty*
each.

quinquaginta, *fifty.*

quis, quid, *interrogative pro-*
noun, who? which? what?

quis, quid, *indefinite pronoun,*
especially after si, ne, nisi, any
one, any thing, somebody, some-
thing.

quisquam, quaequam, quicquam, or quidquam, *any one,*
any thing, something.

quisque, quaeque, quodque,
and *noun,* quicque, *whoever,*
whatever, each, every, with
superlative, *all the.*

quo, *adv., for which reason,*
because, whither, where, for
what purpose, why.

quoad (also **quaad** I. i. 2),
as long as, until, as far as.

quocirca, *conj., to whatever*
place.

quod, *conj., because, that.*

quom, see **cum.**

quoniam, *adv., since then,*
seeing that.

quoque, *conj., also, too.*

quoquo, *adv., to whatever*
place.

quot, *indecl., how many, as*
many as, every.

quotannis, *adv., every year,*
annually.

quotienscumque, *adv., as*
often, so ever as.

quotquot, *indecl., as many*
so ever as.

quotusquisque, *how few,*
sometimes separated as *quotus*
quisque.

R

radicitus, *adv., by the roots,*
completely.

radio, I, *furnish with spokes;*
radiatus, -a, -um, *p.p.p. furn-*
ished with rays, shining.

radius, -ii, *m., staff, rod,*
pointer (III. v. 17).

radix, -icis, *f., root.*

raeda, or **reda,** -ae, *f.,*
travelling-carriage: four-wheeled
wagon (II. vii. 15).

ramulus, -i, *m., dim.* (ramus)
small branch, twig.

ramus, -i, *m., branch, bough.*

rana, -ae, *f., frog.*

raphanus, -i, *m., radish.*

rapum, -i, *n., also* **rapa,** -ae,
f., turnip, rape.

raro, *adv., rarely, seldom.*

rarus, -a, -um, *thin, scattered*
scanty, scarce.

rastellus, -i, *dim.* (rastrum),
hoe, rake, mattock (I. xlix. 1).

rastrum, -i, *n., usually pl.,*
rastri, -orum, *m., toothed hoe,*
rake.

ratio, -onis, *f., reckoning,*
transaction, regard, judgment,
reason, understanding (III. xvi.
4).

ratiocinor, I *dep., reckon,*
consider, reason, argue.

ravus, -a, -um, *greyish-*
yellow, tawny (II. ii. 4).

recens, -entis, *fresh, young,*
recent.

recipero, I, *regain, recover*
one's self (I. xiii. 1).

recipio, -cepi, -ceptum, 3,
take back, regain, receive, accept.

reclino, I, *bend back, lean*
back, recline.

reconcilio, I, *bring together*
again, reconcile.

recte, *adv., in a straight line,*
rightly, correctly.

rectus, -a, -um, *p.p.p.* of
rego, *straight, upright.*

recuso, I, *decline, reject,*
refuse.

reddo, -didi, -ditum, 3, *give*
back, return.

redeo, -ii, -itum, *go* or *come*
back, return.

redigo, -egi, -actum, 3,
drive or lead back.

reduco, also **redduco,** -xi,
-ctum, 3, *bring or conduct back.*

refero, rettuli, relatum, *bear,*
carry, or *bring back, return,*
pay back, relate.

refert, -tulit, impersonal, *it is for one's interest or advantage, it matters.*

reficio, -feci, -fectum, 3, *make anew, restore, refresh.*

refrigero, 1, *make cold, cool off* (II. ii. 11).

regelo, 1, *thaw, warm.*

regie, *adv., royally, splendidly.*

regio, -onis, *f., boundary-line, limit, region, territory.*

regredior, -gressus, 3 *dep., go* or *come back.*

regula, -ae, *f., straight piece of wood, stick, rule, pattern.*

regulus, -i, *m. dim.* (rex), *ruler of a small country, the king-bee* (III. xvi. 18).

reicio, reieci, -iectum, 3, *throw* or *fling back, reject, refuse, despise.*

reiculus, -a, -um, *that is to be rejected, worthless* (II. v. 17).

relaxo, 1, *stretch out, widen, unloose, ease.*

relinquo, -liqui, -lictum, 3, *leave behind, abandon.*

reliquus, -a, -um, *that is left,* or *remains, remaining.*

remaneo, -mansi, -mansum, 2, *remain behind, be left.*

remedium, -ii, *n., cure, remedy.*

remeo, 1, *go* or *come back, return.*

remissio, -onis, *f., slackening, relaxing.*

remitto, -misi, -missum, 3, *send back, loosen relax, restore.*

removeo, -movi, -motum, 2, *move* or *draw back, withdraw.*

remuneror, 1, *dep., repay, recompense.*

renuntio, 1, *report, declare.*

reor, ratus, 2 *dep., reckon, think.*

repentinus, -a, -um, *sudden, unexpected.*

reperio, repperi, repertum, 4, *find, meet with, learn.*

repertor, -oris, *m., discoverer, inventor* (I. ii. 19).

repeto, -ivi, or -ii, -itum, 3, *fall upon again, seek again, fetch, back, repeat.*

repleo, -evi, -etum, 2, *refill, fill up.*

repono, -posui, -positum, 3, *put back, restore.*

reprehendo, -di, -sum, 3, *hold back, seize, pass judgment on.*

repudio, 1, *cast off, scorn.*

requies, -etis, *f., rest, relief.*

requiesco, -evi, -etum, 3, *rest.*

requiro, -sivi, or -sii, -situm, 3, *seek again, search for, need.*

requisitio, -onis, *f., searching, examination, questioning.*

resico, 1, *cut back,* more usually **reseco** (I. xxxi. 2).

resimus, -a, -um, *turned up, bent back.*

resipio, *taste* or *smack of.*

resisto, -stiti, 3, *withstand, resist.*

resono, 1, *resound, re-echo.*

respecto, 1, *look* or *gaze about.*

respondeo, -di, -sum, 2, *answer, agree.*

restibilis, -e, *that is restored, of a field that is sown or tilled every year* (II. vii. 11).

resticula, -ae, *f., dim.* (restis), *small rope.*

restis, -is, *f., rope, cord.*

restituo, -ui, -utum, 3, *set up again, replace, restore.*

rete, -is, *n.* (*f.* III. v. 11), *net.*

retectus, *p.p.p.* retego.

retego, -xi, -ctum, 3, *uncover, open.*

reticulus -i, *m., netting* (III. v. 13).

retineo, -ui, -tentum, 2, *keep back, restrain, detain.*

retinnio, 4, *ring again, resound.*

revertor -versus, 3, *come back, return.*

revivesco (also **revivisco**), -vixi, 3, *come to life again.*

revoco, 1, *call back, recall.*

revolo, 1, *fly back.*

ricinus, -i, *m., vermin that infests sheep, tick.*

ridica, -ae, *f., stake, prop, vine-prop* (I. viii. 4).

rima, -ae, *f., chink, fissure.*

ripa, -ae, *f., bank* of a stream.

rivolus (also **rivulus**), -i, *m. dim.* (rivus), *small stream, rivulet.*

rivus, -i, *m., stream.*

rixo, I, *quarrel, dispute.*

robeus, see **rubeus.**

robigo, -inis, *f., rust, mouldiness on grain* (I. i. 6).

robustus, -a, -um, *of oakwood, hard, strong.*

rogatio, -onis, *f., invitation* (I. ii. 2).

rogo, I, *ask, question.*

ros, roris, *m., dew.*

rosarium, -ii, *n., rosegarden.*

roscidus, -a, -um, (also **ruscidus**), *full of dew, wet with dew.*

rostrum, -i, *n., beak of a bird, snout* or *mouth of an animal.*

rota, -ae, *f., wheel.*

rotundus (also **rutundus**), -a, -um, *round, circular.*

rubens, -entis, *red.*

ruber, -bra, -brum, *red, ruddy.*

rubeus, -a, -um, *red, reddish.*

rubicundus, -a, -um, *red, ruddy.*

rubigo, see **robigo.**

ructor, I *dep., belch* (III. ii. 3).

rudis, -e, *untilled, rough.*

rudis, -is, *f., slender stick* or *rod, goad.*

ruga, -ae, *f., wrinkle, crease.*

ruina, -ae, *f., falling down, downfall, destruction.*

ruminor, I *dep., chew over again, chew the cud, ruminate upon.*

rumis, -is, *f., breast that gives suck, teat.*

rumpo, -rupi, -ruptum, 3, *burst, force open.*

rumpus, -i, *m., vine branch* or *runner* (I. viii. 4).

runco, I, *weed out, clear of weeds* (I. xxx).

rupes, -is, *f., rock.*

rursus, *adv., again.*

rus, ruris, *n., country, lands, country-seat.*

ruscidus, see **roscidus.**

rutundus, see **rotundus.**

S

sacer, sacris, sacre (also sacra, sacrum), *dedicated, consecrated, sacred.*

sacrificor, I *dep., make a sacrifice.*

saepio, -psi, -ptum, 4, *fence in, enclose.*

saeta, -ae, *f., thick, stiff hair* on an animal, *bristle.*

sagino, I, *fatten.*

salax, -acis, *fond of leaping,* especially of male animals, *lustful.*

salictum, -i, *n., willowthicket.*

salio, -ui, 4, *leap, spring.*

saliva, -ae, *f., spittle.*

salix, -icis, *f., willow-tree, willow.*

sallo, -salsum, 3, *salt down.*

salsus, *p.p.p.* of sallo, *salted.*

saltus, -us, *m., forest* or *woodland pasture, mountain valley.*

saluber (or **salubris**), -e, *healthy.*

salubritas, -atis, *f., healthfulness, wholesomeness.*

salus, -utis, *f., health, welfare.*

saluto, I, *greet.*

sancio, -xi, -ctum, 4, *make sacred, decree.*

sane, *adv., certainly.*

sanguis, -inis, *m., blood.*

sanitas, -atis, *f., health.*

sanus, -a, -um, *sound, whole, healthy.*

sapa, -ae, *f., must, new wine boiled thick.*

sapor, -oris, *m., taste.*

sappinus, -i, *f., kind of fir-* or *pine-tree.*

sarcina, -ae, *f., pack., pl., baggage.*

sario, -ui, and -ivi, -itum (also **sarrio**), *weed* (I. xviii. 8).

sarmentum, -i, *n., brushwood.*

sartor (also **sarritor**), -oris, *m., hoer, weeder* (I. xxix. 2).

satio, -onis, *f., sowing, planting.*

satis, *adv., enough, sufficient.*

satisfacio, 3, *give satisfaction.*

satullus, -a, -um, *dim.* (satur), *filled with food* (II. ii. 15).

satur, -ura, -urum, *full of food, that has eaten enough.*

saturo, 1, *fill, satisfy.*

scaber, -bra, -brum, *rough, scurfy.*

scaenicus, -a, -um, *belonging to the stage, dramatic.*

scala, -ae, *f.,* usually *pl.* **scalae,** *flight of steps.*

scapulae, -arum, *f. pl., shoulder-blades.*

scapus, -i, *m., stem, stalk* (I. xxxi. 5).

scientia, -ae, *f., knowledge, skill.*

scilicet, *adv., of course, naturally, undoubtedly.*

scindo, -scidi, -scissum, 3, *cut, tear, break asunder.*

scopus, I. liv. 2.

scorpio, -onis, *m., scorpion, one of the signs of the Zodiac* (I. xxviii. 1).

scriba, -ae, *m., clerk, secretary, official writer.*

scriptor, -oris, *m., secretary, writer.*

scrobis, -is, *m.* or *f., ditch, trench.*

scrofa, -ae, *f., breeding-sow.*

secerno, -crevi, -cretum, 3, *separate, discern.*

secludo, -si, -sum, 3, *shut off, shut in a separate place.*

seclusorium, -ii, *n., coop* (III. v. 5).

seco, -cui, -ctum, 1, *cut, cut up.*

sectio, -onis, *f., cutting, dividing* or *distributing by auction* of confiscated or captured goods (II. x. 4).

sector, 1, *follow after, chase.*

secundus, -a, -um, *following, second, favourable.*

secus, *prep., by, beside, along.*

sedeo, -sedi, -sessum, 2, *sit, remain sitting.*

sedile, -is, *n., seat, bench.*

seditio, -onis, *f., insurrection, civil disorder.*

seditiosus, -a, -um, *seditious.*

seges, -etis, *f., cornfield* (II. *praef.* 4).

segnis, -e, *slow, lingering.*

sella, -ae, *f., seat, stool.*

semen, -inis, *n., seed.*

sementivus, -a, -um, *belonging to seed* or *sowing* (I. ii. 1).

semestris, -e, *of six months, half yearly* (I. ii. 5).

semigraecus, -a, -um, *half-Greek* (II. i. 2).

semimas, -maris, *m., half male, castrated* (III. ix. 3).

seminarium, -ii, *n., nursery garden, seed-plot.*

seminatio, -onis, *f., breeding.*

seminium, -ii, *n., procreation, breeding, stock* (II. i. 14).

semipes, -pedis, *m., half-foot.*

semipiscina, -ae, *f., small fish-pond* (I. xiii. 3).

semita, -ae, *f., path, lane, by-way.*

semivocalis, -e, *half-talking,* applied to cattle (I. xvii. 1).

sempiternus, -a, -um, *everlasting, continual.*

senectus, -utis, *f., old age.*

senescio, -nui, 3, *grow old.*

seni, -ae, -a, *six each.*

separatim, *adv., apart, separately.*

septemtrio, see **septentriones.**

septemtrionalis, -e, *of* or *belonging to the north, northern* (I. 2. 4).

septentriones, *m.,* properly the seven plough-oxen: hence as a constellation *the seven stars near the north pole called also the Wain and the Great* or *Little Bear: the northern regions.*

septingenarius, -a, -um, *belonging to the number seven hundred.*

sequor, secutus, 3 *dep., follow, attend, chase, aim at.*

SELECT VOCABULARY

serenus, -a, -um, *clear: as noun,* **serenum,** -i, *n., fair weather.*

seria, -ae, *f., large earthen jar* (III. ii. 8).

sermo, -onis, *m., talk, conversation.*

sero, *adv., late, at a late hour.*

sero, sevi, satum, 3, *sow, plant.*

serpens, -entis, *f., serpent, snake.*

serpo, -psi, -ptum, 3, *creep.*

serpillum or **serpyllum** (also **serpyllon**), -i, *n., wild-thyme, thyme* (I. xxxv. 2, III. xvi. 13).

serrula, -ae, *f. dim.* (serra), *small saw* (I. l. 2).

servo, I, *preserve, protect.*

servos, also **servus,** -i, *m.,* and **serva,** -ae, *f., slave, servant.*

serus, -a, -um, *late.*

sesquimensis, -is, *m., month and a half* (I. xxvii. 1).

sestertia, *a thousand sesterces* (II. i. 14).

sestertius, -ii, *m., sesterce,* small coin by which the largest sums were reckoned.

sexageni, -ae, a, *sixty each.*

Sextilis, -e, *sixth,* only with mensis, *the month of August,* the sixth month according to the old Roman calendar reckoning from March (I. xxviii. 2).

siccitas, -atis, *f., dryness, drought.*

sicco, I, *make dry, dry up, drain.*

significo, I, *show, indicate.*

signum, -i, *n., mark, sign, military standard.*

silentium, -ii, *n., silence.*

silicula, -ae, *f. dim.* (siliqua), *little husk or pod* (I. xxiii. 3).

siligo, -inis, *f., winter-wheat* (I. xxiii. 2).

silva, -ae, *f., wood, forest, plantation of trees, orchard* (III. v. 12).

silvaticus, -a, -um, *of a wood, wild* (I. xl. 5).

silvestris, -e, *of a wood, wooded.*

similis, -e, *like, resembling.*

similiter, *adv., in the same manner.*

similitudo, -inis, *f., likeness, resemblance.*

simpla, -ae, *f.* (with pecunia understood), *the simple purchase-money* (II. x. 5).

simplex, -icis, *simple, unmixed.*

simul, *adv., at the same time, as soon as.*

simulac, *conj., as soon as.*

simus, -a, -um, *flat* or *snub-nosed.*

sin, *conj., if however, but if.*

singuli, -ae, -a, *one to each, single.*

sinister, -tra, -trum, *left, on the left.*

siquando, *adv., if ever.*

siquidem, *adv., if only, since indeed, since.*

siquis or **siqui,** siqua, siquid, *if any, if anyone,* siqua as *adverb, if in any way, if by any means.*

sirpea (also **scirpea**), -ae, *f., basket work of rushes to form the body of a wagon* (I. xxiii. 5).

sirpiculus (also **scirpiculus**), -i, *m., basket made of rushes.*

sisera, -ae, *f.* (also **sisara**), (III. xvi. 26).

sitiens, -entis, *thirsty.*

situs, *placed, situated, set.*

sive ... sive, *conj., if ... or if.*

socer, -eri, *m., father in law.*

societas, -atis, *f., fellowship, association.*

socius, -i, *m., fellow, comrade, partner.*

sodes, shortened from si audes for audies, *if you will, please.*

soldus, see **solidus.**

solea, -ae, *f., slipper, sandal* (I. xxiii. 6).

soleo, -itus, 2, *be accustomed.*

solidus, -a, -um, *firm, dense, solid.*

solitarius, a, -um, *lonely, solitary.*

solstitialis, -e, *belonging to the summer solstice* (I. xli. 1).

23*

355

solstitium, -ii, *n., the time when the sun seems to stand still, especially the summer solstice* (I. xxxi. 1).

solum, -i, *ground, earth, soil.*

solus, -a, -um, *alone, only.*

solutio, -onis, *f., loosing, payment of money* (II. i. 15).

solvo, solvi, solutum, 3, *loosen, release pay.*

somnus, -i, *m., sleep.*

sonus, -i, *noise, sound.*

spartum, -i, *n., esparto grass, Spanish broom* (I. xxiii. 6).

species, -ei, *f., appearance, shape, form.*

spectaculum, -i, *n., show, sight, spectacle.*

specto, 1 (*freq.*), *watch, observe.*

spero, 1, *hope.*

spina, -ae, *f., thorn.*

spinosus, -a, -um, *thorny.*

spiritus, us, *m., breath.*

spiro, 1, *breathe.*

spissus, -a, -um, *thick, compact, dense.*

spondeo, spopondi, sponsum, 2, *promise, pledge oneself.*

sponte, *abl., of free will, of one's own accord.*

spurcitia, -ae, *f., dirt.*

stabilis, -e, *that stands firm, steady.*

stabulo, 1, *stable, kennel, roost.*

stabulum, -i, *n., dwelling, stable.*

stagnum, -i, *pool, pond.*

stamen, -inis, *n., warp* (I. l. 3).

statim, *adv., immediately.*

statuo, -ui, -utum, 3, *set up, fix in upright position.*

stella, -ae, *f., star.*

stercilinum, -i, *n., dung-pit* (I. xiii. 4).

stercoratio, -onis, *f., dunging, manuring.*

stercoro, 1, *dung, manure.*

stercus, -oris, *n., dung, manure.*

sterilis, -e, *barren.*

sterno, stravi, stratum, 3, *spread out, extend, stretch out.*

stillo, 1, *drip, distil.*

stipulatio, -onis, *f., bargain, stipulation* (legal term II. ix. 7).

stipulor, 1 *dep., bargain, stipulate* (legal term II. v. 10).

stirps, -pis, *f., stock, stem, root.*

stolo, -onis, *m., shoot, twig, sucker.*

stramen, I. l. 3.

stramentum, -i, *n., straw, litter* (I. l. 2).

stratus, -us, *m., spreading, strewing* (I. l. 3).

stria, -ae, *f., furrow, hollow* (I. xxix. 3).

stringo, -inxi, -ictum, 3, *draw tight, tie tight.*

studeo, 2, *be eager, apply one's self to.*

studiosus, -a, -um, *eager, interested in, devoted to.*

studium, -ii, *n., eagerness, study, devotion.*

stylobates, -is, *m., pedestal of a column* or *row of columns* (III. v. 11).

suavis, -e, *sweet, pleasant.*

subalbicans, -antis, *somewhat white* (III. ix. 5).

subcrispus, -a, -um, *somewhat curled* (II. vii. 5).

subduco, -xi, -ctum, 3, *lead away, withdraw.*

subfumigo, 1, *fumigate from below* (III. xvi. 36), *smoke lightly.*

subicio, -ieci, -iectum, 3, *place under, subject.*

subiecto, 1, *place under* (I. lii. 2.)

subigo, -egi, -actum, 3, *bring under, dig up, plough, tame.*

subito, *adv., suddenly.*

sublevo, 1, *raise up, support, assist.*

sublimis, -e, *high, lofty.*

subniger, -gra, -grum, *somewhat black.*

subplementum, -i, *n.* (also **supplementum**), *filling up, supplement.*

subrideo (also **surrideo**), -si, 2, *smile.*

subrumus (also **surrumus**), -a, -um, *under the udder, sucking.*

subseco, -cui, -ctum, I (also **subsico**), *cut away below.*

subsellium, -ii, *n.*, *low bench.*

subsico, see **subseco.**

subsidium, -ii, *n.*, *help, support.*

subsido, -sedi, -sessum, 3, *sit down, settle, remain in a place.*

subsimus, -a, -um, *somewhat snub-nosed.*

substerno, -stravi, -stratum, 3, *strew, scatter, lay under.*

substramen, -inis, *n.*, *litter, what is strewn underneath.*

substruo, -xi, -ctum, 3, *build beneath.*

subter, *adv. and prep.*, *below, beneath, under.*

subtilis, -e, *fine, slender.*

subtus, *adv.*, *below, underneath.*

suburbanus, -a, -um, *situated near the city of Rome, suburban.*

succedo, -cessi, -cessum, 3, *go or come under, approach, succeed.*

succentivus, -a, -um, *sounding to, accompanying of a pipe* (I. ii. 15).

successus, -us, *m.*, *success.*

succido, -cidi, -cisum, 3, *cut from below, cut off or down.*

succino, 3, *sing to, accompany.*

succollo, I, *to take upon the shoulder* (III. xvi. 8).

succumbo, -cubui, -cubitum, 3, *fall down, yield, submit.*

sucus, -i, *m.*, *juice, sap.*

suffragator, -oris, *m.*, *one who votes for another, supporter.*

suffragium, -ii, *n.*, *voting-tablet, ballot.*

suggestum, -i, *n.*, *raised place, platform* (III. v. 16).

sugo, -xi, -ctum, 3, *suck.*

sugrunda, -ae, *f.*, *lower border of a roof, eaves* (III. iii. 5).

suillus, -a, -um, *of swine.*

sulco, I, *furrow, plough.*

sulcus, -i, *m.*, *furrow made by the plough.*

sumen, -inis, *n.*, *breast* (of a woman).

summa, -ae, *f.*, *sum, sum total, summit, chief point.*

summatim, *adv.*, *briefly, slightly.*

summitto, -misi, -missum, 3, *produce, raise, rear* (of farm animals) (II. ii. 18).

summus, -a, -um, *top of, highest part* (I. ii. 10).

sumo, sumpsi, sumptum, 3, *take, lay hold of.*

sumptuosus, -a, -um, *costly, lavish, extravagant.*

sumptus, -us, *m.*, *expense.*

supellex, suppellectilis, *f.*, *household furniture* or *goods.*

supercilium. -ii, *n.*, *eyebrow.*

superior, -ius, *higher, upper.*

supero, I, *rise above, surpass.*

supersum, -fui, *be left, exist still.*

suppedito, I, *have in abundance, afford, supply.*

suppleo, -evi, -etum, 4, *fill up, complete.*

supplicium, -ii, *n.*, *act of worship, sacrifice, punishment.*

suppono, -posui, -positum, 3, *place under.*

supposticius, *substituted, false* (II. viii. 2).

suppressus, *p.p.p.* of supprimo.

supra, *adv. and prep.*, *above, over.*

supprimo, -pressi, -pressum, 3, *press down, check, restrain.*

suptilis, see **subtilis.**

suptus, see **subtus.**

surclus, see **surculus.**

surcularius, -a, -um, *of shoots* or *twigs.*

surculus, -i, *m.*, *young twig* or *branch, shoot.*

surdus, -a, -um, *deaf.*

surgo, surrexi, surrectum, 3, *rise, get up.*

sursum, see **susum.**

sus, suis, *c.*, *swine, pig.*

suscipio, -cepi, -ceptum, 3, *take* or *catch up, support, undertake.*

suspendo, -di, -sum, 3, *hang up.*

sustento, I, *freq.* (sustineo), *uphold, support.*

sustineo, -tinui, -tentum, 2, *hold up, support, check, maintain.*

susum, *adv.,* *upwards, on high.*

T

tabanus, -i, *m.,* *gad-fly* (II. v. 14).

tabella, -ae, *f. dim.* (tabula), *small board attached to the plough* (I. xxix. 2), *writing tablet.*

taberna, -ae, *f.,* *hut, booth, tavern, workshop.*

tabula, -ae, *f.,* *board.*

tabulatum, -i, *n.,* *floor, story, flooring.*

tactus, -us, *m.,* *touch.*

talea, -ae, *f.,* *rod, stick, cutting* or *slip for grafting* (I. xl. 4).

talpa, -ae, *f.,* *mole.*

tamquam, *adv.,* *just as, as if, as it were.*

tango, tetigi, tactum, 3, *touch.*

tantus, -a, -um, *of such size, so great.*

tarde, *adv.,* *late.*

tardus, -a, -um, *slow, late.*

taura, -ae, *f.,* *barren cow* (II. v. 6).

taurus, -i, *m.,* *bull, a constellation of the Zodiac.*

tector, -oris, *m.,* *plasterer.*

tectorium, -ii, *n.,* *covering,* cover, especially of stucco and plaster.

tectorius, -a, -um, *of plaster* or *stucco* (III. xi. 2).

tectum, -i, *n.,* *covering, roof, house, shelter.*

tego, -xi, -ctum, 3, *cover, hide.*

tegulae, -arum, *f. pl.,* *roof-tiles, a tiled roof.*

tempero, I, *temper, govern, rule, be moderate.*

tempestas, -atis, *f.,* *time, weather, storm.*

tempestive, *adv.,* *at the right time.*

tenebrae, -arum, *f. pl.,* *darkness, night, gloom.*

tenebricosus, -a, -um, *full of darkness, gloomy.*

tenellus, -a, -um, *dim.* (tener), *somewhat tender* (I. xli. 1).

teneo, tenui, tentum, 2, *hold, keep.*

tener, -era, -erum, *soft, delicate, tender.*

teneritudo, -inis, *f.,* *softness, tenderness.*

tenuis, -e, *thin, fine.*

tepidus, -a, -um, *lukewarm.*

terni, -ae, a, *three each.*

tero, trivi, tritum, 3, *rub, grind, thresh* (I, xiii, 5).

terrestris, -e, *of the earth* or *land.*

tesserula, -ae, *f.,* *dim.* (tessera), *square piece of stone for paving, voting-tablet.*

testa, -ae, *f.,* *brick, tile.*

testiculus, -i, *m.,* *dim.* (testis), *testicle.*

testimonium, -ii, *n.,* *evidence, proof.*

testis, -is, *c.,* *witness.*

testudo, -inis, *f.,* *tortoise, arch, vault in buildings* (III. v. 1).

textor, -oris, *m.,* *weaver.*

theatridion, *little theatre* (III. v. 13).

theatrum, -i, *n.,* *play-house, theatre.*

tholus, -i, *m.,* *dome, cupola* (III. v. 12).

thymum, -i, *n.,* *thyme.*

tibia, -ae, *f.,* *shin-bone, pipe, flute.*

tibicen, -inis, *m.,* *piper, flute-player.*

tiro, -onis, *m.,* *recruit.*

tolerabilis, -e, *bearable.*

tollo, sustuli, sublatum, 3, *lift up, take away.*

tondeo, totondi, tonsum, 2, *shear, crop, shave.*

tonsura, -ae, *f.,* *shearing.*

torculum, -i, *n.,* *press* (I. lv. 7).

torrens, -entis, *m.,* *torrent.*

torrens, -entis, *burning, hot.*

tortus, -a, -um, *twisted, crooked.*

totidem, *num. adj., in-declinable, just so many.*

totiens, *adv., so often, as many times.*

tracto, I, *freq.* (traho), *draw violently, drag, manage, treat.*

trado, -didi, -ditum, 3, *hand over, deliver, hand down, to posterity.*

tradux, -ucis, *m., vine-branch, vine-layer* (I. viii. 3).

traho, -xi, -ctum, 3, *draw, haul.*

traicio, -ieci, -iectum, 3, *throw or hurl over.*

Transalpinus, -a, -um, *that lies beyond the Alps, Trans-alpine.*

transeo, -ivi or -ii, -itum, *cross over.*

transfero, -tuli, -latum, *carry or bring over, transfer by grafting* (I. xxxix. 3 and I. xl. 4).

translatio, -onis, *f., removing from one place to another, transplanting* (I. xli. 3).

transmarinus, -a, -um, *coming from beyond the sea.*

transversus, -a, -um, *lying across, transverse, crosswise.*

trapetum, -i, *n., olive-mill,* also *pl.,* **trapetes,** -um, *m., accus.* trapetas (I. lv. 5).

traveho, -xi, -ctum, 3 (also **transveho**), *carry or convey over, transport.*

traversus, -a, -um (also **transversus**), *lying across, athwart, crosswise.*

treceni, -ae, a, *num. distrib. adj., three hundred each.*

trecenarius, -a, -um, *belonging to three hundred.*

trepido, I, *be in a state of agitation, be anxious.*

tribulis, is, *m., one of the same tribe with another* (III. ii. 1).

tribulum, -i, *n., threshing-sledge* (I. lii. 1).

tribunus, -i, *m., tribune* (I. ii. 9).

triclinaria, -ium, *n. pl., dining-room* (I. xiii. 7).

triclinium, -ii, *n., couch,*

running round three sides of a table, for reclining at meals.

triduum, -i, *n., the space of three days.*

triennium, -ii, *n., the space of three years.*

trimestris, -e, *of three months.*

trimus, -a, -um, *of three years, three years old.*

trini (also **terni**), *num, dist. adj., three each.*

tripedalis, -e, *of three feet in measure.*

triplex, -icis, *three-fold, triple.*

tritavus, -i, *m., father of an atavus or atavia, pl., remote ancestors* (III. iii. 2).

triticeus, -a, -um, *of wheat.*

triticum, -i, *wheat.*

tritura, -ae, *f., rubbing, threshing of grain* (I. xiii. 5).

triumphus, -i, *m., solemn entrance of a general into Rome after an important victory, triumphal procession.*

tuba, -ae, *f., trumpet.*

tubulus, -i, *m., dim.* (tubus), *small pipe, water-pipe.*

tueor, tuitus, 2 *dep., watch, protect.*

turba, -ae, *f., tumult, crowd, throng.*

turdus, -i, *m., thrush, field-fare.*

turgesco, 3, *begin to swell, swell up.*

turpis, -e, *ugly.*

turris, -is, *f., tower.*

turtur, -uris, *m., turtle-dove.*

tutela, -ae, *f., charge, protection.*

tutus, -a, -um, *safe, out of danger.*

tympanum, -i, *drum, tambourine.*

U

uber, -eris, *n., udder, teat.*

uber, -eris, *full, abundant.*

ubicumque, *adv., wherever.*

ulcus, -eris, *n., sore, ulcer.*

uliginosus, -a, -um, *full of moisture, damp, wet, moist.*

uligo, -inis, *f.,* *moisture* (II. ii. 7).

ulmus, -i, *f., elm-tree.*

ulterior, -ius, *farther, on the farther side.*

ultimus, -a, -um, *farthest, last, farthest part of anything.*

ultra, *adv.* and *prep., adv., on the other side, beyond, besides, prep., on the farther side of, beyond, over, more than.*

ultro, *adv., beyond, to the farther side, voluntarily,* ultro et citro, *back and forth* (III. v. 16).

umbra, -ae, *f., shade.*

umbraculum, -i, *n., shady place, arbor.*

umbrifer, -era, -erum, *shade-giving, casting a shade.*

umbrosus, -a, -um, *shady.*

umerus, -i, m, *shoulder.*

umidus, -a, -um, *moist, damp.*

umor, -oris, *m., liquid, moisture.*

umquam, also **unquam,** *adv., ever.*

unde, *adv., from which place.*

unguen, -inis, *n., ointment, unguent.*

unguentum, -i, *n., ointment, unguent, perfume.*

unguis, -is, *m., finger-nail, claw, hoof, talon.*

ungula, -ae, *f., hoof, claw, talon.*

unguo (also **ungo**), -nxi, -nctum, 3, *smear, anoint.*

universus, -a, -um, *all together, entire, universal.*

urbanus, -a, -um, *of the city* or *town, refined, courteous.*

urgeo, ursi, 2, *press, drive, urge.*

urina, -ae, *f., urine.*

usque, *adv., all the way to, as far as.*

usurpo, 1, *make use of, employ.*

usus, -us, *m., use, employment.*

utensilis, -e, *fit for use.*

uterque, utraque, utrumque, *each* (of two), *both.*

uterus, -i, *m., womb.*

utilis, -e, *useful, advantageous.*

utiliter, *adv., usefully.*

utilitas, -atis, *f., usefulness.*

utique, *adv., certainly, assuredly, especially.*

utor, usus, 3 *dep., with abl., use, make use of.*

utrum, *adv.,* introducing an alternative question followed by *an* beginning the second clause: *whether.*

uva, -ae, *f., grape, bunch of grapes.*

uxor, -oris, *f., wife.*

V

vacca, -ae, *f., cow* (I. xx. 4).

vagina, -ae, *f., sheath, scabbard.*

vagor, 1 *dep., stroll about, wander.*

valde, *adv., strongly, powerfully.*

valeo, 2, *be strong.*

valetudo, -inis, *f., state of health, good health.*

vallis, also **valles,** -is, *f., valley.*

vallus, -i, *m., stake.*

vallus, -i, *f. dim.* (contracted for vannullus, from vannus), *a small winnowing-fan for grain* (I. lii. 2).

vapor, -oris, *m., steam, exhalation.*

vapulo, 1, *be flogged.*

varietas, -atis, *f., difference, diversity.*

varius, -a, -um, *different, various.*

varus, -a, -um, *bent inwards* (II. ix. 4).

vas, vasis, *n., vessel, dish, implement.*

vasculum, -i, *n., dim.* (vas), *small vessel.*

vasto, 1, *lay waste.*

vastus, -a, -um, *empty, desolate, large.*

vatillo, *abl.* (III. vi. 5).

vatius, -a, -um, *bent outwards.*

-ve, *or* (always enclitic).

vectorius, -a, -um, *for carry-ing.*

vectura, -ae, *f., carrying, conveying.*

vegrandis, -e, *not very large, small.*

veha, I. ii. 14.

veho, -xi, -ctum, 3, *bear, convey,* passive, *ride, sail, go.*

vehementer, *adv., eagerly.*

vehiculum, -i, *n., means of transport, carriage, cart.*

velamen, -inis, *n., cover, garment, veil.*

vella, I. ii. 14.

vellico, 1, *pluck, pinch, pull apart.*

vello, vulsi, vulsum, 3, *pluck, pull, tear out.*

vellus, -eris, *n., fleece, wool shorn off.*

velocitas, -atis, *f., swiftness.*

velox, -ocis, *swift.*

vena, -ae, *f., vein.*

venalis, -e, *for sale, purchase-able.*

venaticus, -a, -um, *belong-ing to hunting.*

venatio, -onis, *f., hunting.*

venator, -oris, *m., hunter.*

vendibilis, -e, *that may be sold.*

venditor, -oris, *m., seller, vendor.*

vendo, -didi, -ditum, 3, *sell.*

venenum, -i, *n., drug, poison.*

veneo, -ivi, or -ii, -itum, 4, *be sold,* used as *passive* of vendo.

veneratio, -onis, *f., rever-ence.*

venio, veni, ventum, 4, *come.*

venter, -tris, *m., belly, stom-ach.*

ventilabrum, -i, *n., imple-ment for winnowing grain* (I. lii. 2).

ventilo, 1, *toss, fan.*

ventosus, -a, -um, *windy.*

ventus, -i, *m., wind.*

venus, -eris, *f., love, sexual love.*

venustus, -a, -um, *lovely, fair.*

ver, veris, *n., spring.*

verber, -eris, *n., lash, whip, flogging.*

verbosus, -a, -um, *wordy,*

verbum, -i, *n., word.*

vereor, -itus, 2 *dep., fear, respect.*

vergiliae, -arum, *f. pl., con-stellation that rises at the end of spring* (I. xxviii. 2).

vergo, 3, *bend, incline.*

vermiculus, -i, *m., dim.* (vermis), *small worm, grub.*

vernacula, -ae, *f., female household slave, indigenous* (III. v. 7).

verres, -is, also **verris,** *m., boar.*

versor, 1 *dep., dwell, stay, be engaged in an occupation.*

versus, -us, *m., line of writing, poetry.*

versus, *adv., facing, opposite to.*

verto, -ti, -sum, 3, *turn, turn round or about.*

verus, -a, -um, *true,* **vero,** *in truth.*

vervex, -ecis, *m., wether.*

vescor, 3, *dep., eat., with abl.*

vespa, -ae, *f., wasp.*

vesper, -eris and eri, *m., evening.*

vespertinus, -a, -um, *of evening.*

vestibulum, -i, *n., enclosed space between the entrance of a house and the street; fore-court.*

vestigium, -ii, *n., footstep, footprint.*

vestitus, *p.p.p.* of vestio, *clothed.*

vestitus, -us, *m., clothing.*

veteranus, -a, -um, *old, veteran* (I. xx. 2 of oxen).

veto, -ui, -itum, 1, *forbid.*

vetulus, -a, -um, *dim.* (vetus), *little old, old.*

vetus, -eris, *old.*

vetustas, -atis, *f., old age, long existence.*

viator, -oris, *m., wayfarer, traveller.*

viceni, -ae, -a, *twenty each, num. distrib. adj.*

vicesimus, -a, -um, *twen-tieth.*

vicia, -ae, *f., vetch* (I. xxxi. 5).

vicinia, -ae, *f., neighbourhood, vicinity.*

vicinitas, -atis, *f., neighbourhood, vicinity.*

vicinus, -a, -um, *neighbouring.*

victima, -ae, *f., beast for sacrifice.*

victus, -us, *m., sustenance, victuals.*

vicus, -i, *m., village, country-seat, quarter of a city, street.*

vieo, -etum, 2, *plait, weave.*

vigilo, 1, *keep awake, keep watch.*

vilica, -ae, *f., female overseer.*

vilicus, -i, *m., overseer of a farm, agent, bailiff.*

villaticus, -a, -um, *belonging to a country seat or villa.*

villus, -i, *m., shaggy hair, tuft of hair.*

vimen, -inis, *n., pliant twig, withe, osier.*

vinarius, -a, -um, *adj. of wine.*

vincio, vinxi, vinctum, 4, *bind, tie.*

vinclum (also **vinculum**), -i, *n., band, fetter.*

vinco, vici, victum, 3, *conquer.*

vinctus, -us, *m., binding.*

vindemia, -ae, *f., vintage, grape-gathering.*

vindemiator, -oris, *m., grape-gatherer.*

vindemiatorius, -a, -um, *of the vintage.*

vindico, 1, *claim as one's own, demand.*

vinetum, -i, *n., vineyard.*

vineus, -a, -um, *of wine.*

vinum, -i, *n., wine.*

viola, -ae, *f., violet.*

violarium, -ii, *n., violet bed.*

violo, 1, *treat with violence, injure.*

virga, -ae, *f., slender green branch, rod.*

virgo, -inis, *f., maid, virgin.*

virgula, -ae, *f.* (*dim.* virga), *small twig, small rod, wand.*

virgultum, -i, *n., bush, thicket.*

viridis, -e, *green.*

viritim, *adv., man by man, individually.*

vitalis, -e, *belonging to life, vital.*

viteus, -a, -um, *of the vine* (I. xxxi. 4).

vitiarium, -ii, *n., nursery for vines* (I. xxxi. 2).

vitilis, -e, *platted, interwoven, wicker-work.*

vitiosus, -a, -um, *faulty, bad.*

vitis, -is, *f., vine, grape-vine.*

vitium, -ii, *n., fault, vice.*

vito, 1, *avoid.*

vitulus, -i, *m.,* and **vitula,** -ae, *f., calf.*

viviradix, -icis, *f., set or cutting, having a root, layer, quickset* (I. xxxv. 1).

vivo, vixi, victum, 3, *live.*

vivus, -a, -um, *alive.*

vocabulum, -i, *n., name.*

vocalis, -e, *that can speak.*

vociferor, 1 *dep., cry out, exclaim.*

vocifico, 1, *cry aloud, proclaim.*

vocito, 1 *freq.* (voco), *call, name.*

voco, 1, *call, summon.*

volatura, -ae, *f., flight.*

volito, 1, *freq.* (volo), *fly to and fro, flit about.*

volo, volui, *wish, be willing.*

volo, 1, *fly.*

volucer, -ucris, -ucre, *flying, winged.*

volumen, -inis, *n., roll of writing, volume.*

voluntas, -atis, *f., will, desire.*

voluptas, -atis, *f., pleasure.*

voluto, 1, *freq., roll, turn about.*

volva, -ae, *f., wrapper, covering, womb* (II. i. 19).

vomer, -eris, *m., ploughshare.*

vox, vocis, *f., voice, word.*

vulnero, 1, *wound.*

vulpes (also **vulpis**), -is, *f., fox.*

PRINTED IN GREAT BRITAIN BY UNIVERSITY TUTORIAL PRESS LTD, FOXTON
NEAR CAMBRIDGE